# Dynamics of
# Molecular
# Collisions

## Part B

# MODERN THEORETICAL CHEMISTRY

*Editors:* **William H. Miller,** University of California, Berkeley
**Henry F. Schaefer III,** University of California, Berkeley
**Bruce J. Berne,** Columbia University, New York
**Gerald A. Segal,** University of Southern California, Los Angeles

# Dynamics of
# Molecular
# Collisions

## Part B

Edited by
## William H. Miller
University of California, Berkeley

PLENUM PRESS · NEW YORK AND LONDON

Library of Congress Cataloging in Publication Data

Main entry under title:

Dynamics of molecular collisions.

    (Modern theoretical chemistry; v. 2)
    Includes bibliographical references and index.
    1. Molecular dynamics. 2. Chemical reaction, Conditions and laws of. I. Miller,
William Hughes, 1941-       II. Series.
QD461.D95                     539'.6                     76-12633
ISBN 0-306-33502-6

© 1976 Plenum Press, New York
A Division of Plenum Publishing Corporation
227 West 17th Street, New York, N.Y. 10011

Printed in the United States of America

# Contributors

**R. B. Bernstein,** Chemistry and Physics Departments, The University of Texas, Austin, Texas

**M. S. Child,** Theoretical Chemistry Department, University of Oxford, Oxford, England

**William L. Hase,** Department of Chemistry, Wayne State University, Detroit, Michigan

**P. J. Kuntz,** Hahn–Meitner Institut für Kernforschung, Berlin GmbH, West Germany, and St. Mary's University, Halifax, Nova Scotia, Canada

**R. D. Levine,** Department of Physical Chemistry, The Hebrew University, Jerusalem, Israel

**Philip Pechukas,** Department of Chemistry, Columbia University, New York, New York

**Richard N. Porter,** Department of Chemistry, State University of New York, Stony Brook, New York

**Lionel M. Raff,** Department of Chemistry, Oklahoma State University, Stillwater, Oklahoma

**John C. Tully,** Bell Laboratories, Murray Hill, New Jersey

# Preface

Activity in any theoretical area is usually stimulated by new experimental techniques and the resulting opportunity of measuring phenomena that were previously inaccessible. Such has been the case in the area under consideration here beginning about fifteen years ago when the possibility of studying chemical reactions in crossed molecular beams captured the imagination of physical chemists, for one could imagine investigating chemical kinetics at the same level of molecular detail that had previously been possible only in spectroscopic investigations of molecular stucture. This created an interest among chemists in scattering theory, the molecular level description of a bimolecular collision process. Many other new and also powerful experimental techniques have evolved to supplement the molecular beam method, and the resulting wealth of new information about chemical dynamics has generated the present intense activity in molecular collision theory.

During the early years when chemists were first becoming acquainted with scattering theory, it was mainly a matter of reading the physics literature because scattering experiments have long been the staple of that field. It was natural to apply the approximations and models that had been developed for nuclear and elementary particle physics, and although some of them were useful in describing molecular collision phenomena, many were not. The most relevant treatise then available to students was Mott and Massey's classic *The Theory of Atomic Collisions,*\* but, as the title implies, it dealt only sparingly with the special features that arise when at least one of the collision partners is a *molecule*.

Now, however, chemical applications of scattering theory have evolved considerably, and many new techniques and models have been developed by chemical theorists specifically for treating molecular phenomena. It is interesting, too, that some of these newly developed approaches to scattering are also finding useful application in certain areas of physics, and we are thus able in part to repay our debt to that field.

---

\*N. F. Mott and H. S. W. Massey, 3rd ed., Clarendon Press, Oxford (1965).

This work is a collection of the new theoretical techniques that have been developed in recent years for describing chemical dynamics. The goal has been to assemble a body of material that would be of use for beginning research students in the field. It is expected, therefore, that the reader has a background in quantum mechanics and also some elementary knowledge of scattering theory itself (i.e., elastic scattering by a spherically symmetric potential). The first volume, Part A, deals primarily with quantum mechanical descriptions of molecular collision processes and their application to inelastic and reactive phenomena, while the companion volume, Part B, deals mainly with classical and semiclassical approaches. This division, though somewhat arbitrary and coincidental, is nevertheless symbolic of the field; i.e., the chemical dynamicist must be prepared to think in terms of both classical and quantum mechanics—or perhaps some combination of the two—if he is to make significant progress toward understanding chemical kinetic phenomena at a molecular level.

*W. H. Miller*

# Contents

## Chapter 2.   Features of Potential Energy Surfaces and Their Effect on Collisions

### *P. J. Kuntz*

## Chapter 3.  Dynamics of Unimolecular Reactions
### *William L. Hase*

## Chapter 4.  Semiclassical Methods in Molecular Collision Theory
### *M. S. Child*

*Chapter 5.*   Nonadiabatic Processes in Molecular Collisions
            *John C. Tully*

*Chapter 6.*   Statistical Approximations in Collision Theory
            *Philip Pechukas*

*Chapter 7.* Thermodynamic Approach to Collision Processes
*R. D. Levine and R. B. Bernstein*

# Contents of Part A

# Classical Trajectory Methods in Molecular Collisions

*Richard N. Porter*
*and*
*Lionel M. Raff*

## 1. Introduction

### 1.1. Rationale for the Use of Classical Dynamics

The dynamics of a molecular scattering process is described in exact terms by the solutions to the Schrödinger equation in which the kinetic energy and the electrodynamical interactions of all the nuclei and electrons of the colliding partners are used. If the process to be studied can be assumed to be adiabatic, the Born–Oppenheimer separation can be invoked, and the Schrödinger equation for the scattering is reduced to the problem of nuclear motion on a potential energy surface known as a function of all the internuclear distances. The accuracy of quantum mechanical calculations of the measurable attributes of molecular collisions is limited only by the accuracy of the potential energy surface and by the number of basis functions that can be afforded in terms of computer core storage size and processing time. The technical and economic questions are therefore

  1. How accurate must a calculation be in order to test predictions of a given theory against a given experimental result, and how is this accuracy most efficiently achieved?

*Richard N. Porter* • Department of Chemistry, State University of New York, Stony Brook, New York, and *Lionel M. Raff* • Department of Chemistry, Oklahoma State University, Stillwater, Oklahoma

2.  What calculational expense is commensurate with the scientific value of the result?

Concerted efforts of several groups are rapidly yielding answers to question 1 for nonreactive inelastic scattering and steadily, but much more slowly, for reactive scattering. Answers to question 2 require delicate interplay between scientific judgement and cost–benefit analysis that depends for input on the answers to question 1 and on the scientific richness or technological importance of the phenomenon.

At the state of development of computer technology and numerical analysis in the mid-1970s, the experimental data on molecular collisions are considerably more precise and extensive in scope and detail than quantum mechanical scattering calculations can be. The disparity is only partly due to incomplete knowledge of the potential energy functions. The inevitable question is whether the far more tractable classical mechanical calculations of the attributes of molecular collisions are accurate enough. A somewhat facile answer is that quantal effects are expected to be negligible except for processes involving hydrogen atoms. A deeper insight into the nature of the averaging of primitive quantities to give measurable attributes such as differential cross sections in both the classical and quantal regimes is even more encouraging.[1] The averaging over quantum numbers, such as total angular momentum, that are not measured in the experiment tends to destroy the quantal effects by a superposition that is practically random for most experimental conditions. For experiments that incompletely specify internal states of the colliding molecules, the averaging out of quantal effects is expected to be even more complete. Calculation of thermal averages from classical trajectories is in fact expected to be even more accurate than quantal calculations can be at present, because of the requirement of an impractically large number of partial waves corresponding to the contributing initial and final internal molecular states and the range of orbital angular momentum.

Clear exceptions occur when all channels for a given process are classically closed but energetically open and quantally allowed. Processes with non-negligible contributions from potential surface crossings, tunneling, and scattering into classically forbidden states thus require quantum mechanical elucidation. The extent to which these processes can be treated by semiclassical methods[1–5] has been the subject of intense and fruitful study.*

A quantitative answer to the question of errors introduced by use of classical mechanics must be extracted from comparisons of parallel classical and quantal calculations of a wide variety of attributes of collisions on representative potential energy surfaces. Because treatment of the full dimensionality of the problem induces superposition of more quantal states than does restriction to a lower dimensionality, comparisons of quantal and classical

---

*See Chapter 4 of this volume.

results for the lower dimensionality, although tractable and technically infor-mative, tend to overstate the classical error.

## 1.2. Survey of Applications of Classical Calculations

Some representative classical calculations that have had an impact on ex-perimental reaction dynamics are mentioned briefly so the reader can see some-thing of the scope of the field before we lead him into the details of the method.

### 1.2.1. Total and Differential Cross Sections

The initial determination of the cross section for a chemical reaction and its dependence on the collision energy resulted from an analysis of classical trajectories calculated for the exchange reaction

$$H + H_2 \rightarrow H_2 + H$$

and its isotopic analogues.[6-8] For the reactions[8,9]

$$T + H_2 \rightarrow HT + H$$
$$T + D_2 \rightarrow DT + D$$
$$T + HD \rightarrow HT + D$$
$$T + HD \rightarrow DT + H$$

which are of interest in the study of hot atom processes,[10-14] the total reaction cross section was calculated over the entire range of reactive collision energies. The cross sections for these reactions rise steeply from a threshold energy $E_1$ and exhibit a maximum above which they decay relatively slowly in a concave curve to an effective "antithreshold" $E_2$. Examination of statistically significant samples of trajectories at various energies provides dynamical rationalization of the shape of the cross-section curves.[9]

Although the energy dependences near the maxima of the cross sections for the (H, $H_2$) exchange reactions are yet to be measured, the features of the theoretical curves appear to be quite general for exchange reactions with activation energy. For example, the reaction

$$CH_3I + K \rightarrow CH_3 + KI$$

exhibits a similarly shaped cross section that has been determined in molecular beam experiments.[15,16] Proposals for the mechanism leading to a maximum in the cross section for this reaction have been based both on trajectory studies[17,18] and on simplified general models of molecular collisions.[19] The issue of mechanism is somewhat clouded by the fact that the potential energy surface is much less completely characterized than that for the (H, $H_2$) reaction.

Perhaps the most stringent test of the accuracy of a potential energy function is its ability to reproduce the distributions of the product scattering angle measured in a crossed beam experiment. An excellent example of comparison of theoretical[20] and experimental[21] scattering angle–velocity contour plots is that for the reaction

$$F + D_2 \rightarrow DF + D$$

in which an empirical potential energy function was used in the calculations.

Studies of the mechanisms of inelastic molecular scattering have become particularly interesting because of their relevance to the design of gas phase lasers. Translation–rotation energy transfer has been studied with both accurate *ab initio* potential functions and with simple models in a successful effort to extract the essence of the mechanism for inelastic collisions of this type involving an atom and a diatomic molecule.[22,23] In another important development, inelastic collisions have been analyzed via classical trajectories to determine the roles played by excitation of rotational states and by atomic exchange in the vibrational relaxation,[24,25]

$$Cl + HCl(v) \rightarrow Cl + HCl(v')$$

### 1.2.2. Deposition of Reaction Energy into Product Modes

Chemiluminescence experiments provide information about the distribution among excited vibrational and rotational states of the molecular products of exothermic reactions.[26] Extensive use of classical trajectories has led to the discovery and elucidation of general relationships between features of potential energy surfaces and masses of the atoms, on the one hand, and the deposition of reaction energy into vibrational and rotational modes of the products, on the other. The principle of microscopic reversibility then shows which initial states of the reactant molecules are most effective in the endothermic reverse reactions on a potential energy surface of a given type. These relationships have been summarized[27] and reviewed in detail.[28]*

The results of chemiluminescence experiments and the predictions of classical trajectory calculations have been analyzed by an interesting extension of information theory.[29]† In particular, it is found that plots of $\ln (P_v/P_v^0)$ vs. $v$ (where $P_v$ is the observed population of state $v$ and $P_v^0$ is the population if each product state of the specified energy is equally probable) are in most cases very nearly linear for highly exothermic reactions.

### 1.2.3. Reaction Rate Coefficients and Relaxation Times

Because calculation of reaction rate coefficients and relaxation times involves thermal averaging over large numbers of reactive and inelastic

---

*See also Chapter 2 of this volume.
†See also Chapter 7 of this volume.

trajectories, respectively, these quantities are perhaps least sensitive to errors due to the use of classical mechanics, to inaccurate integration of the equations of motion, and to minor inaccuracies in the potential energy surface. On the other hand, they can be very sensitive to changes in the broader features of the potential energy surface, such as the barrier height and location, the curvature of the reaction path, and the location and depths of wells that may exist in the surface.[28]

By specifying the initial internal states of the reactants in the calculations, one obtains detailed theoretical rate coefficients and relaxation times for reactive and inelastic processes. These quantities can then be averaged over whatever distribution of internal states of reactants is appropriate. In this way, one can generate input to kinetic models of systems such as gas lasers and the upper atmosphere in which some degrees of freedom may have Boltzmann distributions whereas others may not. For some processes, the most reliable values of the rate coefficients are those calculated from classical trajectories. An example is

$$H_2 + F_2 \rightarrow H + HF + F$$

a possibly important branching step in the $(H_2, F_2)$ explosion chain.[30]

One of the first applications of trajectory calculations in chemical kinetics was a study of unimolecular processes in triatomic molecules.[31,32] This program has been expanded to provide insight into assumptions of RRKM theory for isomerization of $CH_3NC$.[33]†

### 1.2.4. Hot Atom Chemistry

Classical trajectories provide estimates of both the reactive and nonreactive cross sections required for the analysis of the thermalization of translationally excited atoms in a reactive Boltzmann gas mixture.[34] Reactions that have been studied extensively include $(T^*, H_2)$,[8] $(F^*, H_2)$,[35] $(T^*, CH_4)$,[36,37] and $(D^*, DBr, H_2)$.[38]

### 1.2.5. Isotope Effects

The most accurate experimental data on kinetic isotope effects are those for the hydrogen isotopes, although in this case the question of mechanism is complicated at low translational energies (or at low temperature) by a contribution from tunneling. Conclusions drawn from analysis of the isotope effect in classical trajectories must therefore be done with caution in the low-temperature regime. At collision energies well above threshold, however, classical trajectory predictions of the kinetic isotope effect offer a sensitive means of evaluating the accuracy of a potential energy surface if the

†See also Chapter 3 of this volume.

experimental data are available.[39] Furthermore, they can provide mechanistic insights not easily obtained by other means, when dynamic effects are important.

### 1.2.6. Investigation of Reaction Mechanisms

If there are several pathways for a chemical reaction, it is usually assumed to proceed by the path that has the lowest activation energy. This assumption is a good one for thermal reactions if the lowest activation energy lies sufficiently far below all the others. Actually, all pathways contribute to the reaction, and if the lowest activation energy is nearly degenerate with one or more of the others, the relative contribution from each of these competing pathways may depend more significantly on dynamic considerations than on energetic ones.

Endothermic reactions provide simple examples. Here the competing pathways involve the different internal states of the reactant molecules, and those excited states having energies below the measured activation energy can be expected to contribute. Trajectory calculations and experimental data show in general that excited vibrational states below the activation energy in fact make the most significant contribution, and in some cases the ground state is relatively "inert."[39]

An example of a more complex question that has been studied by classical trajectories is the relative contribution of atomic and molecular mechanisms to the overall reaction[40,41]

$$H_2 + I_2 \rightarrow 2HI$$

This brief survey suggests only broadly some of the kinds of chemical uses that have been made of classical trajectory calculations. Readers interested in more details of the applications should consult more complete reviews.[28,42,43]

## 2. Review of Classical Hamiltonian Mechanics

The calculation of classical trajectories of the atoms in colliding molecules involves the numerical solution of the appropriate set of equations of motion. In this section we shall briefly review the classical formulation of the mechanical problem for $N$ particles.

According to Newton's second law, the coordinate of the $i$th particle obeys the equation

$$m_i \ddot{\mathbf{r}}_i = \mathbf{F}_i \tag{1}$$

where $\mathbf{F}_i$ is the force acting on particle $i$, and $m_i$ and $\mathbf{r}_i$ are its mass and coordinate, respectively.

Because we generally ignore nonconservative forces in "ideal" molecular collisions, we assume that the forces are given by potential gradients,

$$\mathbf{F}_i = -\nabla_i V \tag{2}$$

where $V$ is the potential energy for the system of particles. Combining Eqs. (1) and (2), we obtain

$$\frac{d}{dt}\left(\frac{\partial T}{\partial \dot{x}_i}\right)+\frac{\partial V}{\partial x_i}=0$$

$$\frac{d}{dt}\left(\frac{\partial T}{\partial \dot{y}_i}\right)+\frac{\partial V}{\partial y_i}=0 \tag{3}$$

$$\frac{d}{dt}\left(\frac{\partial T}{\partial \dot{z}_i}\right)+\frac{\partial V}{\partial z_i}=0$$

where $T$ is the kinetic energy defined by

$$T=\tfrac{1}{2}\sum_{i=1}^{N} m_i(\dot{x}_i^2+\dot{y}_i^2+\dot{z}_i^2) \tag{4}$$

Equations (3) comprise a set of $3N$ second-order differential equations in Cartesian form that could be used to generate the desired trajectories. However, in most cases the constraints imposed upon the initial values of the coordinates and momenta by specification of the initial relative velocity of the colliding molecules and the partitioning of energy among their internal degrees of freedom make a more general formulation desirable.

Let $\{q_i, i = 1, \ldots, 3N\}$ be a set of $3N$ generalized coordinates such that

$$\mathbf{r}_i = \mathbf{r}_i(q_1, q_2, \ldots, q_{3N}, t) \tag{5}$$

In the usual case the coordinate system is fixed so that the explicit time dependence of the rhs (right-hand side) of Eq. (5) disappears. The equations of motion can now be obtained in terms of the $q_i$ by direct substitution of Eq. (5) into (3); however, it is more convenient and instructive to obtain the required equations by making use of the action functional,

$$S[q] \equiv \int_{t_1}^{t_2} L(q, \dot{q})\, dt \tag{6}$$

where $L$ is an arbitrary function of the generalized coordinates $q_i$ and their velocities $\dot{q}_i (i = 1, 2, \ldots, 3N)$ and $t_1$ and $t_2$ are the initial and final times. The *principle of least action* identifies the classical trajectory between the point $\{q_i(t_1)\}$ and the point $\{q_i(t_2)\}$ as the set of functions $\{q_i(t)\}$ for which $S$ is an extremum. To find the conditions for an extremum in $S$, we augment each $q_i(t)$ by the function $\delta q_i(t)$, writing

$$\delta q_i(t) = \varepsilon \sigma_i(t) \tag{7}$$

where $\varepsilon$ is a parameter independent of $i$. We require that the function $\sigma_i(t)$ and its derivative $\dot{\sigma}_i(t)$ be continuous between $t_1$ and $t_2$ and that $\sigma_i(t_1) = \sigma_i(t_2) = 0$. [The latter requirement on the set $\{\sigma_i\}$ assures that the augmented functions $q_i + \delta q_i$ have the specified values $q_i(t_1)$ and $q_i(t_2)$ at the end points.] The functional $S$ is now a function of the parameter $\varepsilon$ whose extremum for a given set $\{\sigma_i\}$ is determined by

$$\frac{d}{d\varepsilon} S[\{q_i + \varepsilon\sigma_i\}] = \sum_i \int_{t_1}^{t_2} \left( \sigma_i \frac{\partial}{\partial q_i} + \dot{\sigma}_i \frac{\partial}{\partial \dot{q}_i} \right) L(q + \varepsilon\sigma, \dot{q} + \varepsilon\dot{\sigma}) \, dt = 0 \qquad (8)$$

As Eq. (8) stands, with the set $\{\sigma_i\}$ specified, it is an equation for $\varepsilon$. What we seek is the set of functions $\{q_i\}$ for which $S$ is stationary for all variations $\{\delta q_i\}$. Thus we allow the set $\{\sigma_i\}$ to be arbitrary and evaluate the functional $dS/d\varepsilon$ at $\{\sigma_i\}$ by setting $\varepsilon = 0$. After integrating the $\dot{\sigma}_i \partial/\partial \dot{q}_i$ terms by parts, we obtain

$$\left. \frac{dS}{d\varepsilon} \right|_{\varepsilon=0} = \sum_i \int_{t_1}^{t_2} \sigma_i \left( \frac{dL}{\partial q_i} - \frac{d}{dt} \frac{\partial L}{\partial \dot{q}_i} \right) dt + \sum_i \left( \sigma_i \frac{\partial L}{\partial \dot{q}_i} \right) \bigg|_{t_1}^{t_2} = 0 \qquad (9)$$

Because the $\sigma_i$ vanish at $t_1$ and $t_2$ but are otherwise arbitrary, Eq. (9) requires that the trajectories $q_i(t)$ satisfy the Lagrange equations of motion*

$$\frac{d}{dt} \frac{\partial L}{\partial \dot{q}_i} - \frac{\partial L}{\partial q_i} = 0 \qquad (i = 1, \ldots, 3N) \qquad (10)$$

where $L(q, \dot{q})$ is the Lagrangian function, chosen to give the observed trajectories, subject to the requirement that $\partial L/\partial \dot{q}_i$ remain finite at the end points. The requirement that Eq. (10) reduce to Newton's law in the form of Eq. (3) when the $q_i$ are Cartesian coordinates shows that for conservative nonrelativistic systems

$$L(q, \dot{q}) = T - V \qquad (11)$$

Equation (10) with $L$ defined by Eq. (11) is a set of $3N$ second-order differential equations that is applicable to any coordinate system. Although these equations could be used to calculate the trajectories $\{q_i(t)\}$, it is generally more convenient to transform to a set of $6N$ first-order equations. To do this, we use the Hamiltonian formalism.

The Hamiltonian is the function

$$H(q, p) \equiv \sum_{i=1}^{3N} p_i \dot{q}_i - L(q, \dot{q}) \qquad (12)$$

where the momenta $\{p_i\}$ are new variables whose relation to $\{q_i\}$, $\{\dot{q}_i\}$ will be determined later. We find the equations of motion again from the requirement that $S$ be stationary. In view of Eq. (12), we now write Eq. (6) in the form

$$S[q, p] = \int_{t_1}^{t_2} \left[ \sum_{i=1}^{3N} p_i \dot{q}_i - H(q, p) \right] dt \qquad (13)$$

To find the functions $q_i(t)$ and $p_i(t)$ that make $S$ an extremum, we proceed as in

*The left-hand side of Eq. (10) is the functional derivative $\delta S/\delta q_i$ (see Reference 44).

Eq. (9), obtaining

$$\frac{d}{d\varepsilon}S[\{q_i + \varepsilon\sigma_i\}, \{p_i + \varepsilon\tau_i\}]_{\varepsilon=0} = \sum_{i=1}^{3N} \int_{t_1}^{t_2} \left[\tau_i\left(\dot{q}_i - \frac{\partial H}{\partial p_i}\right) - \sigma_i\left(\dot{p}_i + \frac{\partial H}{\partial q_i}\right)\right] dt = 0 \quad (14)$$

where $\{\sigma_i\}$ is a set of functions with continuous derivatives $\{\dot{\sigma}_i\}$ between $t_1$ and $t_2$ such that $\sigma_i(t_1) = \sigma_i(t_2) = 0$ and $\{\tau_i\}$ is a set of functions continuous between $t_1$ and $t_2$. Because Eq. (14) must hold for any set of functions $\{\sigma_i\}$ and $\{\tau_i\}$ satisfying these conditions, Hamilton's equations

$$\dot{q}_i = \frac{\partial H}{\partial p_i}$$

$$(i = 1, \ldots, 3N) \quad (15)$$

$$\dot{p}_i = -\frac{\partial H}{\partial q_i}$$

must be satisfied by the trajectories $\{q_i(t)\}$, $\{p_i(t)\}$.

The momenta $p_i$ are yet to be defined. Differentiation of Eq. (12) gives

$$dH = \sum_{i=1}^{3N} \left[\left(p_i - \frac{\partial L}{\partial \dot{q}_i}\right) d\dot{q}_i + \dot{q}_i\, dp_i - \frac{\partial L}{\partial q_i}\, dq_i\right] \quad (16)$$

Comparing Eqs. (15) and (16), we find that for trajectories

$$\dot{p}_i = \frac{\partial L}{\partial q_i} \quad (17)$$

whereupon we conclude from Eq. (10) that $p_i$ and $\partial L/\partial \dot{q}_i$ can differ at most by a constant. The choice

$$p_i \equiv \frac{\partial L}{\partial \dot{q}_i} \quad (18)$$

allows the dependence of $H$ on the $\dot{q}_i$ to vanish entirely [as we have already assumed in Eq. (14)], making $H$ a constant of the motion because

$$\frac{dH}{dt} = 0 \quad (19)$$

Furthermore, if $V$ is independent of the velocities $\dot{q}_i$, Eqs. (11), (12), and (18) show that

$$H = \sum_{i=1}^{3N} \frac{\partial T}{\partial \dot{q}_i}\dot{q}_i - T + V \quad (20)$$

But if the transformations given by Eq. (5) are independent of $t$, the kinetic energy $T$ is a homogeneous function of the $\dot{q}_i$ of second degree, for Eq. (4) transforms into

$$T = \frac{1}{2}\sum_{i=1}^{N}\sum_{j=1}^{3N}\sum_{k=1}^{3N} m_i\left(\frac{\partial \mathbf{r}_i}{\partial q_j}\frac{\partial \mathbf{r}_i}{\partial q_k}\right)\dot{q}_j\dot{q}_k \quad (21)$$

Thus, from Euler's theorem

$$\sum_{i=1}^{3N} \frac{\partial T}{\partial \dot{q}_i} \dot{q}_i = 2T \tag{22}$$

and Eq. (20) becomes

$$H = T + V \equiv E \tag{23}$$

where $E$ is the energy. The $6N$ first-order differential equations given by Eq. (15) with $H$ given by Eq. (23) are in the form generally employed in the numerical calculation of molecular trajectories.

We now require explicit procedures for finding transformations between the variables $\{q_i\}$, $\{p_i\}$ and a new set $\{Q_j\}$, $\{P_j\}$ such that the latter also satisfies equations of the form of Eq. (15). Such transformations are called contact or canonical transformations.[45] In our applications, we know the $Q_j$ as functions of the $q_i$ and wish to find the $p_i$ as functions of the $Q_j$ and $P_j$ so that the transformed Hamiltonian

$$K(Q, P) \equiv H[q(Q), p(Q, P)] \tag{24}$$

can be found. For $\{Q_j\}$ and $\{P_j\}$ to obey equations analogous to Eq. (15), the action

$$R[Q, P] \equiv \int_{t_1}^{t_2} \left[ \sum_{j=1}^{3N} P_j \dot{Q}_j - K(Q, P) \right] dt \tag{25}$$

must be stationary with respect to variations in $\{Q_j\}$ and $\{P_j\}$. Because $S[q, p]$ of Eq. (13) must also be stationary, it follows that the difference

$$S[q, p] - R[Q, P] = \int_{t_1}^{t_2} \left[ \sum_i p_i dq_i - \sum_j P_j dQ_j - (H - K) dt \right] \tag{26}$$

is stationary. In general, the integral on the rhs of Eq. (26) is path dependent. But we seek a family of sets $\{Q_j\}$, $\{P_j\}$ that renders the integral independent of path. Thus the bracketed expression is an exact differential, because the variations in the functions vanish at $t_1$ and $t_2$. We therefore write

$$dF_1 = \sum_i p_i dq_i - \sum_j P_j dQ_j - (H - K) dt \tag{27}$$

The form of the transformation we require is obtained if we define a new function $F_2$ such that

$$F_2 = \sum_j Q_j P_j + F_1 \tag{28}$$

Then from Eqs. (28) and (27) we have

$$dF_2 = \sum_j Q_j dP_j + \sum_i p_i dq_i + (K - H) dt \tag{29}$$

from which it follows that

$$Q_j = \frac{\partial F_2}{\partial P_j} \tag{30}$$

$$p_i = \frac{\partial F_2}{\partial q_i} \tag{31}$$

$$K = H + \frac{\partial F_2}{\partial t} \tag{32}$$

If we now take $F_1$ to be identically zero so that

$$F_2 = \sum_{j=1}^{3N} Q_j P_j \tag{33}$$

we obtain

$$Q_j = Q_j(q_i) \tag{34}$$

$$p_i = \sum_{j=1}^{3N} P_j \frac{\partial Q_j}{\partial q_i} \tag{35}$$

$$K(Q_j, P_j) = H[q_i(Q_j), p_i(Q_j, P_j)] \tag{36}$$

where the $Q_j$ are assumed to be functions of the $q_i$ only. Such transformations are known as point contact transformations.

The working procedure for setting up the differential equations for a classical trajectory calculation is therefore

1. Write the Hamiltonian in Cartesian coordinates.
2. Transform to the desired set of coordinates, $Q_j$, via a point contact transformation.
3. Invert Eq. (34) to obtain the Cartesian coordinates as functions of the $Q_j$.
4. Use Eq. (35) to obtain the Cartesian momenta as functions of the momenta, $P_j$, and the Cartesian coordinates.
5. Use Eq. (36) and the inverse coordinate transformation to obtain the new Hamiltonian $K$ as a function of the $Q_j$ and $P_j$.
6. Use Eq. (15) to set up the $6N$ first-order differential equations in terms of $Q_j$, $P_j$, $\dot{Q}_j$, and $\dot{P}_j$.

Numerical techniques commonly employed to effect the integration of the equations are discussed in Section 6.

## 3. Classical Scattering, Cross Sections, and Rate Coefficients

In this section the equations required to compute bimolecular reaction rate coefficients and cross sections from the results of trajectory calculations are developed.

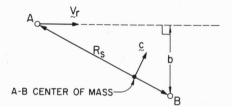

A-B CENTER OF MASS

Fig. 1. Kinematics of the collision of
two particles.

### 3.1. Two-Particle Collisions

Consider first the collision of two point masses, A and B, with masses $m_A$ and $m_B$, at relative velocity $\mathbf{V}_r$, center of mass velocity $\mathbf{c}$, and impact parameter $b$, defined to be the perpendicular distance from point B to the extension of $\mathbf{V}_r$, drawn from A. These variables are illustrated in Fig. 1, where $R_s$ is the initial A–B separation for the initial state of the collision. If $f_A(\mathbf{v}_A)$ and $f_B(\mathbf{v}_B)$ are the normalized probability density functions giving the fraction of A and B molecules moving with laboratory velocities in the range $d\mathbf{v}_A$ at $\mathbf{v}_A$ and $d\mathbf{v}_B$ at $\mathbf{v}_B$, then the fraction of (A, B) pairs that are found to be simultaneously moving with velocities in the above ranges is

$$\delta = f_A(\mathbf{v}_A)f_B(\mathbf{v}_B)\, d\mathbf{v}_A\, d\mathbf{v}_B \tag{37}$$

These pairs exhibit a relative velocity

$$\mathbf{V}_r = \mathbf{v}_A - \mathbf{v}_B$$

For (A, B) pairs moving with relative velocity magnitude $V_r$ and restricted to pass at impact parameters in the range $b$ to $b + db$, the reaction volume $\tau$ swept out per unit time is

$$\tau = V_r(2\pi b\, db) \tag{38}$$

and the total number $dN$ of such (A, B) pairs per unit volume is

$$dN = \delta N_A N_B \tau \tag{39}$$

where $N_A$ and $N_B$ are the number densities of A and B molecules, respectively.

Combining Eqs. (37), (38), and (39), we obtain $dZ_{AB}(V_r, \mathbf{v}_A, \mathbf{v}_B, b)$, the A–B collision frequency per unit volume in the ranges $db$ at $b$, $d\mathbf{v}_A$ at $\mathbf{v}_A$, and $d\mathbf{v}_B$ at $\mathbf{v}_B$:

$$dZ_{AB}(V_r, \mathbf{v}_A, \mathbf{v}_B, b) = N_A N_B f_A(\mathbf{v}_A)f_B(\mathbf{v}_B)V_r \times 2\pi b\, db\, d\mathbf{v}_A\, d\mathbf{v}_B \tag{40}$$

Because the attributes of the collision are independent of the motion of the center of mass of the A–B pair, it is convenient to transform Eq. (40) so that the independent velocity variables are $\mathbf{c}$ and $\mathbf{V}_r$. The required transformation equations are easily deduced from the Newton velocity vector diagram shown in Fig. 2; they are

$$\mathbf{v}_A = \mathbf{c} + m_B \mathbf{V}_r / M$$

$$\mathbf{v}_B = \mathbf{c} - m_A \mathbf{V}_r / M \tag{41}$$

where $M = m_A + m_B$. The Jacobian of this transformation is easily seen to be unity. The transformed equation for the A–B collision frequency is therefore

$$dZ_{AB}(\mathbf{V}_r, \mathbf{c}, b) = N_A N_B F_{AB}(\mathbf{V}_r, \mathbf{c}) V_r \times 2\pi b \, db \, d\mathbf{V}_r \, d\mathbf{c} \tag{42}$$

where $F_{AB}$ is the joint probability density defined by

$$F_{AB}(\mathbf{V}_r, \mathbf{c}) \, d\mathbf{V}_r \, d\mathbf{c} = f_A(\mathbf{v}_A) f_B(\mathbf{v}_B) \, d\mathbf{v}_A \, d\mathbf{v}_B \tag{43}$$

To proceed further we must now consider the actual experimental situation for which we wish to calculate the collision frequency. Specifically, we must either deduce or measure the form of $f_A(\mathbf{V}_r, \mathbf{c})$ and $f_B(\mathbf{V}_r, \mathbf{c})$ appropriate to the experimental system in question. For systems in translational equilibrium at temperature $T$,

$$f_A(\mathbf{v}_A) = (m_A / 2\pi kT)^{3/2} \exp(-m_A v_A^2 / 2kT)$$

$$f_B(\mathbf{v}_B) = (m_B / 2\pi kT)^{3/2} \exp(-m_B v_B^2 / 2kT) \tag{44}$$

Direct substitution of Eqs. (41) and (44) into Eq. (42) yields

$$dZ_{AB}(\mathbf{V}_r, \mathbf{c}, b) = \frac{N_A N_B (m_A m_B)^{3/2}}{(2\pi kT)^3} \exp\left(-\frac{Mc^2 + \mu_{AB} V_r^2}{2kT}\right) V_r \times 2\pi b \, db \, d\mathbf{V}_r \, d\mathbf{c} \tag{45}$$

where $\mu_{AB}$ is the A–B reduced mass,

$$\mu_{AB} = m_A m_B / (m_A + m_B) \tag{46}$$

Let us now define $P_\Lambda(V_r, b)$ to be the trajectory-computed probability of occurrence of some arbitrary chemical process $\Lambda$ upon A–B collision in the ranges $dV_r$ at $V_r$ and $db$ at $b$. The differential reaction rate for this process may then be written

$$dR_\Lambda(\mathbf{V}_r, \mathbf{c}, b) = P_\Lambda(\mathbf{V}_r, b) \, dZ_{AB}(\mathbf{V}_r, \mathbf{c}, b) \tag{47}$$

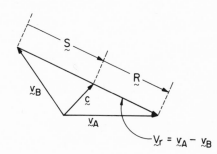

Fig. 2. Newton diagram for conversion to the center of mass system.

Because we expect $P_\Lambda$ to be independent of **c** and of the spherical polar direction angles, $\theta$ and $\phi$, of $\mathbf{V}_r$, we integrate over these variables at once:

$$dR_\Lambda(V_r, b) = \frac{N_A N_B (m_A m_B)^{3/2}}{(2\pi kT)^3} \exp\left(-\frac{\mu_{AB} V_r^2}{2kT}\right) V_r^3 \times 2\pi b\, db\, dV_r P_\Lambda(V_r, b)$$

$$\times \int_0^{2\pi} \int_0^\pi \sin\theta\, d\theta\, d\phi \int_{-\infty}^\infty \int_{-\infty}^\infty \int_{-\infty}^\infty \exp\left[-\frac{M(c_x^2 + c_y^2 + c_z^2)}{2kT}\right] dc_x\, dc_y\, dc_z \tag{48}$$

This yields

$$dR_\Lambda(V_r, b) = N_A N_B \left(\frac{2}{\pi}\right)^{1/2} \left(\frac{\mu_{AB}}{kT}\right)^{3/2} \exp\left(-\frac{\mu_{AB} V_r^2}{2kT}\right) V_r^3$$
$$\times 2\pi b P_\Lambda(V_r, b)\, db\, dV_r \tag{49}$$

If the impact parameter is treated as a continuous variable (see Section 5 for further discussion of this point), the total rate for process $\Lambda$ is the integral

$$R_\Lambda = N_A N_B \left(\frac{2}{\pi}\right)^{1/2} \left(\frac{\mu_{AB}}{kT}\right)^{3/2} \int_0^\infty \int_0^\infty \exp\left(-\frac{\mu_{AB} V_r^2}{2kT}\right) P_\Lambda(V_r, b) V_r^3 \times 2\pi b\, db\, dV_r \tag{50}$$

The cross section for process $\Lambda$ is defined to be

$$S_\Lambda(V_r) = \int_0^\infty P_\Lambda(V_r, b) \times 2\pi b\, db \tag{51}$$

so that

$$R_\Lambda = N_A N_B \left(\frac{2}{\pi}\right)^{1/2} \left(\frac{\mu_{AB}}{kT}\right)^{3/2} \int_0^\infty S_r(V_r) \exp\left(-\frac{\mu_{AB} V_r^2}{2kT}\right) V_r^3\, dV_r \tag{52}$$

## 3.2. Atom–Diatomic Molecule Collisions

Consider the atom–diatomic collision $A + BC$. The set of internal variables required to describe the BC molecule consists of two spherical polar orientation angles, $\theta$ and $\phi$; the numbers $v$ and $J$ that define the internal action* of the molecule; the BC interatomic separation, $R_{BC}$; and an angle $\eta$ specifying the BC rotation plane. Because the relative translational motion of the A–BC system is separable from the internal motions, Eq. (45) is applicable to the A–BC system. Consequently, the A–BC collision frequency in the ranges $dV_r$ at $V_r$, $db$ at $b$, $d\theta$ at $\theta$, $d\phi$ at $\phi$, $dR_{BC}$ at $R_{BC}$, and $d\eta$ at $\eta$ for internal

---

*The numbers $v$ and $J$ are the vibrational and rotational quantum numbers, respectively, when the radial action variable $(v + \frac{1}{2})h \equiv \oint p_r\, dr$ and the rotational angular momentum $J(J+1)\hbar^2 \equiv L^2$ are quantized. Otherwise $v$ and $J$ are continuous variables.

vibration–rotation states $v$ and $J$ has the form

$$dZ_{AB}(V_r, b, \theta, \phi, R_{BC}, \eta, v, J) = N_A N_B \left(\frac{2}{\pi}\right)^{1/2} \left(\frac{\mu_{A,BC}}{kT}\right)^{3/2} \exp\left(-\frac{\mu_{A,BC} V_r^2}{2kT}\right)$$

$$\times V_r^3 \, dV_r \times 2\pi b \, db \, I_{BC}(\theta, \phi, R_{BC}, \eta, v, J) \, d\theta \, d\phi \, dR_{BC} \, d\eta \, dv \, dJ \quad (53)$$

where $\mu_{A,BC}$ is the A–BC reduced mass and $I_{BC}(\theta, \phi, R_{BC}, \eta, v, J)$ is the normalized probability density function for the internal BC variables. The distributions for $\phi$ and $\eta$ are independent of the values of the other variables; that for $R_{BC}$ depends on the values of $v$ and $J$, whereas the distribution for $\theta$ depends on $R_{BC}$, $v$, and $J$. Because the latter dependence, which results from rotation–vibration coupling, is very weak when $J$ is not too large, we ignore it and write $I_{BC}$ in the factored form

$$I_{BC}(\theta, \phi, R_{BC}, \eta, v, J) = \Theta(\theta)\Phi(\phi)G(R_{BC}; v, J)H(\eta)F(v, J) \quad (54)$$

The form chosen for each of the above distribution functions depends on the experimental conditions. For example, in an idealized molecular beam experiment employing perfect internal state selection in the incident BC beam, $F(v, J)$ would be represented by the delta function product

$$F(v, J) = \delta(v - v_0)\delta(J - J_0) \quad (55)$$

where $v_0$ and $J_0$ represent the selected $v$ and $J$ states, respectively.

In the more usual case the distribution functions will be those for an isolated system in thermal Boltzmann equilibrium. In the classical case, we have

$$F(v, J) = (2J + 1) \exp(-E_{vJ}/kT)/Q_{v-r} \quad (56)$$

where $E_{vJ}$ is the vibration–rotation energy of state $(v, J)$ and $Q_{v-r}$ the classical vibrational–rotational partition function,

$$Q_{v-r} = \int_0^\infty \int_0^\infty (2J + 1)e^{-E_{vJ}/kT} \, dv \, dJ \quad (57)$$

In the quasi-classical sampling, the numbers $v$ and $J$ are restricted to integral values by inserting the factor

$$\sum_{i=0}^\infty \sum_{j=0}^\infty \delta(v - i)\delta(J - j)$$

on the rhs of Eq. (56) and into the integrand of Eq. (57).

The probability for spatial orientation of BC in the range $d\theta$ at $\theta$ and $d\phi$ at $\phi$ is proportional to the solid angle subtended; thus

$$\Theta(\theta)\Phi(\phi) \, d\theta \, d\phi = \left[\frac{1}{2}\sin\theta \, d\theta\right]\left[\frac{1}{2\pi} \, d\phi\right] \quad (58)$$

The density function for the internuclear distance $G(R_{BC}; v, J)$ depends on the BC internuclear potential. We shall therefore defer discussion of the explicit form of this function to Section 5.5.

Because the orientation of the rotation planes in an ensemble of BC molecules in field-free space will be isotropically distributed, we have

$$H(\eta) = [(1/2\pi)\, d\eta] \tag{59}$$

If $P_\Lambda(V_r, b, \theta, \phi, R_{BC}, \eta, v, J)$ is the trajectory-determined probability for the occurrence of process $\Lambda$ upon A–BC collision in the ranges $dV_r$ at $V_r$, $db$ at $b$, $d\theta$ at $\theta$, $d\phi$ at $\phi$, $dR_{BC}$ at $R_{BC}$, and $d\eta$ at $\eta$ for BC molecules in internal state $(v, J)$, the differential rate for the process is

$$dR_\Lambda(V_r, b, \theta, \phi, R_{BC}, \eta, v, J)$$

$$= N_A N_B \left(\frac{2}{\pi}\right)^{1/2} \left(\frac{\mu_{A,BC}}{kT}\right)^{3/2} \left[\frac{(2J+1)\exp(-E_{vJ}/kT)}{Q_{v-r}}\, dv\, dJ\right]\left[\frac{1}{2}\sin\theta\, d\theta\right]$$

$$\times \left[\frac{1}{2\pi}\, d\phi\right]$$

$$\times [G(R_{BC}; v, J)\, dR_{BC}]\left[\frac{1}{2\pi}\, d\eta\right]\left[\exp\left(-\frac{\mu_{A,BC}V_r^2}{2kT}\right) V_r^3\, dV_r\right]$$

$$\times [2\pi b\, db] P_\Lambda(V_r, b, \theta, \phi, R_{BC}, \eta, v, J) \tag{60}$$

and the total rate is

$$R_\Lambda = \frac{N_A N_B (2/\pi)^{1/2}(\mu_{A,BC}/kT)^{3/2}}{Q_{v-r}} \int_0^\infty dv \int_0^\infty dJ \int_{V_r=0}^\infty \int_{b=0}^\infty$$

$$\times \int_{\theta=0}^\pi \int_{\phi=0}^{2\pi} \int_{R_{BC}=\rho_-}^{\rho_+} \int_{\eta=0}^{2\pi} (2J+1)\exp\left(-\frac{E_{vJ}}{kT}\right)\left[\frac{1}{2\pi}\, d\eta\right]$$

$$\times [G(R_{BC}; v, J)\, dR_{BC}]$$

$$\times \left[\frac{1}{2\pi}\, d\phi\right]\left[\frac{1}{2}\sin\theta\, d\theta\right][2\pi b\, db]\exp\left(-\frac{\mu_{A,BC}V_r^2}{2kT}\right) V_r^3\, dV_r$$

$$\times P_\Lambda(V_r, b, \theta, \phi, R_{BC}, \eta, v, J) \tag{61}$$

where $\rho_+$ and $\rho_-$ represent the BC bond length at the outer and inner classical turning points, respectively.

The cross section $S_\Lambda(V_r, v, J)$ for process $\Lambda$ is

$$S_\Lambda(V_r, v, J) = \int_{b=0}^\infty \int_{\theta=0}^\pi \int_{\phi=0}^{2\pi} \int_{R_{BC}=\rho_-}^{\rho_+} \int_{\eta=0}^{2\pi} \left[\frac{1}{2\pi}\, d\eta\right]$$

$$\times [G(R_{BC}; v, J)\, dR_{BC}]\left[\frac{1}{2\pi}\, d\phi\right]\left[\frac{1}{2}\sin\theta\, d\theta\right]$$

$$\times [2\pi b\, db] P_\Lambda(V_r, b, \theta, \phi, R_{BC}, \eta, v, J) \tag{62}$$

If one assumes that rotational and vibrational energies are separable, then a rotationally averaged cross section may be defined:

$$S_\Lambda(V_r, v) = \int_0^\infty dJ(2J+1) \exp\left(-\frac{E_J}{kT}\right) S_\Lambda(V_r, v, J)/Q_r \qquad (63)$$

where $Q_r$ is the rotational partition function,

$$Q_r = \int_0^\infty dJ(2J+1)e^{-E_J/kT} \qquad (64)$$

and $E_J$ is evaluated at the equilibrium internuclear distance $(R_{BC})_0$:

$$E_J = J(J+1)\hbar^2/2\mu(R_{BC})_0 \qquad (65)$$

The quasi-classical selection of $J$ is effected by insertion of the factor

$$\sum_{j=0}^\infty \delta(J-j)$$

into Eqs. (63) and (64). Combining Eqs. (61)–(63), we obtain

$$R_\Lambda = \frac{1}{Q_v} \int_0^\infty dv \exp\left(-\frac{E_v}{kT}\right) \left[ N_A N_B \left(\frac{2}{\pi}\right)^{1/2} \left(\frac{\mu_{A,BC}}{kT}\right)^{3/2} \right.$$
$$\left. \times \int_0^\infty S_\Lambda(V_r, v) \exp\left(-\frac{\mu_{A,BC}V_r^2}{2kT}\right) V_r^3\, dV_r \right] \qquad (66)$$

where $Q_v$ is the vibrational partition function,

$$Q_v = \int_0^\infty dv e^{-E_v/kT} \qquad (67)$$

It should be noted that the quantity within the brackets in Eq. (66) can be identified with a rate corresponding to reaction of BC molecules in vibrational state $v$. Thus,

$$R_\Lambda = \frac{1}{Q_v} \int_0^\infty dv \exp\left(-\frac{E_v}{kT}\right) R_\Lambda(v) \qquad (68)$$

and the overall rate becomes a Boltzmann average of rates characteristic of specified BC vibrational states. For quasi-classical selection, the factor

$$\sum_{i=0}^\infty \delta(v-i)$$

is inserted into the integrands of Eqs. (66)–(68).


### 3.3. Diatomic–Diatomic Molecule Collisions

We now extend the formalism to diatomic–diatomic four-body reactions. In the collision of AB with CD we require 12 internal variables consisting of the set of four spherical polar orientation angles, $\theta_1, \theta_2, \phi_1, \phi_2$; the AB and CD internuclear separations, $R_1$ and $R_2$; two rotation planes designated by $\eta_1$ and $\eta_2$; and the two sets of vibration–rotation state action variables, $(v_1, J_1)$ and $(v_2, J_2)$.

Because there is initially no interaction between AB and CD, the normalized probability density functions for the internal variables will be identical to those obtained in the atom–diatomic molecule case. The reaction rate for process $\Lambda$ may therefore be written

$$
R_\Lambda = \frac{N_{AB} N_{CD}}{(Q_{v-r})_1 (Q_{v-r})_2} \left(\frac{2}{\pi}\right)^{1/2} \left(\frac{\mu_{AB-CD}}{kT}\right)^{3/2} \int_{V_r=0}^{\infty} \int_{b=0}^{\infty} \int_{\theta_1=0}^{\pi} \int_{\phi_1=0}^{2\pi} \int_{R_1=\rho_{1-}}^{\rho_{1+}}
$$

$$
\times \int_{\eta_1=0}^{2\pi} \int_0^{\infty} dv_1 \int_0^{\infty} dJ_1 \int_{\theta_2=0}^{\pi} \int_{\phi_2=0}^{2\pi} \int_{R_2=\rho_{2-}}^{\rho_{2+}} \int_{\eta_2=0}^{2\pi} \int_0^{\infty} dv_2
$$

$$
\times \int_0^{\infty} dJ_2 \left[\frac{1}{2} \sin\theta_1 \, d\theta_1\right]
$$

$$
\times \left[\frac{1}{2\pi} d\phi_1\right] [G_1(R_1; v_1, J_1) \, dR_1] \left[\frac{1}{2\pi} d\eta_1\right] (2J_1 + 1) \exp\left(-\frac{E_{v_1 J_1}}{kT}\right)
$$

$$
\times \left[\frac{1}{2} \sin\theta_2 \, d\theta_2\right]
$$

$$
\times \left[\frac{1}{2\pi} d\phi_2\right] [G_2(R_2; v_2, J_2) \, dR_2] \left[\frac{1}{2\pi} d\eta_2\right]
$$

$$
\times (2J_2 + 2) \exp\left(-\frac{E_{v_2 J_2}}{kT}\right) \exp\left(-\frac{\mu_{AB-CD} V_r^2}{2kT}\right) V_r^3
$$

$$
\times 2\pi b P_\Lambda(V_r, b, \theta_1, \theta_2, \phi_1, \phi_2, R_1, R_2, \eta_1, \eta_2, v_1, v_2, J_1, J_2) \, db \, dV_r \tag{69}
$$

provided both the relative translational motion and the internal vibration–rotation of AB and CD exhibit Boltzmann equilibrium at temperature $T$. The appropriate insertions of delta functions are made in the quasi-classical case.

## 3.4. Atom–Triatomic Molecule Collisions

The collision of an atom A with the triatomic molecule BCD is a considerably more difficult problem, as are all trajectory computations involving polyatomic species.

The internal vibration–rotation states of BCD require the specification of a set of five action variables: $v_1, v_2, v_3, J$, and $K$. The first three correspond to the level of vibrational excitation in each of the three normal vibrational modes, whereas the last two specify the rotational state. The ranges of $J$ and $K$ are $0 \le J < \infty$, $-J \le K \le J$. If BCD is either a prolate or oblate symmetric top, the $\pm K$ states for a given $J$ are degenerate. In the case of an asymmetric top this degeneracy is lifted.

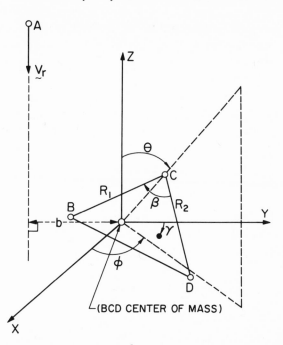

Fig. 3. Some of the parameters defining the collision of an atom and a triatomic molecule.

The spatial orientation of BCD is determined by the specification of $R_1$, $R_2$, $\beta$, $\theta$, $\phi$, and $\gamma$. The first three variables yield the relative BCD positions, wheras $\theta$, $\phi$, and $\gamma$ give the orientation of BCD with respect to atom A and the initial relative velocity vector $V_r$. Figure 3 illustrates each of these spatial variables.

If the distribution of internal energy states is Boltzmann, and if we further assume that the normal mode vibrations are independent and not coupled with the rotational levels, the overall vibrationally and rotationally averaged cross section for process $\Lambda$ takes the form

$$S_\Lambda(V_r) = \int_{b=0}^{\infty} \int_{\theta=0}^{\pi} \int_{\phi=0}^{2\pi} \int_{R_1 R_2 \beta} \int_{\eta=0}^{2\pi} \int_{\gamma=0}^{2\pi} \int_0^\infty dv_1 \int_0^\infty dv_2 \int_0^\infty dv_3$$

$$\times \int_0^\infty dJ \int_{-J}^{J} dK \left[\frac{1}{2} \sin\theta\, d\theta\right]\left[\frac{1}{2\pi} d\phi\right]$$

$$\times [G(R_1 R_2 \beta; v_1, v_2, v_3, J, K)\, dR_1\, dR_2\, d\beta]\left[\frac{1}{2\pi} d\gamma\right]\left[\frac{1}{2\pi} d\eta\right]$$

$$\times \left[\exp\left(-\frac{E_{v_1}}{kT}\right)\right] Q_1^{-1}\left[\exp\left(-\frac{E_{v_2}}{kT}\right)\right] Q_2^{-1}$$

$$\times \left[\exp\left(-\frac{E_{v_3}}{kT}\right)\right] Q_3^{-1}\left[\exp\left(-\frac{E_{JK}}{kT}\right)\right] Q_r^{-1} g_{JK}$$

$$\times P_\Lambda(V_r, b, \theta, \phi, R_1, R_2, \beta, \eta, \gamma, v_1, v_2, v_3, J, K) \times 2\pi b\, db \qquad (70)$$

where $Q_1$, $Q_2$, and $Q_3$ are the vibrational partition functions for each normal mode, $Q_r$ is the rotational partition function, $\eta$ is the angle specifying the initial plane of rotation, and $g_{JK}$ is the rotational degeneracy.

In the above equation, the function $G(R_1, R_2, \beta; v_1, v_2, v_3, J, K)$ represents the normalized probability density function for the internal variables, $R_1$, $R_2$, and $\beta$. Because BCD exhibits simultaneous vibration in three normal modes, each with three distinct frequencies (except for $D_{3h}$ molecules) and phase angles, it is simpler to select initial states of BCD numerically than to represent $G(R_1, R_2, \beta; v_1, v_2, v_3, J, K)$ in analytical form. Discussion of this point is deferred until Section 5.3.

Systems of even greater complexity have already received attention. The reader is referred to References 36 and 37 for more detailed accounts of such systems.

## 4. Statistical Averaging

Because the reaction probability $P_\Lambda$ cannot, in general, be obtained in closed analytic form, the multidimensional integrals for the rate or cross section that have been developed in Section 3 must be evaluated numerically. In principle, the numerical integration could be accomplished by any number of procedures, e.g., Simpson's rule, Gaussian quadrature, etc. However, from a practical point of view such procedures are not feasible, because for a $k$-dimensional integral, the number of points at which the integrand must be evaluated is on the order of $N^k$, where $N$ is usually on the order of 10. Each evaluation requires the computation of a complete trajectory. Thus, for such methods the computer time requirement becomes prohibitive.

To circumvent these difficulties the multidimensional integrals are usually evaluated by a Monte Carlo procedure.[46] This method converges at a rate that is approximately independent of the dimensionality of the integral.

Consider the following multidimensional integral:

$$I = \int_{x_k} \cdots \int_{x_2} \int_{x_1} F(\mathbf{x}) w(\mathbf{x})\, dx \tag{71}$$

where $F(\mathbf{x})$ is a multivariable function of the $k$-tuple $\mathbf{x} = (x_1, x_2, \ldots, x_k)$ and $w(\mathbf{x})$ is a weight function. We first perform a transformation of variables that maps $\mathbf{x} \to \boldsymbol{\xi} \equiv (\xi_1, \xi_2, \ldots, \xi_k)$ such that $0 \leq \xi_i \leq 1$ for all $i$. The transformed integral has the form

$$I = \int_0^1 \cdots \int_0^1 F(\boldsymbol{\xi}) w(\boldsymbol{\xi}) |J|\, d\boldsymbol{\xi} = \int_0^1 \cdots \int_0^1 F(\boldsymbol{\xi})\, d\boldsymbol{\xi} \tag{72}$$

where $J$ is the Jacobian of the transformation and is equal to $[w(\boldsymbol{\xi})]^{-1}$.

The Monte Carlo approximant, $I'$, is

$$I \simeq I' = \frac{1}{N} \sum_{i=1}^{N} F(\boldsymbol{\eta}_i) \tag{73}$$

where $\boldsymbol{\eta}_i$ is a $k$-tuple whose elements are a particular set $(\xi_1^{(i)}, \xi_2^{(i)}, \ldots, \xi_k^{(i)})$ with each $\xi_j^{(i)}$ chosen randomly from a uniform distribution.

The Monte Carlo error can be estimated from the variance $\sigma^2$ of $F(\boldsymbol{\eta})$ from the calculated mean $I'$:

$$\sigma^2 = \sum_{i=1}^{N} \frac{[F(\boldsymbol{\eta}_i) - I']^2}{N(N-1)} \tag{74}$$

For large $N$ this yields

$$\sigma^2 = \frac{1}{N} \left\{ \frac{1}{N} \sum_{i=1}^{N} [F^2(\boldsymbol{\eta}_i)] - \left[ \frac{1}{N} \sum_{i=1}^{N} F(\boldsymbol{\eta}_i) \right]^2 \right\}$$

$$= [\langle F^2(\boldsymbol{\eta}) \rangle - \langle F(\boldsymbol{\eta}) \rangle^2]/N \tag{75}$$

There is about a 68% probability that the actual error $\varepsilon$, where

$$\varepsilon = |I - I'| \tag{76}$$

is such that $\varepsilon \leq \sigma$.

Equation (75) shows that $\sigma$ is independent of $k$ and decreases as $N^{-1/2}$. This is a rather slow rate of convergence and means that $N$ must be increased fourfold in order to double the accuracy of a Monte Carlo integration. Consequently, the statistical error in many reported trajectory studies is on the order of 10%.

Suzukawa et al.[47] have reported a detailed study of the use of a nonrandom, Diophantine integration procedure for the evaluation of rate coefficients and cross sections. This procedure also converges at a rate that is essentially independent of the dimensionality of the integral and may possess a number of advantages over the more standard Monte Carlo method.

The Diophantine approximant is

$$I \simeq I'' = \frac{1}{N} \sum_{i=0}^{N-1} F(\boldsymbol{\eta}_i) \tag{77}$$

where

$$\boldsymbol{\eta}_i \equiv i \cdot \boldsymbol{\alpha} \quad (\text{mod } 1) \tag{78}$$

and $\boldsymbol{\alpha}$ is an arbitrary $k$-tuple whose elements $\alpha_j$ are linearly independent with respect to the rational numbers $r_j$, that is,

$$\sum_{j=1}^{k} r_j \cdot \alpha_j \neq 0 \quad \text{unless } \boldsymbol{\alpha} = 0 \tag{79}$$

Thus $\alpha$ in general consists of a set of irrational numbers. Such a procedure is termed "open" Diophantine integration because the set of integration points $\eta_i$ never repeats. If $\alpha$ is constructed from a set of rational numbers that are linearly independent for integer $r_j$ in Eq. (79), the procedure is termed "closed" Diophantine integration because the set of integration points will then repeat when $N$ exceeds the common denominator of the $\alpha_j$. The form $g_i/N$ is therefore taken for $\alpha_i$, with $g_i$ a positive integer less than $N$.

In principle $\alpha$ should be some optimal set that gives the best approximant to the integral of some "worst possible function" as defined by the behavior of its Fourier coefficients. In practice, however, such a set is usually not available. Fortunately, Haselgrove[48] has shown that for most $\alpha$ the error is not substantially worse than that which would occur using an optimal set. Tables of $\alpha$ vectors for closed Diophantine integration with $2 \leq k \leq 12$ have been reported by Conroy[49] and by Cheng et al.[50] Mickish employed a set made up from the square roots of prime numbers selected so that the first integration point, $\eta_1$, had its components approximately evenly distributed on the interval 0 to 1.[51]

The analysis of the error associated with Eqs. (77) and (78) is more involved than for the Monte Carlo procedure because the integration points are not random variables. It can be shown that if the Fourier coefficients of the $k$-tuple expansion of $F(\eta)$,

$$F(\eta) = \sum_{n_1} \cdots \sum_{n_k} a(n_1 \cdots n_k) e^{2\pi i n \cdot \eta} \tag{80}$$

satisfy the inequality

$$|a(n_1 \cdots n_k)| \leq M_t |n_1 n_2 \cdots n_k|^{-t} \tag{81}$$

for $t > 1$ for some $M_t$ with zero factors removed on the right, then the Diophantine error approaches zero at the same rate as $1/N$ does. In many cases it can also be shown that an upper limit to the error $\varepsilon$ is of the order

$$\varepsilon \leq (\ln N)^k / N$$

Clearly, the Diophantine procedure is superior to a Monte Carlo integration in that it offers the possibility of more rapid convergence, with the error decreasing as $N^{-1}$ compared with $N^{-1/2}$ in the Monte Carlo case. The results reported by Suzukawa et al.[47] for the (H, H$_2$) reaction system support this observation.

Unfortunately, Diophantine integration contains some disadvantages when compared to the Monte Carlo method. In addition to the difficulties associated with obtaining a reasonably good lattice vector $\alpha$ along with a reliable error estimate, further problems exist if one employs a "closed" procedure. In such a case it is necessary to determine in advance the total number of integration points required. Should this number $N_1$ be judged to be insufficient after integration, a new lattice vector for $N_2$ points must be selected and all $N_2$ trajectories computed. In the case of a Monte Carlo integration, only $N_2 - N_1$ additional points need to be computed. Thus, the effort required to increase the accuracy of a closed Diophantine integration is greater than that for a Monte Carlo procedure. It would appear that the use of an open integration would remove this difficulty, but certain advantages of methodology of optimizing the set would be lost.

## 5. Selection of Initial States

The Monte Carlo or Diophantine integration procedures described in Section 4 require the evaluation of an integrand $F(\boldsymbol{\eta})$ at a series of integration points denoted by the $k$-dimensional lattice vector $\boldsymbol{\eta}_i$. In either case it is necessary to convert the elements of this lattice vector into a specification of the initial states of the trajectory so that Eqs. (15) can be integrated to yield the corresponding required reaction probability by the Monte Carlo or Diophantine procedure. Throughout this section we shall assume that the lattice vector elements $\xi_1, \xi_2, \xi_3, \ldots, \xi_k$ (where $0 \le \xi_i \le 1$) of $\boldsymbol{\eta}_i$ have been chosen randomly from a uniform distribution (Monte Carlo) or have been generated by Eq. (78) (Diophantine).

We shall now describe some commonly used procedures for converting the $\xi_i$ into an appropriate set of initial values of the coordinates and momenta for a trajectory.

### 5.1. Relative Translation

As pointed out in Section 3, the form of the integral over the relative speed $V_r$ depends on the experiment in question. We shall consider three common cases: thermal bulk systems, velocity-selected molecular beams, and hot atoms or a velocity-selected molecular beam in a thermal medium.

For systems in Maxwell–Boltzmann equilibrium, the calculation of the reaction rate requires that the cross section be integrated over the relative speed. This integral has the form [see, for example, Eq. (52)]

$$I = \int_0^\infty S_\Lambda(V_r, v) \left[ \exp\left(-\frac{\mu V_r^2}{2kT}\right) \right] V_r^3 \, dV_r \tag{82}$$

provided the impact parameter is treated as a continuous variable. For systems with potential energy barriers to reaction, the classical reaction cross section is zero below a certain minimum relative speed $V_M$. We can therefore rewrite Eq. (82) as

$$I = \int_{V_M}^\infty S_\Lambda(V_r, v) \left[ \exp\left(-\frac{\mu V_r^2}{2kT}\right) \right] V_r^3 \, dV_r \tag{83}$$

Let $\xi(V_r)$ be the cumulative distribution function

$$\xi(V_r) \equiv K(V_M)^{-1} \int_{V_M}^{V_r} \left[ \exp\left( -\frac{\mu x^2}{2kT} \right) \right] x^3 \, dx \tag{84}$$

where the normalizing constant $K(V_M)$ is

$$K(V_M) = \frac{kT}{\mu} \left[ \exp\left( -\frac{\mu V_M^2}{2kT} \right) \right] \left( V_M^2 + \frac{2kT}{\mu} \right) \tag{85}$$

Thus $\xi(V_r)$ is a monotonic function with limits

$$\xi(V_M) = 0, \qquad \xi(\infty) = 1 \tag{86}$$

Transforming to $\xi$ as the integration variable, from Eqs. (83) and (84), we obtain

$$I = K(V_M) \int_0^1 S[V_r(\xi), v] \, d\xi \tag{87}$$

where $V_r(\xi)$ is the solution to Eq. (84) taken as a transcendental equation in the unknown $V_r$.

The procedure for selection of the initial relative speed $V_r$ in the case of thermal equilibrium is therefore:

1. Determine $V_M$ from a consideration of the potential energy barrier and the range of internal energies to be sampled.
2. Choose the number $\xi_i$ in the interval $(0, 1)$ by one of the methods of Section 4.
3. Calculate $V_r$ as the solution to the equation

$$\left[ \exp\left( -\frac{\mu V_r^2}{2kT} \right) \right] \left( V_r^2 + \frac{2kT}{\mu} \right) = \xi_i \left[ \exp\left( -\frac{\mu V_M^2}{2kT} \right) \right] \left( V_M^2 + \frac{2kT}{\mu} \right) \tag{88}$$

The numerical solution to Eq. (88) may be effected by a number of standard techniques, such as Newton–Raphson.

For reactions occurring in a crossed molecular beam experiment in which both beams are velocity selected, one need not average the rate over the relative translational speed. In such a case the normalized probability density functions in Eq. (40) are the delta functions,

$$f_A(\mathbf{v}_A) = \delta(\mathbf{v}_A - \mathbf{v}_A^0)$$
$$f_B(\mathbf{v}_B) = \delta(\mathbf{v}_B - \mathbf{v}_B^0) \tag{89}$$

where $\mathbf{v}_A^0$ and $\mathbf{v}_B^0$ are the selected velocities of the A and B beams, respectively. For the differential rate Eqs. (42) and (47) therefore give

$$dR_\Lambda(b) = N_A N_B V_r \times 2\pi b \, db P_\Lambda(V_r, b) \tag{90}$$

where $V_r$ is given by

$$V_r = (v_A^{02} + v_B^{02} - 2v_A^0 v_B^0 \cos \gamma)^{1/2} \qquad (91)$$

with $\gamma$ being the beam crossing angle. Thus the integral over $V_r$ disappears as we previously indicated, and all trajectories are run with $V_r$ as calculated in Eq. (91).

For the case in which a velocity-selected molecular beam interacts with a thermal medium, the normalized probability density functions in Eq. (40) are

$$f_A(\mathbf{v}_A) = \delta(\mathbf{v}_A - \mathbf{v}_A^0)$$
$$f_B(\mathbf{v}_B) = (m_B/2\pi kT)^{3/2} \exp(-m_B v_B^2/2kT) \qquad (92)$$

where we choose $v_A^0$ to be directed along the $z$ axis. Equations (92) also can be used to describe the situation for hot atoms, A, reacting with a thermal medium, B, if the energy imparted to A by the hot atom production process is large relative to the average energy of any thermalized A atoms in the system and the barrier to the process $\Lambda$ is large compared to $kT$.

Substitution of Eq. (92) into Eq. (40) followed by transformation to relative velocity yields

$$dZ_{AB}(\mathbf{V}_r, b) = N_A N_B \left(\frac{m_B}{2\pi kT}\right)^{3/2}$$

$$\times \exp\left[-\frac{m_B}{2kT}(v_A^{02} + V_r^2 - 2\mathbf{v}_A^0 \cdot \mathbf{V}_r)\right] V_r \times 2\pi b \, db \, d\mathbf{V}_r \qquad (93)$$

Writing $\mathbf{V}_r$ in terms of its magnitude and spherical polar direction angles and integrating over $\theta$ and $\phi$, we obtain

$$dZ_{AB}(V_r, b) = N_A N_B \left(\frac{m_B}{2\pi kT}\right)^{1/2} \left(\frac{V_r^2}{v_A^0}\right)$$

$$\times \left\{\exp\left[-\frac{m_B}{2kT}(v_A^0 - V_r)^2\right] - \exp\left[-\frac{m_B}{2kT}(v_A^0 + V_r)^2\right]\right\} dV_r$$
$$\times 2\pi b \, db \qquad (94)$$

so that the total rate expression for process $\Lambda$ analogous to Eq. (52) is

$$R_\Lambda = N_A N_B \left(\frac{m_B}{2\pi kT}\right)^{1/2} (v_A^0)^{-1} \int_0^\infty S_\Lambda(V_r) V_r^2$$

$$\times \left\{\exp\left[-\frac{m_B}{2kT}(v_A^0 - V_r)^2\right] - \exp\left[-\frac{m_B}{2kT}(v_A^0 + V_r)^2\right]\right\} dV_r \qquad (95)$$

If the reaction cross section is known to be zero below $V_r = V_1$, the lower limit in Eq. (95) may be replaced by $V_1$; furthermore, it is also likely that there will be an effective upper limit $V_2$. The cumulative distribution function in this case is

$$\xi(V_r) = K(V_1, V_2)^{-1} \int_{V_1}^{V_r} x^2 \left\{ \exp\left[ -\frac{m_B}{2kT}(v_A^0 - x)^2 \right] \right.$$
$$\left. - \exp\left[ -\frac{m_B}{2kT}(v_A^0 + x)^2 \right] \right\} dx \tag{96}$$

where $K(V_1, V_2)^{-1}$ is chosen so that $\xi(V_2) = 1$. Evaluation of the integral in Eq. (96) gives

$$\xi(V_r) = K(V_1, V_2)^{-1} \left[ \frac{\pi^{1/2}}{2a^3} (a^2 v_A^{02} + 1) \{ \text{erf}[a(V_r - v_A^0)] - \text{erf}[a(V_r + v_A^0)] \right.$$

$$- \text{erf}[a(V_1 - v_A^0)] + \text{erf}[a(V_1 + v_A^0)] \} - \left( \frac{V_r + v_A^0}{2a^2} \right) \exp[-a^2(V_r - v_A^0)^2$$

$$+ \left( \frac{V_r - v_A^0}{2a^2} \right) \exp[-a^2(V_r + v_A^0)^2] + \left( \frac{V_1 + v_A^0}{2a^2} \right) \exp[-a^2(V_1 - v_A^0)^2]$$

$$- \left( \frac{V_1 - v_A^0}{2a^2} \right) \exp[-a^2(V_1 + v_A^0)^2] \right] \tag{97}$$

where $a \equiv (m_B/2kT)^{1/2}$ and

$$\text{erf}(x) \equiv \frac{2}{\pi^{1/2}} \int_0^x e^{-t^2} \, dt \tag{98}$$

Substitution of Eq. (97) into the integral of Eq. (95) yields an integral of the form

$$I = K(V_1, V_2) \int_0^1 S_\Lambda[\xi(V_r)] \, d\xi \tag{99}$$

The procedure for the integration of Eq. (95) is therefore identical to that for the case of thermal equilibrium for both A and B, except that the initial value of $V_r$ is taken to be the solution to Eq. (97) with $\xi = \xi_i$.

In the event that $v_A^0$ is very large relative to the average thermal velocity of B, the relative speed $V_r$ approaches $v_A^0$ in almost all collisions, so that Eq. (94) reduces to

$$dZ_{AB}(b) \simeq N_A N_B (m_B/2\pi kT)^{1/2} v_A^0 \times 2\pi b \, db \tag{100}$$

and the differential rate is of the form of Eq. (90).

## 5.2. Impact Parameter

If it is assumed that the impact parameter behaves as a classical continuous variable, the cross section is given by the integral

$$S_\Lambda(V_r, \{q_i\}) = 2\pi \int_0^\infty b P_\Lambda(V_r, b, \{q_i\}) \, db \tag{101}$$

where $\{q_i\}$ represents the set of internal variables appropriate for the system under consideration. Because the reaction probability will always approach zero as the impact parameter approaches some sufficiently large value $b_M$, it is possible to write Eq. (101) in the form

$$S_\Lambda = 2\pi \int_0^{b_M} bP(V_r, b, \{q_i\}) \, db \qquad (102)$$

In practice, the value of $b_M$ must be determined empirically by a series of preliminary trajectory computations.

A commonly employed transformation to the integration variable $\xi$ is

$$b = b_M \xi_i^{1/2} \qquad (103)$$

so that Eq. (102) becomes

$$S_\Lambda = \pi b_M^2 \int_0^1 P_\Lambda(V_r, \xi, \{q_i\}) \, d\xi \qquad (104)$$

In many scattering problems, however, Eq. (103) turns out to be a relatively poor choice in that it heavily weights the larger impact parameters. Because reactive scattering tends to occur at small values of the impact parameter, use of Eq. (103) results in low statistical accuracy. A better transformation is the linear one

$$b = b_M \xi_i \qquad (105)$$

which gives

$$S_\Lambda = 2\pi b_M^2 \int_0^1 \xi P_\Lambda(V_r, \xi, \{q_i\}) \, d\xi \qquad (106)$$

Rather than select $b$ from the continuum of possible values between 0 and $b_M$, it is sometimes desirable to treat the impact parameter as a quasi-classical variable by quantizing the initial orbital angular momentum at a given relative velocity. Such a procedure often results in a more rapid convergence of the integrals. Accordingly, for A–B collisions we write

$$b = [l(l+1)]^{1/2} \hbar / \mu_{AB} V_r \qquad (107)$$

This transformation converts Eq. (102) into the form

$$S_\Lambda = \frac{\pi \hbar^2}{\mu_{AB}^2 V_r^2} \int_0^{l_M} (2l+1) P_\Lambda(V_r, l, \{q_i\}) \, dl \qquad (108)$$

where $l_M$ is the nearest integral solution to

$$b_M = [l_M(l_M+1)]^{1/2} \hbar / \mu_{AB} V_r \qquad (109)$$

The initial orbital angular momentum quantum numbers are selected for each integration point as the nearest integer to $l = l_M \xi_i$ so that the integral over the impact parameter becomes

$$S_\Lambda = \frac{\pi \hbar^2 l_M}{\mu_{AB}^2 V_r^2} \int_0^1 (2 l_M \xi + 1) P_\Lambda(V_r, \xi, \{q_i\}) \, d\xi \tag{110}$$

It is important to note that a quasi-classical selection of impact parameters alters the form of the integrals over the relative speed, because a factor of $V_r^2$ appears in the denominator of Eq. (110). In this case Eq. (83) becomes

$$I = \frac{\pi \hbar^2}{\mu_{AB}^2} \int_{V_M}^\infty dV_r V_r \exp\left(-\frac{\mu V_r^2}{2kT}\right) l_M \int_0^1 d\xi (2 l_M \xi + 1) P_\Lambda(V_r, \xi, \{q_i\}) \tag{111}$$

and the required transformation for Monte Carlo or Diophantine integration analogous to Eq. (84) is

$$\xi(V_r) = K(V_M)^{-1} \frac{kT}{\mu}\left[\exp\left(-\frac{\mu V_M^2}{2kT}\right) - \exp\left(-\frac{\mu V_r^2}{2kT}\right)\right] \tag{112}$$

with

$$K(V_M) = \frac{kT}{\mu} \exp\left(-\frac{\mu V_M^2}{2kT}\right) \tag{113}$$

The initial value of $V_r$ is therefore given by the simple closed formula

$$V_r = (V_M^2 - 2kT \ln \xi_i/\mu)^{1/2} \tag{114}$$

because the distributions of $1 - \xi$ and of $\xi$ are identical. Corresponding formulas for the case of velocity-selected molecular beams and for the interaction of the velocity-selected beam with a thermal medium can be derived.

## 5.3. Internal States

### 5.3.1. Rotation

If separability of rotational–vibrational motion is assumed, the Monte Carlo or Diophantine sum over the rotational states for a linear molecule has the form given by Eq. (63),

$$S_\Lambda(V_r, v) = Q_r^{-1} \sum_{J=0}^\infty (2J+1) \exp\left(-\frac{E_J}{kT}\right) S_\Lambda(V_r, v, J) \tag{115}$$

The simplest assumption is that the rotational states are given by the rigid rotor quantization

$$E_J = J(J+1)\hbar^2/2I \tag{116}$$

with the molecular moment of inertia $I$ taken as constant.

If the rotational levels are closely spaced, the sum in Eq. (115) may be replaced by an integral with $J$ regarded as a continuous variable:

$$S_\Lambda(V_r, v) = Q_r^{-1} \int_0^\infty (2J+1) \exp\left[-\frac{J(J+1)\hbar^2}{2IkT}\right] S_\Lambda(V_r, v, J) \, dJ \quad (117)$$

where

$$Q_r = \int_0^\infty (2J+1) \exp\left[-\frac{J(J+1)\hbar^2}{2IkT}\right] dJ = \frac{2IkT}{\hbar^2} \quad (118)$$

The cumulative distribution function $\xi(J)$ is

$$\xi(J) = Q_r^{-1} \int_0^J (2x+1) \exp\left[-\frac{x(x+1)\hbar^2}{2IkT}\right] dx$$

$$= 1 - \exp\left[-\frac{J(J+1)\hbar^2}{2IkT}\right] \quad (119)$$

Substition of Eqs. (118) and (119) into Eq. (117) yields

$$S_\Lambda(V_r, v) = \int_0^1 S_\Lambda[V_r, v, J(\xi)] \, d\xi \quad (120)$$

Thus the initial value of $J$ is the nearest integer to

$$J = \tfrac{1}{2}\{-1 + [1 - 8IkT \ln(1-\xi_i)/\hbar^2]^{1/2}\} \quad (121)$$

If we are dealing with a molecule such as $H_2$ whose rotational levels are too widely spaced to be regarded as continuous, Eqs. (118) and (119) are replaced by the summations

$$Q_r = \sum_{J=0}^\infty (2J+1) \exp\left[-\frac{J(J+1)\hbar^2}{2IkT}\right] \quad (122)$$

$$\xi(J) = Q_r^{-1} \sum_{J'=0}^J (2J'+1) \exp\left[-\frac{J'(J'+1)\hbar^2}{2IkT}\right] \quad (123)$$

The initial $J$ value corresponds to the nearest integral solution to Eq. (123), with $\xi = \xi_i$.

Equation (115) may be more accurately evaluated by employing the quantization obtained for a rotating Morse oscillator. If the classical Hamiltonian for the Morse oscillator is truncated at the quadratic potential terms, the vibration–rotation energy levels are given by

$$E_{vJ} = BJ(J+1) - D[J(J+1)]^2 + \hbar\omega_0(v+\tfrac{1}{2}) - \frac{\hbar^2\omega_0^2}{4D_e}(v+\tfrac{1}{2})^2 - A(v+\tfrac{1}{2})J(J+1)$$

$$\begin{aligned} B &= \hbar^2/2\mu r_e^2 \\ D &= \hbar^4/4\mu^2\alpha^2 r_e^6 D_e \\ A &= 3[1-(\alpha r_e)^{-1}]\hbar^3\omega_0^3/4\mu\alpha r_e^3 D_e \end{aligned} \quad (124)$$

The parameters $D_e$, $\alpha$, and $r_e$ appearing in Eq. (124) are the usual Morse parameters, i.e.,

$$V_{\text{Morse}} = D_e\{1 - \exp[-\alpha(r - r_e)]\}^2 \tag{125}$$

and $\mu$ and $\omega_0$ are the reduced mass and classical frequency, respectively.

For the uncoupled case we ignore the coupling term in Eq. (124). This yields the "pure rotational" energy

$$E_J = BJ(J+1) - D[J(J+1)]^2 \tag{126}$$

If the temperature and molecular moment of inertia are such that Eq. (122) converges rapidly, i.e., $J_{\text{max}} \lesssim 50$, then the quadratic approximation used in the derivation of Eq. (124) will be acceptably accurate, and Eq. (126) may be substituted for $E_J$ in equations analogous to (122) and (123) to obtain a more accurate sum over the internal rotational states.

The integration over the internal states of a nonlinear polyatomic molecule is more complicated. The difficulty arises, in part, because the rotational energy levels of molecules that have no threefold or higher axis of symmetry (asymmetric top molecules) cannot be represented by an explicit analytic formula. For rigid symmetric top molecules containing a threefold or higher axis, the rotational levels can be represented by

$$E_{J,K} = BJ(J+1) + (A - B)K^2 \tag{127}$$

with

$$B = \hbar^2/2I_B \quad \text{and} \quad A = \hbar^2/2I_A$$

where $I_B$ represents the two equal moments of inertia and $I_A$ the third moment of inertia. The specification of the rotational energy level now requires two quantum numbers, $J$ and $K$, such that $-J \le K \le J$, because $K$ gives the component of angular momentum in the direction of the threefold axis. The Monte Carlo or Diophantine sum over the rotational states therefore becomes a double sum of the form

$$S_\Lambda(V_r, v) = Q_r^{-1} \sum_{J=0}^{\infty} \sum_{K=-J}^{+J} (2J+1) \exp\left(-\frac{E_{JK}}{kT}\right) S_\Lambda(V_r, v, J, K) \tag{128}$$

where

$$Q_r = \sum_{J=0}^{\infty} \sum_{K=-J}^{+J} (2J+1) \exp\left(-\frac{E_{JK}}{kT}\right) \tag{129}$$

Obviously states with equal $J$ and $|K|$ will be degenerate.

Because $A$ and $B$ may be of comparable size, the ordering of energy levels from Eq. (127) is not clear. In a given case, however, it is always possible to order the energy levels. These ordered levels, each associated with a particular

$J$ and $K$, can now be designated with a single quantum number, $M$, and Eqs. (128) and (129) can be written in the form

$$S_\Lambda(V_r, v) = Q_M^{-1} \sum_{M=0}^{\infty} g_M \exp\left(-\frac{E_M}{kT}\right) S_\Lambda(V_r, v, M) \qquad (130)$$

where

$$Q_M = \sum_{M=0}^{\infty} g_M \exp\left(-\frac{E_M}{kT}\right) \qquad (131)$$

These equations are of the same form as Eq. (115), except for generalization of the degeneracy factor, so that the previously described methods may be employed to convert the elements of $\boldsymbol{\eta}_i$ to an initial $M$ state and consequently to an initial $(J, K)$ pair.

In the special case where the molecule has two or more threefold axes of symmetry, $I_A = I_B$, and the molecule becomes a spherical top. In such a case Eq. (127) reduces to

$$E_J = BJ(J+1) \qquad (132)$$

Because there are now $2J+1$ possible values of $K$ for each $J$ state, each of which will have degeneracy $2J+1$ for the spherical top, we set $g_J = (2J+1)^2$. Using this value for the degeneracy, the Monte Carlo or Diophantine sum over the rotational states for a spherical top molecule is given by Eq. (115). The elements of $\boldsymbol{\eta}_i$ may be converted to an initial $J$ state using the analogues of Eqs. (122) and (123) with $g_J = (2J+1)^2$.

Approximate formulas giving the energy levels for nonrigid symmetric top molecules are available, but at the present time no calculations have been reported in which such formulas have been employed.

The specification of the initial direction of the rotational angular momentum vector $\mathbf{L}$ in the case of a linear molecule is relatively simple. For such molecules $\mathbf{L}$ is always perpendicular to the internuclear axis so that its direction may be specified by the angle $\eta$ that $\mathbf{L}$ makes with some convenient reference plane. The integral over $\eta$ has the form

$$I = \int_0^{2\pi} \frac{1}{2\pi} P_\Lambda(V_r, b, \theta, \phi, R_{BC}, \eta, v, J) \, d\eta \qquad (133)$$

The elements of the lattice vector for Monte Carlo or Diophantine integration may be converted into initial values of $\eta$ by the transformation

$$\eta = 2\pi \xi_i \qquad (134)$$

which yields an integral of the form required by Eq. (72), i.e.,

$$I = \int_0^1 P_\Lambda(V_r, b, \theta, \phi, R_{BC}, \xi, v, J) \, d\xi \qquad (135)$$

In the case of symmetric or spherical top molecules, the magnitude of **L** and its $z$ component are given by

$$L = [J(J+1)]^{1/2}\hbar \tag{136}$$

and

$$L_z = K\hbar \tag{137}$$

where $J$ and $K$ are the rotational quantum numbers in Eq. (127). If we take the threefold symmetry axis to be the $z$ axis, then **L** must be oriented such that the angle $\gamma$ between **L** and the $z$ axis is

$$\gamma = \arccos\{K/[J(J+1)]^{1/2}\} \tag{138}$$

The complete specification of the initial direction of **L** therefore requires the selection of only one additional angle of rotation, $\eta$, about the $z$ axis. This selection can be carried out in the manner described by Eqs. (134) and (135).

Before closing this section, we note that statistical weights deriving from the degeneracies of the nuclear spin states associated with the molecular rotational states have not been included in our discussion. Where they arise (as, for example, in $H_2$) they should be incorporated in an obvious way.

### 5.3.2. Vibration

The summation over the vibrational states of a diatomic molecule BC has the form

$$R_\Lambda = Q_v^{-1} \sum_{v=0}^{\infty} \exp\left(-\frac{E_v}{kT}\right) R_\Lambda(v) \tag{139}$$

The vibrational energy $E_v$ is generally calculated in one of three ways:

1. Harmonic approximation:

$$E_v = (v + \tfrac{1}{2})\hbar\omega_0 \tag{140}$$

2. Morse oscillator approximation:

$$E_v = (v + \tfrac{1}{2})\hbar\omega_0 - \frac{\hbar^2\omega_0^2}{4D_e}(v + \tfrac{1}{2})^2 \tag{141}$$

3. A series expansion using the spectroscopically measured frequencies, anharmonicity terms, and rotation–vibration interaction.

The sum in Eq. (139) can be evaluated directly after computation of $R_\Lambda(v)$, the rate from initial vibrational state $v$. We may, however, carry out a Monte Carlo or Diophantine sum by taking the initial vibrational quantum state $v$ to be the nearest integral solution to

$$\sum_{v'=0}^{v} \exp\left(-\frac{E_{v'}}{kT}\right) = Q_v \xi_i \tag{142}$$

This transforms Eq. (139) into the form

$$R_\Lambda = \int_0^1 R_\Lambda(\xi)\, d\xi \tag{143}$$

as required.

For a nonlinear polyatomic system containing $N$ atoms, Eq. (139) becomes a $(3N-6)$-dimensional summation. In principle it is possible to calculate the energy states of each of the $3N-6$ normal vibrations by any of the methods listed above. In practice, however, only method (1) has thus far been used.

## 5.4. Initial Orientation

For a diatomic molecule the integrals over the spherical polar orientation angles, $\theta$ and $\phi$, have the form

$$I = \int_0^\pi \int_0^{2\pi} P_\Lambda(V_r, b, \theta, \phi, R_{BC}, \eta, v, J)\left[\frac{1}{2\pi}\, d\phi\right]\left[\frac{1}{2}\sin\theta\, d\theta\right] \tag{144}$$

The elements of the lattice vector for Monte Carlo or Diophantine integration may be converted to initial values of $\theta$ and $\phi$ by taking

$$\phi = 2\pi\xi_i \tag{145}$$

and

$$\xi_j = \tfrac{1}{2}(1+\cos\theta) \tag{146}$$

Substitution of Eqs. (145) and (146) into (144) yields the required form:

$$I = \int_0^1 \int_0^1 P_\Lambda(V_r, b, \xi_j, \xi_i, R_{BC}, \eta, v, J)\, d\xi_i\, d\xi_j \tag{147}$$

## 5.5. Vibrational Phase

The integral over the vibrational phase angle is equivalent to one over the initial internuclear separation; this integral has form

$$I = \int_{\rho_-}^{\rho_+} G(R_{BC})P_\Lambda(V_r, b, \theta, \phi, R_{BC}, \eta, v, J)\, dR_{BC} \tag{148}$$

for a diatomic molecule BC.

In principle, this integral may be transformed to the required form by making use of the cumulative distribution function $F(R_{BC})$, where

$$F(R_{BC}) = \int_{\rho_-}^{R_{BC}} G(R'_{BC})\, dR'_{BC} \Big/ \int_{\rho_-}^{\rho_+} G(R'_{BC})\, dR'_{BC}$$

Substitution into Eq. (148) yields

$$I = \int_0^1 P_\Lambda(V_r, b, \theta, \phi, F, \eta, v, J) \, dF \times \int_{\rho_-}^{\rho_+} G(R_{BC}) \, dR_{BC} \qquad (149)$$

The element $\xi_i$ of the lattice vector can now be converted to an initial B—C separation from the solution to

$$F(R_{BC}) = \xi_i \qquad (150)$$

Equation (149) now becomes

$$I = \int_0^1 P_\Lambda(V_r, b, \theta, \phi, \xi, \eta, v, J) \, d\xi \times \int_{\rho_-}^{\rho_+} G(R_{BC}) \, dR_{BC} \qquad (151)$$

as required.

In practice the analytic form of $F(R_{BC})$ depends on the diatomic internuclear potential being used in the trajectory study. If only low-lying vibrational states need be considered, little error is introduced by using a harmonic internuclear potential. In such a case

$$G(R_{BC}) \, dR_{BC} = dX/\pi(X_+^2 - X^2)^{1/2} \qquad (152)$$

with

$$X = R_{BC} - R_e, \qquad X_+ = \rho_+ - R_e \qquad (153)$$

where $R_e$ is the BC equilibrium separation. This yields the cumulative distribution function

$$F(R_{BC}) = \frac{1}{\pi} \sin^{-1}\left(\frac{X}{X_+}\right) + \frac{1}{2} \qquad (154)$$

from which the initial BC distance can be obtained from Eq. (150).

If higher vibrational states need to be considered, a more realistic form of the internuclear potential is required. For a rotating–vibrating Morse oscillator, a good approximation to the cumulative distribution function is[52]

$$F(R_{BC}) = \frac{1}{2} - \frac{1}{\pi} \sin^{-1}\left[\frac{b\gamma + 2a}{\gamma(b^2 - 4ac)^{1/2}}\right] \qquad (155)$$

where

$$\gamma = \exp(-\alpha X)$$

$$a = E_{vJ} - D_e - AL^2$$

$$b = 2D_e - BL^2$$

$$c = -D_e + CL^2$$

$$A = \frac{1}{2\mu_{BC}r_e^2}\left[1 - \frac{3}{\alpha r_e}\left(1 - \frac{1}{\alpha r_e}\right)\right] \qquad (156)$$

$$B = \frac{2}{\mu \alpha r_e^3} \left( 1 - \frac{3}{2\alpha r_e} \right)$$

$$C = \frac{1}{2\mu \alpha r_e^3} \left( 1 - \frac{3}{\alpha r_e} \right)$$

$$L^2 = J(J+1)\hbar^2$$

The parameters $\alpha$, $r_e$, and $D_e$ are the Morse potential energy parameters appearing in Eq. (125), and $E_{vJ}$ is the rotation–vibration energy given by Eq. (124). Inversion of Eq. (155) gives the initial value of $R_{BC}$:

$$R_{BC} = r_e + \frac{1}{\alpha} \ln \left\{ \left( -\frac{b}{2a} \right) \left[ 1 - \left( 1 - \frac{4ac}{b^2} \right)^{1/2} \cos(\pi \xi_i) \right] \right\} \tag{157}$$

If $J$ is set to zero, Eqs. (155) and (157) reduce exactly to those previously given by Slater[53] for the nonrotating Morse oscillator.

In the case of polyatomic molecules, and perhaps for some diatomic molecules, the above procedures cannot be used, because the radial density function $G(R_{BC})$ is generally not available in closed form. The integral over vibrational phase is, however, equivalent to one over the vibrational period. Because the distribution of the time $t$ is "flat," regardless of the form of the radial density function, $G(R_{BC})$ is related to the time density function by

$$G(R_{BC}) \, dR_{BC} = \frac{2dt}{\tau} \tag{158}$$

where $\tau$ is the vibrational period. If we take the zero of time to be the point at which $R_{BC} = \rho_-$, Eq. (148) becomes

$$I = \int_0^{\tau/2} P_\Lambda(V_r, b, \theta, \phi, t, \eta, v, J) \frac{2dt}{\tau} \tag{159}$$

The cumulative time distribution function is

$$\xi(t) = \int_0^t \frac{2dt}{\tau} \bigg/ \int_0^{\tau/2} \frac{2dt}{\tau} = \frac{2t}{\tau} \qquad \text{(mod 1)} \tag{160}$$

If we take

$$\xi(t) = \xi_i \tag{161}$$

then Eq. (159) takes the form required for Monte Carlo or Diophantine integration.

The initial internuclear separations may be obtained from Eq. (161) either by varying the initial reaction radius[7] or by a "stored structure" method.[40] In the first case, the internuclear separations are all set equal to either their equilibrium values or their turning point values, and the reaction radius vector $\mathbf{R}_s$ between the centers of mass of the colliding molecules is replaced by

$$\mathbf{R}_s' = \mathbf{R}_s - \mathbf{V}_r \tau \xi_i / 2 \tag{162}$$

Thus the time required for the colliding system to reach the reaction radius $\mathbf{R}_s$ is $\xi_i \tau/2$, i.e., the $t$ of Eq. (160).

The above procedure is inconvenient for cases in which $\tau$ is large. In such an event, $R'_s$ becomes unreasonably large, and a great deal of computer time must be consumed integrating the equations of motion to bring the colliding species together. In this case the stored structure method may be employed. In this procedure, the equations of motion for the isolated molecule are integrated, and the coordinates and conjugate momenta of all atoms are stored at equally spaced time intervals $\Delta t$ with $0 \le t \le \tau/2$. An initial time is then chosen via Eq. (160), and the stored structure nearest this selected value of $t$ is employed as the initial configuration of the system.

It should be noted that even the stored structure method may lead to difficulties in the case of polyatomic reactants. Such molecules usually have extremely long "periods," because the vibrational frequencies are generally incommensurate. Thus, the equations of motion must be integrated for long periods of time before a vibrational configuration will approximately repeat itself. In such an event, it will be necessary to integrate over only a subset of vibrational phase space and assume that the stored structures of that subset are representative of the entire set. For a spherical top molecule such as $CH_4$, this appears to be a reasonable assumption, but for other systems it may not be sufficiently accurate.

### 5.6. Calculation of Initial Coordinates and Momenta

Once the initial orientation angles, angular momentum direction, initial vibration–rotation states, and vibrational phases are chosen, it is generally necessary to convert these values into Cartesian position coordinates and their conjugate momenta for the diatomic or polyatomic molecule in question. In the case of a diatomic molecule, BC, this conversion is relatively simple.

Let $x$, $y$, and $z$ be the initial Cartesian position components of B relative to C with conjugate momenta $P_x$, $P_y$, and $P_z$. If BC is aligned along the $z$ axis with $\mathbf{L}$ pointed in the positive $x$ direction, the coordinates and their conjugate momenta are given by

$$x_0 = y_0 = 0$$
$$z_0 = R_{BC}$$
$$P_x^0 = 0 \tag{163}$$
$$P_y^0 = [J(J+1)]^{1/2} \hbar / R_{BC}$$
$$P_z^0 = \pm \{2\mu_{BC}[E_v - V(R_{BC})]\}^{1/2}$$

where $E_v$ is the vibrational energy given by Eq. (140) or (141) or an appropriate series expansion and $V(R_{BC})$ is the vibrational potential energy. If a Morse potential is being employed, then $V(R_{BC})$ is represented by Eq. (125) with $r$ replaced by $R_{BC}$. The selection of the plus or minus sign on $P_z^0$ determines whether the vibrational phase is in the part of the period between 0 and $\tau/2$ or between $\tau/2$ and $\tau$. The molecule and its angular momentum may now be oriented in the required directions by successive operation with rotation matrices:

$$\begin{pmatrix} x \\ y \\ z \end{pmatrix} = \mathbf{R}_z(\phi)\mathbf{R}_y(\theta)\mathbf{R}_z(\eta) \begin{pmatrix} x_0 \\ y_0 \\ z_0 \end{pmatrix} \tag{164}$$

and

$$\begin{pmatrix} P_x \\ P_y \\ P_z \end{pmatrix} = \mathbf{R}_z(\phi)\mathbf{R}_y(\theta)\mathbf{R}_z(\eta) \begin{pmatrix} P_x^0 \\ P_y^0 \\ P_z^0 \end{pmatrix} \tag{165}$$

where

$$\mathbf{R}_z(\beta) = \begin{pmatrix} \cos\beta & \sin\beta & 0 \\ -\sin\beta & \cos\beta & 0 \\ 0 & 0 & 1 \end{pmatrix} \tag{166}$$

and

$$\mathbf{R}_y(\beta) = \begin{pmatrix} \cos\beta & 0 & -\sin\beta \\ 0 & 1 & 0 \\ \sin\beta & 0 & \cos\beta \end{pmatrix} \tag{167}$$

A pictorial illustration of the effect of these operations has been given.[40]

The situation is considerably more complicated for polyatomic molecules. In general, it is simplest to employ a Cartesian coordinate system in the laboratory frame of reference. We therefore require the laboratory Cartesian coordinates and conjugate momenta for each atom in the system.

Assuming that the stored structure technique is being used for vibrational phase averaging, it is most convenient to start the symmetric top molecule in its equilibrium geometry with its threefold symmetry axis aligned along the $z$ axis. In this orientation all the vibrational energy is initially present as kinetic energy, and $\mathbf{L}$ makes an angle $\gamma$ [given by Eq. (138)] with the $z$ axis. Because all the vibrational energy is initially kinetic for this orientation, the total vibrational energy of the $i$th normal mode is given by

$$E_{vJ}^i = \sum_{n=1}^N [(P_{xn}^i)^2 + (P_{yn}^i)^2 + (P_{zn}^i)^2]\frac{1}{2M_n} \tag{168}$$

where $E_{vJ}^i$ is given by either Eq. (140) or (141), $M_n$ is the mass of atom $n$, and

$P^i_{xn}$, $P^i_{yn}$, and $P^i_{zn}$ are the $x$, $y$, and $z$ momentum components, respectively, for atom $n$ due to its vibration in the $i$th normal mode. The ratio of momentum components may be expressed in terms of the normal mode Cartesian displacement vectors. If $C^i_{xj}$, $C^i_{yj}$, and $C^i_{zj}$ are the $x$, $y$, and $z$ normal mode Cartesian displacement vectors, respectively, for the $j$th atom vibrating in the $i$th normal mode, the ratio of momentum components is

$$P^i_{yk}/P^i_{xj} = M_k C^i_{yk}/M_j C^i_{xj} \tag{169}$$

Substitution of Eq. (169) into (168) yields

$$E^i_{vJ} = \frac{(P^i_{xj})^2}{2M^2_j(C^i_{xj})^2} \sum_{n=1}^{N} M_n[(C^i_{xn})^2 + (C^i_{yn})^2 + (C^i_{zn})^2] \tag{170}$$

Combination of Eqs. (168)–(170) with $E^i_{vJ}$ for each of the $3N-6$ normal modes gives the initial conjugate momenta for vibration.

If the coordinate system is oriented so that $\mathbf{L}$ initially lies in the $zx$ plane, a single rotation will align $\mathbf{L}$ with the $z$ axis. The new Cartesian coordinates of atom $i$ in the rotated system are $x'_i$, $y'_i$, $z'_i$:

$$\begin{pmatrix} x'_i \\ y'_i \\ z'_i \end{pmatrix} = \mathbf{R}_y(\gamma) \begin{pmatrix} x_i \\ y_i \\ z_i \end{pmatrix} \tag{171}$$

where $\mathbf{R}_y(\gamma)$ is the rotation matrix given by Eq. (167). In this rotated system the additional linear momentum of atom $i$ due to rotation is

$$P'_i = M_i v'_i = M_i b_i \omega \tag{172}$$

$$b_i = (x'^2_i + y'^2_i)^{1/2} \tag{173}$$

$$\omega = \frac{L}{I_L} = \frac{[J(J+1)]^{1/2}\hbar}{\sum\limits_{i=1}^{N} M_i b^2_i} \tag{174}$$

The Cartesian components of $P'_i$ are

$$P'_{x_i} = P'_i y'_i / b_i$$

$$P'_{y_i} = P'_i x'_i / b_i \tag{175}$$

$$P'_{z_i} = 0$$

In the nonrotated coordinate system the momentum components are therefore

$$\begin{pmatrix} P_{x_i} \\ P_{y_i} \\ P_{z_i} \end{pmatrix} = \mathbf{R}_y(-\gamma) \begin{pmatrix} P'_{x_i} \\ P'_{y_i} \\ P'_{z_i} \end{pmatrix} \tag{176}$$

After assignment of the initial position coordinates and momenta from Eqs. (169), (170), and (176), the equations of motion for the isolated polyatomic molecule can be integrated to obtain the stored structure required by Eq. (161). After selection of an initial structure, the molecule is appropriately oriented with use of Eqs. (164) and (165). This completes the selection of the initial phase point for the trajectory.

## 6. Integration of the Equations of Motion

The trajectories for a system of $N$ atoms are obtained by integrating the $6N$ Hamilton equations from a set of initial points (chosen by methods discussed in Section 5) in the reactant region of phase space, until the trajectories pass into one or another region of phase space that is identified with product states. The equations have the general form

$$\frac{d\mathbf{y}}{dt} = \mathbf{f}[\mathbf{y}(t)] \tag{177}$$

where $\mathbf{y}$ and $\mathbf{f}$ are the $6N$-tuples that have as their elements

$$y_i = Q_i, \qquad f_i = \frac{\partial H}{\partial P_i}; \qquad 1 \le i \le 3N$$

$$y_i = P_i, \qquad f_i = -\frac{\partial H}{\partial Q_i}; \qquad 3N < i \le 6N \tag{178}$$

There is a variety of methods available for integration of systems of coupled first-order differential equations. They can generally be classified as single-step and multiple-step methods.

### 6.1. Single-Step Methods

If $\mathbf{y}$ is known at some time $t_\nu$, then $\mathbf{y}$ can be approximated at $t_{\nu+1} \equiv t_\nu + h$ by a truncated Taylor series:

$$\mathbf{y}(t_\nu + h) = \mathbf{y}(t_\nu) + h\mathbf{f}[\mathbf{y}(t_\nu)] + \sum_{k=1}^{r} \frac{h^{k+1}}{(k+1)!} \left\{ \frac{d^k}{dt^k} \mathbf{f}[\mathbf{y}(t)] \right\}_{t=t_\nu} \tag{179}$$

Direct use of Eq. (179) requires evaluation of derivatives of $\mathbf{f}$ with respect to the components of $\mathbf{y}$. This can be avoided by replacing the integral

$$\mathbf{y}(t_\nu + h) = \mathbf{y}(t_\nu) + \int_{t_\nu}^{t_\nu + h} \mathbf{f}[\mathbf{y}(t)] \, dt \tag{180}$$

by the finite quadrature formula

$$\mathbf{y}(t_\nu + h) = \mathbf{y}(t_\nu) + h \sum_{i=0}^{r-1} w_i \mathbf{f}[\mathbf{y}(t_\nu + \zeta_i h)] \tag{181}$$

in which the range of the quadrature points $\zeta_i$ is $(0, 1)$. The value of **y** at the intermediate times $t_\nu + \zeta_i h$ can be similarly approximated by

$$\mathbf{y}(t_\nu + \zeta_i h) = \mathbf{y}(t_\nu) + h \sum_{j=0}^{i-1} w_{ij} \mathbf{f}[\mathbf{y}(t_\nu + \zeta_j h)] \tag{182}$$

For a quadrature of degree $r$, there are thus $\frac{1}{2}r(r+3)$ parameters $\zeta_i$, $w_i$, and $w_{ij}$ to be determined. In the Runge–Kutta method,[54] these parameters are chosen so that the truncation error is the largest possible order in $h$, for a given value of $r$, in the limiting case of a single differential equation. This is accomplished by equating coefficients of powers of $h$ in the one-dimensional version of Eq. (179) with (1) those obtained in an expansion of the one-dimensional version of Eq. (181) directly in powers of $h$ and (2) those obtained by first substituting the one-dimensional version of Eq. (182) successively into Eq. (181) until all the $\zeta_i$ have been eliminated for a given value of $r$ and then expanding the resulting expression in powers of $h$. The conditions on the parameters obtained in this way by equating the three forms of the coefficients of the first four powers of $h$ are

$$\sum_{i=0}^{r-1} w_i = 1 \tag{183}$$

$$\sum_{j=0}^{i-1} w_{ij} = \zeta_i, \qquad i \geq 1 \tag{184}$$

$$\sum_{i=0}^{r-1} w_i \zeta_i^p = (p+1)^{-1}, \qquad 1 \leq p \leq (r-1), \qquad r \geq 2 \tag{185}$$

$$\sum_{i=0}^{r-1} w_i \sum_{j=0}^{i-1} w_{ij} \zeta_i = \tfrac{1}{6}, \qquad r \geq 3 \tag{186}$$

$$\sum_{i=0}^{r-1} w_i \sum_{j=0}^{i-1} w_{ij} (2\zeta_i \zeta_j + \zeta_j^2) = \tfrac{1}{3}, \qquad r \geq 4 \tag{187}$$

and

$$\sum_{i=0}^{r-1} w_i \sum_{j=0}^{i-1} w_{ij} \sum_{k=0}^{j-1} w_{jk} \zeta_k = \tfrac{1}{24}, \qquad r \geq 4 \tag{188}$$

The use of parameters that satisfy these conditions thus allows an approximation for $\mathbf{y}(t_\nu + h)$ to be obtained by $r$ evaluations of $\mathbf{f}(\mathbf{y})$ within the interval $(t_\nu, t_\nu + h)$ such that the result is identical to that given by the Taylor series expansion [Eq. (179)] truncated to the $r$th power of $h$ in the case of a single differential equation or in the limit of a decoupled set of equations. The truncation error is therefore of order $r+1$ in $h$.

Equations (183)–(188) and the additional ones obtained by equating coefficients of higher powers of $h$ provide $\frac{1}{2}r(r+1)$ relations among the parameters, leaving $r$ degrees of freedom in their choice. One may eliminate

Table 1. *Parameters for the Fourth-Order Runge–Kutta Method*[a]

| $i$ | 0 | 1 | 2 | 3 |
|---|---|---|---|---|
| $\zeta_i$ | 0 | $\frac{1}{3}$ | $\frac{2}{3}$ | 1 |
| $w_i$ | $\frac{1}{8}$ | $\frac{3}{8}$ | $\frac{3}{8}$ | $\frac{1}{8}$ |
| $w_{i0}$ | | $\frac{1}{3}$ | $-\frac{1}{3}$ | 1 |
| $w_{i1}$ | | | 1 | $-1$ |
| $w_{i2}$ | | | | 1 |

[a] Nondegenerate quadrature points.

the indeterminacy by using the equally spaced quadrature points

$$\zeta_i = \frac{i}{r-1} \tag{189}$$

For $r = 4$, this choice leads to the parameter values given in Table 1. Four new evaluations of $\mathbf{f}$ are required for each step.

If instead of solving a differential equation of the form of Eq. (177) we were integrating an explicit function of $t$, say $f(t)$, the time spent in the $f(t)$ subroutine (or the number of points required to be tabulated) can be reduced by one fourth if the two interior quadrature points are chosen to be degenerate. Then Eqs. (183)–(188) can be satisfied for any choice of $w_i$ and $w_2$ such that

$$w_1 + w_2 = \tfrac{2}{3}$$

For the degenerate choice $w_1 = w_2 = \frac{1}{3}$, the results are those of Table 2. In solving Eq. (177) by the Runge–Kutta method, we obtain $r$ distinct values of $\mathbf{y}$ even in the limit that not all quadrature points $\zeta_j$ are distinct, as can be seen from the form of Eq. (182).

Going one step further, Gill[55] improved the efficiency of the fourth-order Runge–Kutta method when used in digital computer programs by using the weights

$$w_1 = \frac{1}{3}\left(1 - \frac{1}{\sqrt{2}}\right)$$

$$w_2 = \frac{1}{3}\left(1 + \frac{1}{\sqrt{2}}\right)$$

from which one obtains the parameter values given in Table 3. By this choice Gill reduced the number of storage registers required for the calculation, an important consideration for the early machines with severely limited storage capacity. In addition, Gill modified the method to include a correction for the accummulated roundoff error.[55]

The Runge–Kutta methods are self-starting and are easy to program. They do not, however, make use of information from prior steps in the trajectory.

*Table 2. Parameters for the Fourth-Order*
*Runge–Kutta Method*[a]

| $i$ | 0 | 1 | 2 | 3 |
|---|---|---|---|---|
| $\zeta_i$ | 0 | $\frac{1}{2}$ | $\frac{1}{2}$ | 1 |
| $w_i$ | $\frac{1}{6}$ | $\frac{1}{3}$ | $\frac{1}{3}$ | $\frac{1}{6}$ |
| $w_{i0}$ | | $\frac{1}{2}$ | 0 | 0 |
| $w_{i1}$ | | | $\frac{1}{2}$ | 0 |
| $w_{i2}$ | | | | 1 |

[a] Degenerate quadrature points.

For these reasons, it is generally agreed that a given accuracy is often efficienctly achieved if a Runge–Kutta routine is used to start the trajectory, after which the integration is carried out by a multiple-step method such as the ones described in Section 6.2.

## 6.2. Multiple-Step Methods

Multiple-step methods use values of **y** at previous steps and therefore require fewer evaluations of **f** for a given step size. The integration of Eq. (177) takes the form

$$\mathbf{y}(t_\nu + h) = \mathbf{y}(t_{\nu-p}) + h \int_0^{p+1} \mathbf{f}[\mathbf{y}(t_{\nu-p} + \zeta h)]\, d\zeta$$

$$\approx \mathbf{y}(t_{\nu-p}) + h \sum_{i=1}^{n} \alpha_i \mathbf{f}[\mathbf{y}(t_{\nu-n+i})] \tag{190}$$

The weights $\alpha_i$ can be found either by forcing agreement with a truncated Taylor series in the manner described in Section 6.1 or more elegantly by use of finite difference operator algebra.[56] Two useful results are the five-point

*Table 3. Parameters for the Fourth-Order*
*Runge–Kutta–Gill Method*

| $i$ | 0 | 1 | 2 | 3 |
|---|---|---|---|---|
| $\zeta_i$ | 0 | $\frac{1}{2}$ | $\frac{1}{2}$ | 1 |
| $w_i$ | $\frac{1}{6}$ | $\frac{1}{3}(1-\sqrt{\frac{1}{2}})$ | $\frac{1}{3}(1+\sqrt{\frac{1}{2}})$ | $\frac{1}{6}$ |
| $w_{i0}$ | | $\frac{1}{2}$ | $-\frac{1}{2}+\sqrt{\frac{1}{2}}$ | 0 |
| $w_{i1}$ | | | $1-\sqrt{\frac{1}{2}}$ | $-\sqrt{\frac{1}{2}}$ |
| $w_{i2}$ | | | | $1+\sqrt{\frac{1}{2}}$ |

formulas[55,56]

$$\mathbf{y}_{\nu+1} = \mathbf{y}_\nu + \frac{h}{720}(1901\mathbf{f}_\nu - 2774\mathbf{f}_{\nu-1} + 1361\mathbf{f}_{\nu-2} - 1274\mathbf{f}_{\nu-3} + 251\mathbf{f}_{\nu-4})$$

$$+ \frac{95}{288} h^6 \mathbf{f}^{(5)}(\bar{t}), \qquad t_{\nu+1} < \bar{t} < t_{\nu-4} \tag{191}$$

$$\mathbf{y}_{\nu+1} = \mathbf{y}_{\nu-5} + \frac{3h}{10}(11\mathbf{f}_\nu - 14\mathbf{f}_{\nu-1} + 26\mathbf{f}_{\nu-2} - 14\mathbf{f}_{\nu-3} + 11\mathbf{f}_{\nu-4})$$

$$+ \frac{41}{140} h^7 \mathbf{f}^{(6)}(\bar{t}), \qquad t_\nu < \bar{t} < t_{\nu-5} \tag{192}$$

where we use the notation

$$\mathbf{y}_{\nu+j} \equiv \mathbf{y}(t_{\nu+j}) \equiv \mathbf{y}(t_\nu + jh) \tag{193}$$

$$\mathbf{f}_{\nu+j} \equiv \mathbf{f}[\mathbf{y}(t_{\nu+j})] \tag{194}$$

$$\mathbf{f}^{(n)}(\bar{t}) \equiv \left\{ \frac{d^n}{dt^n} \mathbf{f}[\mathbf{y}(t)] \right\}_{t=\bar{t}} \tag{195}$$

The numerical coefficients of the error terms can be made smaller by better than an order of magnitude if the "open" quadrature in Eq. (190) is replaced by the "closed" form[54,56]

$$\mathbf{y}_{\nu+1} \simeq \mathbf{y}_{\nu-p} + h \sum_{i=2}^{n+1} \beta_i \mathbf{f}_{\nu-n+i} \tag{196}$$

Two special cases of interest are

$$\mathbf{y}_{\nu+1} = \mathbf{y}_\nu + \frac{h}{720}(251\mathbf{f}_{\nu+1} + 646\mathbf{f}_\nu - 264\mathbf{f}_{\nu-1} + 49\mathbf{f}_{\nu-2} - 19\mathbf{f}_{\nu-3})$$

$$- \frac{3}{160} h^6 \mathbf{f}^{(5)}(\bar{t}), \qquad t_{\nu+1} < \bar{t} < t_{\nu-3} \tag{197}$$

$$\mathbf{y}_{\nu+1} = \mathbf{y}_{\nu-3} + \frac{2h}{45}(7\mathbf{f}_{\nu+1} + 32\mathbf{f}_\nu + 12\mathbf{f}_{\nu-1} + 32\mathbf{f}_{\nu-2} + 7\mathbf{f}_{\nu-3})$$

$$- \frac{8}{945} h^7 \mathbf{f}^{(6)}(\bar{t}), \qquad t_{\nu+1} < \bar{t} < t_{\nu-3} \tag{198}$$

Although these formulas are more accurate than the open formulas, their right-hand sides contain $\mathbf{y}_{\nu+1}$ as the argument of $\mathbf{f}_{\nu+1}$. An iterative solution for $\mathbf{y}_{\nu+1}$ can be obtained in principle from either Eq. (197) or (198) if an initial approximation to $\mathbf{y}_{\nu+1}$ is first obtained from an open formula such as Eq. (191) or (192). In practice, the closed formula is usually used only once, because iteration is expensive and it may not in fact converge at each point on a trajectory. These "predictor–corrector" methods thus require only two evaluations of $\mathbf{f}(\mathbf{y})$ for each time step, whereas the fourth-order Runge–Kutta method requires four evaluations.

The use of Eqs. (191) and (197) as predictor and corrector, respectively, is known as a five-point Adams–Moulton method, whereas use of Eqs. (192) and (198) is a five-point Milne method. The Adams–Moulton method is generally preferred for trajectory calculations even though its truncation error is of lower order in $h$, because the Milne method is inherently unstable,[57] that is, as the number of steps becomes large, the error becomes larger with each step. The stability properties of an Adams–Moulton method depend on both the step size and the form of the functions $\mathbf{f}(\mathbf{y})$ (i.e., the gradients of the potential energy surface). Brumer[58] has shown that for atom–diatomic molecule energy transfer calculations, the fourth-order Runge–Kutta method requires actually fewer evaluations of $\mathbf{f}(\mathbf{y})$ in the course of a trajectory than a fifth-order Adams–Moulton method when $h$ is chosen in each case to give stable solutions. The higher-order predictor–corrector methods tend to require smaller step sizes for stability. For exchange reactions on a symmetric potential surface with relatively low activation energy $(H+H_2)$, Brumer[58] shows that stability of the solution becomes a problem only near turning points. Long-lived collision complexes that require a large number of integration steps require a careful choice of a stable integrator. Brumer and Karplus[59] found Gear's hybrid method[60] to be acceptable in these cases.

Hamming[57] has given a stable fourth-order corrector of the form

$$\mathbf{y}_{\nu+1} = \sum_{i=0}^{2} \beta_i \mathbf{y}_{\nu-i} + h \sum_{i=0}^{2} \gamma_i \mathbf{f}(\mathbf{y}_{\nu+1-i}) + Ah^5 f^{(4)}(\bar{t}) \qquad (199)$$

With the six coefficients, it is possible to fit Eq. (199) to a fifth-order Taylor expansion in $h$, but the result, like Milne's method, is unstable for all choices of $h$. Hamming therefore fitted to a fourth-order Taylor expansion and used the extra degree of freedom to force stability for a wide range of $h$ by setting $\beta_1 = 0$. The other coefficients are $\beta_0 = \frac{9}{8}$, $\beta_2 = -\frac{1}{8}$, $\gamma_0 = \frac{3}{8}$, $\gamma_1 = \frac{3}{4}$, and $\gamma_2 = -\frac{3}{8}$. The coefficient of the error term is $A = -\frac{1}{40}$. The error term is necessarily larger than for Milne's three-term formula. Unfortunately, attempts to extend Hamming's method to obtain a stable method of higher order with a lower truncation error than the corresponding Adams–Moulton method appear to be fruitless.

### 6.3. Variable Time Steps

If a predictor–corrector method with the same order of the truncation error for each formula is used, the difference between the predicted and corrected value of $\mathbf{y}_{\nu+1}$ can be used to estimate the absolute error. For example, in the fifth-order Adams–Moulton method if we let $\mathbf{y}_{\nu+1}^*$ be the value predicted by Eq. (191) and $\mathbf{y}_{\nu+1}$ be the corrected value given by Eq. (197), then, because the error terms are omitted in the calculations, we have

$$\varepsilon^*_{\nu+1} \equiv \mathbf{y}_{\nu+1} - \mathbf{y}^*_{\nu+1} = \tfrac{3}{160} h^6 \mathbf{f}^{(5)}(\bar{t}) + \tfrac{95}{288} h^6 \mathbf{f}^{(5)}(\bar{t}^*) \tag{200}$$

If we assume that $\mathbf{f}^{(5)}(\bar{t}) = \mathbf{f}^{(5)}(\bar{t}^*)$, we obtain for the error of the corrected value

$$\varepsilon_{\nu+1} = \tfrac{3}{160} h^6 \mathbf{f}^{(5)}(\bar{t}) = -\tfrac{27}{502} \varepsilon^*_{\nu+1} \tag{201}$$

In this way the estimated error can be monitored at each step; if it exceeds predetermined limits, the step size can be adjusted for the proper compromise between accuracy and efficiency.

When the step size is decreased, the values of $\mathbf{y}$ at the new intervals backward from $\mathbf{y}_\nu$ must be obtained by an accurate interpolation; alternatively, the solution can be started again at $\mathbf{y}_\nu$ or at some earlier point by Runge–Kutta. The speed disadvantage of using too high an order predictor–corrector in a variable time step routine is an obvious one.

## 6.4. Checks on Trajectory Accuracy

No one integrator that has been used extensively in trajectory calculations is without disadvantages for at least a few applications. It is therefore necessary to have means of monitoring the accuracy of any given set of trajectories. Besides the method described in Section 6.3 for evaluating the accuracy of a trajectory, several other methods may be used. A crude indication of error is instability of the constants of the motion, such as total energy and angular momentum. Variation of the total energy is an indication of roundoff error or, in extreme cases, of programming blunders. On the other hand, constancy of the energy is not necessarily an indication of accuracy of the trajectory.

A standard check on trajectory accuracy is to run the trajectory backward from final to initial point and compare with the original initial values of the coordinates and momenta. This is accomplished by reversing the signs of the momenta and the time step $h$ at the end of $s$ steps and then integrating an additional $s$ steps. The tolerance depends on the purpose of the calculation. It might be argued that if a large number of trajectories are to be averaged, individual trajectory accuracy is not critical. If, on the other hand, thresholds for a given mechanism are to be calculated, or if trajectories are to be found that connect specified initial and final states as in semiclassical calculations, high accuracy is desirable.

For high efficiency in a given calculation the step size and the type and order of the integrator should be chosen so that the values of the quantities being calculated are just within tolerance of the converged value estimated by extrapolation of successively more accurate calculations. Schreiber[61] has made a study of convergence limits and integrator tolerances for several reaction attributes, including product energy and scattering angle distributions.

## 7. Analysis of Final States

In Section 4 we reviewed the use of statistical analysis of trajectories to obtain cross sections and rates for various molecular processes. This requires a classification of trajectories as to their initial and final states. Because Hamilton's equations constitute an initial value problem (as opposed to boundary value problems), we *choose* the initial states and *analyze* the final states. The choice of initial states was discussed in Section 5. In this section we shall describe the classification of the final states.

### 7.1. Discrimination of Products

The end point of a trajectory occurs when it enters a region of phase space designated as reactant or product space. The boundaries of these regions can be defined by the separation of groups of atoms from one another and by the internal energies of the separated groups. The critical separation $R_{si}^0$ for the designation of product configuration $i$ (including the reactant configuration as, say, $i = 0$) should be chosen large enough to reduce the residual interaction between the groups to an acceptably small value; each $R_{si}^0$ for a given system must usually be found by a numerical search, including averaging over the internal coordinates of the separated groups. In practice, the separations $R_{si}$ are evaluated at each step (or at most at each interval of a few steps in duration) and tested to determine whether $R_{si} \leq R_{si}^0$. For "mechanically balanced" systems in which the centers of mass of the groups lie near their geometrical centers, $R_{si}$ can be taken as the center of mass to center of mass distance between the pair of groups. For "mechanically imbalanced" systems such as D + HI, the distance from D to the center of mass of HI can be large whereas the D–H distance is small. The most reliable method is therefore to test all interatomic distances. This can be done very efficiently at each step, because these distances are required in the calculation of the potential gradients. The interatomic distances for three- and four-atom systems have been given elsewhere[7,40] in terms of the generalized coordinates in which the equations of motion are written. For simplicity and brevity, we shall restrict our discussion to cases of no more than two atoms in a separated group.

Once the product molecules have been determined by testing interatomic distances, it is necessary to determine whether the molecules are in bound, quasi-bound, or dissociative states. If the internal energy $E_{vri}$ of molecule $i$ is below the dissociation limit, molecule $i$ is in a bound state.

If $E_{vri}$ is above the dissociation limit and either

1. $R_i > R_{ib}$, where $R_i$ is the internuclear distance of molecule $i$ and $R_{ib}$ is the position of the rotational barrier at the maximum,

$$V_{0i} = V_i(R_{ib}) + \frac{L_i^2}{2\mu_i R_{ib}^2} \tag{202}$$

where $\mathbf{L}_i$ is the rotational angular momentum of molecule $i$, or

2. $R_i < R_{ib}$ and $E_{vri} > V_{0i}$, then molecule $i$ is in a dissociative state.

If $R_i < R_{ib}$ and $E_{vri} < V_{0i}$, molecule $i$ is in a quasi-bound state. The evaluation of $E_{vri}$ and $\mathbf{L}_i$ is discussed in Section 7.4.


## 7.2. Relative Translation

The final relative translational energy between group $i$ and group $j$ is

$$E'_T = \tfrac{1}{2}\mu_{ij}\mathbf{V}_{ij}^2 \tag{203}$$

where $\mu_{ij}$ is the reduced mass of groups $i$ and $j$ and $\mathbf{V}_{ij}$ is the relative velocity between their centers of mass,

$$\mathbf{V}_{ij} = \dot{\mathbf{R}}_{ij} \tag{204}$$

With most reasonable choices of coordinate systems, $\mathbf{R}_{ij}$ and $\mathbf{V}_{ij}$ are simple functions of the coordinates $Q_k$ and the momenta $P_k$, respectively.


## 7.3. Scattering Angle

The angle $\theta_c$ into which products are scattered in the center of mass system is defined by

$$\cos\theta_c = \frac{\mathbf{V}_{ij}\cdot\mathbf{V}_R}{|\mathbf{V}_{ij}|\cdot|\mathbf{V}_R|} \tag{205}$$

where $\mathbf{V}_R$ is the initial relative velocity between reactants. For beam experiments in which one of the reactants and one of the products is an atom, the directions of $\mathbf{V}_R$ and of $\mathbf{V}_{ij}$ are usually chosen so that $\cos\theta_c = -1$ when the product atom moves after the collision in a direction collinear with the motion of the reactant atom before the collision.

### 7.4. Internal Energy

#### 7.4.1. Rotation

The rotational angular momentum $\mathbf{L}_i$ of product molecule $i$ is given by

$$\mathbf{L}_i = \mu_i \mathbf{R}_i \times \dot{\mathbf{R}}_i \tag{206}$$

where $\mathbf{R}_i$ is the internuclear distance and $\dot{\mathbf{R}}_i$ is its velocity. The vector $\mathbf{L}_i$ is a constant of the motion if groups $i$ and $j$ are sufficiently separated that their interaction is negligible, but the rotational energy, given to a first approximation by

$$E_r = \frac{L_i^2}{2\mu_i R_i^2} \tag{207}$$

is not conserved, because $R_i$ is variable. If one is required to separate out the rotational energy, one may use an appropriate average value for $R_i^{-2}$ in Eq. (207). We shall see, however, in the next section that the action variables $v_i$ and $J_i$ for classical vibration and rotation are well defined, even in the realistic nonseparable case.

#### 7.4.2. Vibration

The radial action variable $v_i$ that goes over to the vibrational quantum number in the quantal regime is uniquely defined by the phase integral

$$v_i = -\frac{1}{2} + \frac{2\mu_i}{h} \int_{R_{i<}}^{R_{i>}} \dot{R}_i \, dR_i \tag{208}$$

where $R_{i>}$ and $R_{i<}$ are the outer and inner turning points, respectively. The value of $v_i$ for a given trajectory is thus easily calculated by continuing the trajectory through half a vibrational period (as determined by testing for the turning points) in order to numerically evaluate the integral in Eq. (208). It is then appropriate to designate the final vibration state of molecule $i$ by the value of the conserved quantity $v_i$ and the final rotation state by the conserved quantity $J_i$, defined as

$$J_i = -\tfrac{1}{2} + \tfrac{1}{2}(1 + 4L_i^2/\hbar^2)^{1/2} \tag{209}$$

The vibration–rotation energy is then the total internal energy of molecule $i$ obtained as the sum of the kinetic and potential energies calculated in an appropriate internal coordinate system.

In classifying trajectories, the usual practice is to count the number of trajectories whose final states fall within certain ranges of the variables. In doing this, it is appropriate to choose the boxes for $v_i$ and $J_i$ to have their boundaries at the half-integer values; in this way one can calculate, for example, the classical approximation to the state-to-state differential cross section for

$$A + BC(v, J) \rightarrow AB(v', J') + C$$

namely,

$$\frac{d^2 S_{v'J'V' \leftarrow cJV}(\theta_c)}{d\theta_c dV'} = \frac{\pi \hbar^2 \sin \theta_c}{\mu_{A,BC}^2 V_r^2} \sum_{i=1}^{N} l_{Mi} (2l_i + 1) P_i(AB, v', J', V', \theta_c) \qquad (210)$$

where $P_i$ is the probability ($=0$ or $1$) that trajectory $i$ leads to product AB with the scattering angle in the range $d \cos \theta_c$ at $\cos \theta_c$, the final relative speed in the range $dV'$ at $V'$, and the vibrational and rotational quantum numbers in the ranges $(v' - \frac{1}{2}v' + \frac{1}{2})$ and $(J' - \frac{1}{2}, J' + \frac{1}{2})$, respectively.

## 8. Consecutive Collisions

Classical mechanical studies of the mechanisms of thermolecular reactions such as recombinations

$$A + B + M \rightarrow AB + M$$

and exchange reactions of the type

$$2NO + O_2 \rightarrow 2NO_2$$

$$I + I + H_2 \rightarrow 2HI$$

require the extension of the methods we have described to include consecutive collisions. The idea is to choose a pair of reactants as the initial colliding partners and then to allow the third reactant to collide with that collision complex. The stored structure method of phase averaging described in Section 5.5 provides the means of sampling the initial state of the second collision. For details of the treatment of such consecutive collisions, the reader is referred to Reference 40.

## 9. Prognosis for Classical Molecular Scattering Calculations

To what degree will the methods of this chapter be used in the 1970s and 1980s? During the 1970s a simple extrapolation shows rapidly increasing use of classical trajectory calculations in close connection with experiments in chemical dynamics. During this same period, simple models that reduce the number of variables or simplify features of the potential energy surface and reaction criteria will continue to develop. Detailed trajectory studies will therefore provide quantitative standards by which the efficacy of the simple models are tested. Semiclassical calculations will undoubtedly continue to make use of trajectories for calculating the $S$ matrix from the classical action. The impetus for treatment of larger systems and biologically important reactions will encourage the development of ways of combining trajectories with the formalisms of quantum mechanics and nonequilibrium statistical mechanics in some direct way, so that the dimensionality can be reduced with minimal loss of dynamical information.

In the 1980s, classical trajectories alone and in the combinations we have mentioned will perhaps share their places with accurate quantal methods in the chemist's "tool case"[61] for molecular collision calculations. Breakthroughs in digital computer circuitry and in information storage and retrieval technology, or elimination of noise in analog circuits, may make the trajectory calculations of the 1980s more pleasurable. In any case, the chemical theorist can expect to be increasingly concerned with trajectories, whether his contribution is in devising potential energy surfaces, contributing to the extension of trajectory applications, pondering the chemical import of the results of trajectory analysis, or constructing schemes by which trajectories may ultimately be replaced.

## *References*

1. W. H. Miller, Classical $S$ matrix for rotational excitation: Quenching of quantum effects in molecular collisions, *J. Chem. Phys.* **54**, 5386–5397 (1971).
2. R. A. Marcus, Extension of the WKB method to wave functions and transition probability amplitudes ($S$-matrix) for inelastic or reactive collisions, *Chem. Phys. Lett.* **7**, 525–532 (1970).
3. W. H. Miller, Semiclassical theory of atom–diatom collisions: Path integrals and the classical $S$-matrix, *J. Chem. Phys.* **53**, 1949–1959 (1970).
4. R. A. Marcus, Theory of semiclassical transition probabilities for inelastic and reactive collisions. V. Uniform approximation in multidimensional systems, *J. Chem. Phys.* **57**, 4903–4909 (1972).
5. W. H. Miller, Classical-limit quantum mechanics and the theory of molecular collisions, *Adv. Chem. Phys.* **25**, 69–177 (1974).
6. M. Karplus, R. N. Porter, and R. D. Sharma, Dynamics of reactive collisions: The $H + H_2$ exchange reaction, *J. Chem. Phys.* **40**, 2033–2034 (1964).
7. M. Karplus, R. N. Porter, and R. D. Sharma, Exchange reactions with activation energy. I. Simple barrier potential for H, $H_2$), *J. Chem. Phys.* **43**, 3259–3287 (1965).
8. M. Karplus, R. N. Porter, and R. D. Sharma, Energy dependence of cross sections for hot tritium reactions with hydrogen and deuterium molecules, *J. Chem. Phys.* **45**, 3871–3873 (1966).
9. M. Karplus, L. G. Pedersen, R. N. Porter, R. D. Sharma, and D. L. Thompson, unpublished.
10. A. Kuppermann and J. M. White, Energy threshold for $D + H_2 \rightarrow DH + H$ reaction, *J. Chem. Phys.* **44**, 4352–4354 (1966).
11. D. Seewald, M. Gersh, and R. Wolfgang, Exchange between atomic and molecular hydrogen at energies above threshold, *J. Chem. Phys.* **45**, 3870–3871 (1966).
12. C. C. Chou and F. S. Rowland, Exchange reactions of 2.8-eV tritium atoms with isotopic molecular hydrogen—$H_2$, $D_2$, and HD, *J. Chem. Phys.* **46**, 812–813 (1967).
13. C. C. Chou and F. S. Rowland, Threshold energy for substitution of T for D in $CD_4$, *J. Chem. Phys.* **50**, 2763–2764 (1969).

14. C. C. Chou and F. S. Rowland, Reactions of 2.8-eV tritium atoms with methane, *J. Chem. Phys.* **50**, 5133–5140 (1969).
15. M. E. Gersh and R. B. Bernstein, Measurement of the energy dependence of the cross section for the reaction $K + CH_3I \rightarrow KI + CH_3$ from 0.1–1 eV, *J. Chem. Phys.* **55**, 4661 (1971).
16. M. E. Gersh and R. B. Bernstein, Translational energy dependence of the reaction cross section for $K + CH_3I \rightarrow KI + CH_3$ from 0.1 to 1 eV (c.m.), *J. Chem. Phys.* **56**, 6131–6146 (1972).
17. D. L. Bunker and E. A. Goring, $Rb + CH_3I$: Empirically determined potential and predicted cross sections for reactive scattering, *Chem. Phys. Lett.* **15**, 521–523 (1972).
18. R. A. LaBudde, P. J. Kuntz, R. B. Bernstein, and R. D. Levine, Classical trajectory study of the $K + CH_3I$ reaction, *J. Chem. Phys.* **59**, 6286–6298 (1973).
19. R. M. Harris and D. R. Herschbach, Comment in *Discuss. Faraday Soc.* **55**, 121–123 (1973).
20. J. T. Muckerman, Classical dynamics of the reaction of fluorine atoms with hydrogen molecules. II. Dependence on the potential energy surface, *J. Chem. Phys.* **56**, 2997–3006 (1972).
21. T. P. Schafer, P. E. Siska, J. M. Parson, F. P. Tully, Y. C. Wong, and Y. T. Lee, Crossed Molecular beam study of $F + D_2$, *J. Chem. Phys.* **53**, 3385–3387 (1970).
22. R. A. LaBudde and R. B. Bernstein, Classical study of rotational excitation of a rigid rotor: $Li^+ + H_2$, *J. Chem. Phys.* **55**, 5499–5516 (1971).
23. R. A. LaBudde and R. B. Bernstein, Classical study of rotational excitation of a rigid rotor: $Li^+ + H_2$. II. Correspondence with quantal results, *J. Chem. Phys.* **59**, 3687–3691 (1973).
24. D. L. Thompson, On a classical trajectory study of energy transfer in some atom–diatomic molecule systems, *J. Chem. Phys.* **56**, 3570–3580 (1972).
25. I. W. M. Smith and P. M. Wood, Vibrational relaxation in atom-exchange reactions: A classical, Monte-Carlo, trajectory study, *Mol. Phys.* **25**, 441–454 (1973).
26. K. G. Anlauf, P. J. Kuntz, D. H. Maylotte, P. D. Pacey, and J. C. Polanyi, Energy distribution among reaction products. Part 2. $H + X_2$ and $X + HY$, *Discuss. Faraday Soc.* **44**, 183–193 (1967).
27. J. C. Polanyi, Some concepts in reaction dynamics, *Acc. Chem. Res.* **5**, 161–168 (1972).
28. J. C. Polanyi and J. L. Schreiber, The dynamics of bimolecular reactions, *in: Physical Chemistry—An Advanced Treatise*, (H. Eyring, W. Jost, and D. Henderson, eds.), Vol. 6A, "Kinetics of Gas Reactions," Chap. 6, pp. 383–487, Academic Press, Inc., New York (1974).
29. R. D. Levine and R. B. Bernstein, Energy disposal and energy consumption in elementary chemical reactions: The information theoretic approach, *Acc. Chem. Res.* **7**, 393–400 (1974).
30. D. L. Thompson and D. R. McLaughlin, A quasiclassical trajectory study of the $H_2 + F_2$ reaction, *J. Chem. Phys.* **62**, 4284–4299 (1975).
31. D. L. Bunker, Monte-Carlo calculation of triatomic dissociation rates. I. $N_2O$ and $O_3$, *J. Chem. Phys.* **37**, 393–403 (1962).
32. D. L. Bunker, Monte-Carlo calculations. IV. Further studies of unimolecular dissociation, *J. Chem. Phys.* **40**, 1946–1957 (1964).
33. D. L. Bunker and W. L. Hase, On non-RRKM unimolecular kinetics: Molecules in general, and $CH_3NC$ in particular, *J. Chem. Phys.* **59**, 4621–4632 (1973).
34. R. N. Porter, Theoretical studies of hot-atom reactions. I. General formulation, *J. Chem. Phys.* **45**, 2284–2291 (1966).
35. J. T. Muckerman, Classical dynamics of the reaction of fluorine atoms with hydrogen molecules. III. The hot-atom reactions of $^{18}F$ with HD, *J. Chem. Phys.* **57**, 3388–3396 (1972).
36. L. M. Raff, Theoretical investigations of the reaction dynamics of polyatomic systems: Chemistry of the hot atom ($T^* + CH_4$) and ($T^* + CD_4$) systems, *J. Chem. Phys.* **60**, 2220–2244 (1974).
37. T. Valencich and D. L. Bunker, Trajectory studies of hot atom reactions. II. An unrestricted potential for $CH_5$, *J. Chem. Phys.* **61**, 21–29 (1974).
38. H. Y. Su, J. M. White, L. M. Raff, and D. L. Thompson, Abstraction versus exchange in the reaction of H with DBr, *J. Chem. Phys.* **62**, 1435–1433 (1975).
39. R. N. Porter, L. B. Sims, D. L. Thompson, and L. M. Raff, Classical dynamical investigations of reaction mechanism in three-body hydrogen–halogen systems, *J. Chem. Phys.* **58**, 2855–2869 (1973).

40. L. M. Raff, D. L. Thompson, L. B. Sims, and R. N. Porter, Dynamics of the molecular and atomic mechanisms for the hydrogen–iodine exchange reactions, *J. Chem. Phys.* **56**, 5998–6027 (1972).

41. J. B. Anderson, Mechanism of the bimolecular (?) hydrogen–iodine reaction, *J. Chem. Phys.* **61**, 3390–3393 (1974).

42. D. L. Bunker, Classical trajectory methods, *Methods Comput. Phys.* **10**, 287–325 (1971).

43. R. N. Porter, Molecular trajectory calculations, *Annu. Rev. Phys. Chem.* **25**, 317–355 (1974).

44. V. Volterra, *Theory of Functionals and of Integral and Integro-Differential Equations*, Dover Publications, Inc., New York (1959).

45. H. Goldstein, *Classical Mechanics*, Addison-Wesley Publishing Company, Inc., Reading, Mass. (1950), Chap. 8, pp. 237–272.

46. Y. A. Schreider, *The Monte-Carlo Method*, Pergamon Press, Inc., Elmsford, N.Y. (1966).

47. H. H. Suzukawa, Jr., D. L. Thompson, V. B. Cheng, and M. Wolfsberg, Empirical testing of the suitability of a nonrandom integration method for classical trajectory calculations: Comparisons with Monte-Carlo techniques, *J. Chem. Phys.* **59**, 4000–4008 (1973).

48. C. B. Haselgrove, A method of numerical integration, *Math. Comput.* **15**, 323–337 (1961).

49. H. Conroy, Molecular Schrödinger equation. VIII. A new method for the evaluation of multidimensional integrals, *J. Chem. Phys.* **47**, 5307–5318 (1967).

50. V. B. Cheng, H. H. Suzukawa, Jr., and M. Wolfsberg, Investigations of a nonrandom numerical method for multidimensional integration, *J. Chem. Phys.* **59**, 3992–3999 (1973).

51. D. J. Mickish, *Ab Initio* Calculations on the Li and $H_3$ Systems Using Explicitly Correlated Wave Functions and Quasirandom Integration Techniques, Ph.D. thesis, Oklahoma State University, Stillwater, Oklahoma (1970).

52. R. N. Porter, L. M. Raff, and W. H. Miller, Quasiclassical selection of initial coordinates and momenta for a rotating Morse oscillator, *J. Chem. Phys.* **63**, 2214–2218 (1975).

53. N. B. Slater, Vibrational characteristics of quasi-harmonic systems related to diatomic molecules, *Proc. Leeds Philos. Lit. Soc. Sci. Sect.* **8**, 93–108 (1959).

54. E. Isaacson and H. B. Keller, *Analysis of Numerical Methods*, John Wiley & Sons, Inc., New York (1966), Chap. 8, pp. 364–441.

55. S. Gill, A process for the step-by-step integration of differential equations in an automatic digital computing machine, *Proc. Cambridge Philos. Soc.* **47**, 96–108 (1951).

56. F. B. Hildebrand, *Introduction to Numerical Analysis*, McGraw-Hill Book Company, New York (1956).

57. R. W. Hamming, Stable predictor corrector methods for ordinary differential equations, *J. Assoc. Comput. Mach.* **6**, 37–47 (1959).

58. P. Brumer, Stability concepts in the numerical solution of classical atomic and molecular scattering problems, *J. Comput. Phys.* **14**, 391–419 (1974).

59. P. Brumer and M. Karplus, Collision complex dynamics in alkali halide exchange reactions, *Discuss. Faraday Soc.* **55**, 80–92 (1973).

60. C. W. Gear, Hybrid methods for initial value problems in ordinary differential equations, *SIAM J. Num. Anal.* **2B**, 69–86 (1965).

61. J. L. Schreiber, Classical Trajectory Studies of Chemical Reactions, Ph.D. thesis, University of Toronto, Toronto (1973).

# 2

# Features of Potential Energy Surfaces and Their Effect on Collisions

## P. J. Kuntz

## 1. Introduction

### 1.1 Molecular Collisions

In this chapter we shall discuss molecular rearrangement collisions that take place on a single potential energy surface. Nonreactive collisions, particularly those involving vibrational energy transfer, are treated in Chapter 4 of Part A; multipotential processes are treated in Chapter 5 of Part B. Because most of the work devoted to the correlation of collision phenomena with features of potential energy functions has been done within the framework of classical mechanics, the major part of the discussion will utilize the language appropriate to such a description of the motion. For most chemical systems, such a treatment is entirely adequate.

Experimental work over the past decade has done a great deal to characterize exoergic elementary reactions by the specificity of their energy disposal and endoergic reactions by the selectivity of their energy requirements.[1] Thus, many exoergic reactions do not distribute energy among the products' degrees of freedom in a statistical way but tend to populate certain states rather than others.[2,3] Likewise, endoergic reactions are able to utilize internal energy much more effectively than relative kinetic energy.[4-6] An example of the former phenomenon is the reaction $H + Cl_2 \rightarrow HCl + Cl$, which is exothermic

*P. J. Kuntz* • Hahn–Meitner Institut für Kernforschung, Berlin GmbH, West Germany, and St. Mary's University, Halifax, Nova Scotia, Canada

by $\sim 45$ kcal/mole; enough energy is released by the reaction to populate the product HCl molecules up to vibrational level $v' = 6$. A statistical distribution of this energy consistent with the total energy available would put most of the energy in the lowest vibrational levels of HCl, whereas the experimental distribution peaks at around $v' = 3$.[7] An example of selectivity is provided by the endothermic reaction $HCl + K \rightarrow H + KCl$, which is found to have a much larger cross section for HCl in state $v' = 1$ than $v' = 0$.[4,5]

The experimentally observable quantities of interest in reactive molecular collisions are obtained chiefly from molecular beam[8] and chemiluminescence experiments.[9] The chemiluminescence work measures the intensity of radiation emitted by the product molecules, and from this intensity it is possible to estimate the rate of formation of product molecules in their various states (usually vibrational-rotational states). The molecular beam technique measures the flux intensity of molecules scattered outward from a small zone wherein the molecular collisions occur. From this technique one obtains distributions of the angle at which product molecules are scattered relative to the reagent beam direction as well as the velocity distribution of the molecules. From combinations of these and other related techniques, it is possible to obtain estimates of reactive cross sections, differential cross sections for formation of molecules with specific amounts of internal energy and scattered into specific angular regions, and rate constants for the formation of molecules in specific vibrational–rotational levels. With more specialized techniques it is possible to study the effect of orienting the reagent molecules [10] and state selecting the reagent molecules.[11] In favorable cases, it is possible to obtain some of this information at different energies of collision.[12] The reaction $K + ICH_3 \rightarrow KI + CH_3$ is one of these cases.

## 1.2. Potential Energy Surfaces

The goal of a theory of molecular collisions is to understand and correlate observable phenomena in terms of consistent fundamental concepts. The most basic concept is the Schrödinger equation for all the nuclei and electrons that make up the collision system. This equation is the starting point for rigorous theoretical work, but it is not a very appropriate basis for discussion of collision processes, because the interest there lies in the interactions between the atoms and molecules themselves, and not between each electron and nucleus that make them up. A more useful point of reference is the potential energy function for the nuclei. It is obtainable from the Schrödinger equation (in principle) and is designed to represent interactions at the atomic and molecular level.

Forgoing the rigor of the Schrödinger equation for the convenience of the potential function is done only at a price. There arise two related complications. The reagents and products can exist in many different electronic states and can interact, therefore, via many different potentials. It is also possible for the collision process itself to induce a transition from one potential to another, because the whole concept of potential energy rests upon an approximation that is not always valid throughout a collision. The notion of the potential energy can still be preserved in these cases, provided that the simultaneous influence of several potentials is considered. In such cases, the electronic structure of the collision system itself is a feature that has a profound influence on the collision process. Systems where such considerations are important are treated in Chapter 5.

In attempting to correlate features of a potential function with collision phenomena, one is immediately faced with the problem of characterizing the function. *Ab initio* potentials are usually served up as rather indigestible tables of numbers. Occasionally these tables are garnished with arbitrary functional expressions that fit the main course quite well (i.e., they are also indigestible), and although these expressions are palatable for calculation, they provide little food for thought about what sort of dynamics the potential will produce. Much more suggestive are plots of the potential function considered as a hypersurface in many dimensions, or, more accurately, as two-dimensional cuts through the hypersurfaces. These plots characterize the potential in terms of its shape. In this chapter, we shall attempt to comprehend collision dynamics in terms of concepts more closely related to surface topology than to the mathematical description of the surface. This is clearly a step in the qualitative direction, but it is rewarded by achieving a more widely applicable viewpoint, from which it is possible to predict the approximate nature of collision dynamics on new surfaces that are computed.

## 2. Potential Energy Surfaces and Dynamics

### 2.1. Separation of Electronic and Nuclear Motion

The method of arriving at a potential energy function from the Schrödinger equation involves treating the motions of the nuclei and electrons separately. Because the mass of the electron is so much smaller than that of the nuclei, this is often a good approximation, and its application to the Schrödinger equation produces two equations, an electronic one and a nuclear one. The electronic equation is solved for fixed positions of the nuclei to give a set of electronic eigenvalues and eigenfunctions for the molecular system. The eigenvalues and eigenfunctions so obtained depend on the coordinates of the nuclei in a parametric fashion, and the set of eigenvalues constitute the potential energy functions for the system. The separation procedure is outlined here only briefly, because detailed derivations exist elsewhere.[13-15]

The Hamiltonian operator for all particles consists of kinetic energy operators for the electrons and nuclei plus the potential energy of interaction:

$$H = T_e + T_N + V(r, R) \tag{1}$$

where $r$ and $R$ denote the collection of electronic and nuclear coordinates, respectively. The Schrödinger equation is

$$H\Psi(r, R) = E\Psi(r, R) \tag{2}$$

If the coordinates $R$ are held fixed, one can treat the motion of the electrons alone:

$$[T_e + V(r, R)]\phi_n(r, R) = U_n(R)\phi_n(r, R) \tag{3}$$

The subscript $n$ labels the various electronic eigenvalues $U_n(R)$. The nuclear equation is obtained from Eq. (2) by expanding $\Psi(r, R)$ in terms of the electronic eigenvalues:

$$\Psi(r, R) = \sum_n X_n(R)\phi_n(r, R) \tag{4}$$

Substituting Eq. (4) into Eq. (2), multiplying by $\phi_n^*(r, R)$, and integrating over $r$, one obtains

$$[T_N + \tilde{U}_n(R) - E]X_n(R) + \sum_{j \neq n} C_{nj}X_j(R) = 0 \tag{5}$$

where the $C_{nj}$ are sums of integrals over $r$ involving the $\phi_n$; the diagonal term, $C_{nn}$, is absorbed in $\tilde{U}_n$:

$$\tilde{U}_n(R) = U_n(R) + C_{nn}(R) \tag{6}$$

If the coupling terms in Eq. (5) are neglected ($C_{nj} = 0$), $\tilde{U}_n(R)$ plays the role of the potential function for the nuclei, and $\tilde{U}_n(R) \approx U_n(R)$.

## 2.2. Classical Dynamics

A rigorous treatment of the nuclear motion involves the solution of the nuclear wave equation (5). For many chemical systems, however, a classical treatment of the motion is adequate. This is done by using $U_n(R)$ as the potential energy function in the classical equations of motion (Newton's, Hamilton's, or Lagrange's equations), i.e., the rate of change of the momentum with time is equal to minus the gradient of the potential function, $-\nabla U_n(R)$. This law implies that the influence of the potential function on the dynamics is greater the smaller the momentum, because the perturbations in momentum due to the change in potential (i.e., the forces) are relatively larger. The largest influence of the potential will be in those regions where all the momenta are simultaneously small, i.e., in regions of low kinetic energy. These regions are the "critical states" through which the collision passes.

The classical description is quite different from the quantum one, for it allows one to describe the coordinates and momenta simultaneously as a function of time, and one can follow the path of the system as it goes from reagents to products during a collision. These paths, or trajectories, provide a "motion picture" of the collision process. A single trajectory, however, cannot represent the results of any real collision event. This can be done only by computing a large number of trajectories to obtain distributions of post-collision properties of interest (e.g., energy or angular distributions). Such postcollision distributions are obtained by averaging the properties of single trajectories over the (known) precollision distributions, which are determined by the experiment being simulated. In fact, the trajectory calculation really does nothing more than transform one distribution function (precollision, or reagent, distribution) into another one (postcollision, or product, distribution), and this transformation is determined by the potential energy function. The methods of computing these trajectories appear in Chapter 1 and are given in other reviews.[16,17]

It is important to bear in mind that when collision properties (e.g., average vibrational excitation of a product molecule) are discussed in relation to features of the potential function, it is always with reference to certain reagent distributions of initial variables. These distributions, along with the atomic masses, the total energy, and the reagent energy partitioning, are additional factors that influence the outcome of molecular collisions. The features of the potential surface do not have an influence that is independent of these distributions, but on the contrary, these various factors are intimately related.

## 3. Potential Energy Representations

### 3.1. Introduction

There are two aspects of surface representation that are relevant here: graphical and functional. The former is necessary for a description of the surface in topological terms. The latter is necessary because of practical problems associated with dynamical calculations, namely, there are invariably too few points at which the potential function is known from *ab initio* studies to allow a numerical calculation.

Most of the systematic dynamical studies on which this study is based utilize not *ab initio* surfaces but rather model surfaces generated by simple expressions of the potential as functions of internuclear distances. Surface topology is changed by manipulating the parameters in the functional expressions. The conclusions drawn about the effect of surface features on collisions are therefore open to criticism that the manipulation of functional parameters to alter a particular surface feature will, in fact, change the entire surface, and, therefore, the change in dynamics that results cannot be attributed to the alteration of the feature on which attention has been focused. It is often, however, the wrong approach to attribute dynamical effects to local features of a surface, because a given trajectory is in general governed by the cumulative influence of the potential in all the regions it traverses. A more useful goal is to identify those topological features that can serve as useful diagnostic indicators of the dynamical behavior on the entire surface. To achieve this goal, the model potentials used must possess the essential attributes of real potentials. Thus, the nature of the model potentials has an important bearing on the conclusions drawn, and so a somewhat detailed account is given of the more important ones in Section 3.4.

### 3.2. Graphical Representations of Potentials

*3.2.1. Contour Maps*

The potential energy functions needed to describe molecular collisions are in general multidimensional. The potential for an $N$-atom system is a function of $3N-6$ independent variables, usually taken to be the internuclear distances, $R_1, R_2, \ldots, R_{3N-6}$.

The most common manner of representing these multidimensional functions on paper is by a series of contour maps. Such a map is constructed by taking two variables on which the potential depends and representing them as distances along axes $x$ and $y$ in the plane of the paper. The potential function is then considered as a function only of $x$ and $y$, with all the other variables held fixed at some specified values. The function $U(x, y)$ can then be computed at any point in the $xy$ plane, and contours are drawn by joining all points $(x, y)$ for which $U(x, y)$ has the same value. The map consists of several contours corresponding to a series of values of the potential function, $U(x, y) = U_1$, $U(x, y) = U_2$, etc.

If the function $U(x, y)$ were plotted along the $z$ axis, it would form a surface over the $xy$ plane—hence, the expression "potential energy surface." Of course, the potential $U$ depends on more than two variables in general, so that the two-dimensional (2D) surface is really a cut through a multidimensional hypersurface. As the number of atoms in the system increases, one requires more and more such cuts to adequately picture the complete hypersurface.

As an example, consider one of the simplest rearrangement processes, a triatomic transfer reaction, $A + BC \rightarrow AB + C$. The AB, BC, and CA interatomic distances are denoted by $r_1, r_2$, and $r_3$, respectively, and the potential function by $U(r_1, r_2, r_3)$. The various asymptotic limits are $U_1(r_1) \equiv U(r_1, \infty, \infty)$ for AB + C, $U_2(r_2) \equiv U(\infty, r_2, \infty)$ for A + BC, $U_3(r_3) \equiv U(\infty, \infty, r_3)$ for CA + B, and $U_\infty \equiv U(\infty, \infty, \infty)$ for A + B + C. Because there are three independent variables, $r_1$, $r_2$, and $r_3$, a series of 2D maps with some quantity held fixed is needed. Two types of surfaces are commonly used: the fixed molecule surface (FMS) and fixed angle surface (FAS).

### 3.2.2. Fixed Molecule Surface for A + BC

Here, the BC molecule is situated along the $x$ axis with its midpoint at the origin and its bond distance fixed at $r_2 = R_M$. The values of $r_1$ and $r_3$ for any point in the $xy$ plane are determined by geometry, and the potential function is computed as $U[r_1(x, y), R_M, r_3(x, y)]$.

Examples are shown in Fig. 1 for the F + H$_2$ system with the H$_2$ molecule at the equilibrium distance (0.74 Å) and at an extended distance (1.0 Å). The contours are in electron volts (eV) and are referred to the reagent asymptotic region as zero $[U_2(r_2^0) = 0]$. At the H$_2$ equilibrium distance the potential is everywhere repulsive, whereas at extended geometries the FH$_2$ molecule is lower in energy than the reagents, the collinear geometry being preferred over a very bent one. The shape of FMS shown is quite typical of LEPS (London–Eyring–Polanyi–Sato) surfaces for exothermic reaction (see Section 3.4.3).

The FMS representation is best suited to describing the early part of a reactive collision and to displaying the orientational dependence of the potential. It cannot conveniently describe the whole rearrangement process in a single map.

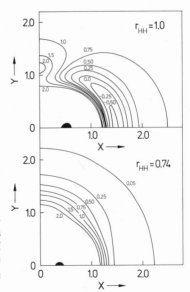

Fig. 1. Fixed molecule contour maps for an LEPS representation of the F + H$_2$ system. The HH molecule lies on the $x$ axis with the midpoint at the origin and the H atom at the position of the solid semicircle. Contours are in electron volts relative to the energy of F + H$_2$ at the H$_2$ equilibrium distance (0.74 Å).

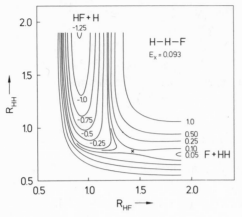

Fig. 2. Collinear (i.e., $\theta = 180°$) fixed angle contour map for an LEPS representation of $F + H_2 \rightarrow FH + H$. Contours are in electron volts (again relative to $F + H_2$) and the barrier is indicated by an x.

### 3.2.3. Fixed Angle Surfaces for $A + BC$

This representation focuses attention on the transition from reagents, through the transition region, to the product region and, as such, is better suited to a description of the rearrangement process. The $x$ and $y$ axes are chosen to represent the $r_1$ and $r_2$ distances, and the $r_3$ distance is obtained from these distances and the value of the ABC angle, fixed at a value $\theta$. The surface is normally drawn so that the reagent region lies in the lower right-hand corner and the product region in the upper left. FAS maps are normally drawn for $90° \leq \theta \leq 180°$, and they include the important collinear surface ($r_3 = r_1 + r_2$) when $\theta = 180°$.

Two examples of FAS representations of the LEPS surface for $F + H_2 \rightarrow HF + H$ are shown in Figs. 2 and 3. The contours are in electron volts, and the zero is again $U_2(r_2^0) = 0$. The reaction is exothermic by 1.34 eV and has barrier heights of 0.093 eV for $\theta = 180°$ and 0.98 eV for $\theta = 90°$. The positions of the barriers are shown by an x. The lower barrier height in the linear geometry is typical of the LEPS function. This is shown more clearly in Fig. 4,

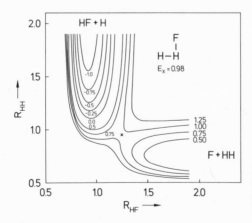

Fig. 3. Fixed angle ($\theta = 90°$) contour map (LEPS) for $F + H_2 \rightarrow FH + H$. Contours are in electron volts, and the barrier is marked by an x.

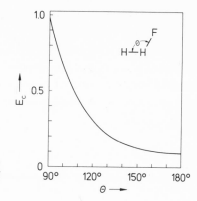

Fig. 4. Fixed angle surface barrier heights as a function of H–H–F angle for the LEPS representation of $F + H_2 \rightarrow FH + H$.

where barrier heights for a series of FAS cuts are plotted against the ABC angle.

Several terms used to describe FAS cuts are defined now for later reference. The lower right-hand region leads to the asymptotic region for the reagents; it is called the entrance channel, or entrance valley. Similarly, the upper left-hand region is the exit valley. A cut through the entrance (exit) valley along a line parallel to the exit (entrance) valley has the shape of the potential curve for the reagent (product) molecule; these cuts both terminate in the region corresponding to dissociation of the diatoms: $A + B + C$. The straight line $r_2 = r_2^0$ along the floor of the entrance channel is called the entry line; the corresponding line $r_1 = r_1^0$ is the exit line. The region near the intersection of the two lines is the corner of the surface. Another curve of use is the minimum reaction path (MRP). This is the path of steepest descent leading from the lowest point in the entrance channel to the lowest point in the exit channel (or vice versa if the reaction is endothermic); if there is a barrier, the starting point is taken at the crest of the barrier, and the portions going to the exit and entrance valleys are both obtained by steepest descent.

FAS representations are often used to characterize the entire potential. This is especially so in the case of collinear FAS maps, for if the velocities of the three atoms lie in the collinear direction, they will never depart from that direction, so that a collinear map can actually support a real dynamical path, a collinear trajectory.

## 3.3. Functional Representations

### 3.3.1. General Requirements

Closely connected with the problem of graphical representation of surfaces is that of functional representation. To solve the equations of motion, one must be able to find quickly and easily the first derivatives of the potential

$U(R)$ at any point $R$. In favorable cases, the potential energy function will be available in the form of a small set of (*ab initio*) values of $U(R)$ at various points in the interaction region. These values are the result of the lengthy quantum calculations necessary to solve the electronic Schrödinger equation, and as such they will rarely be spaced closely enough for direct use. Some interpolation method is therefore needed.

The method used ought to have several properties that make it suitable for dynamical calculations:

1. It should adequately describe the potential in the asymptotic regions. This ensures that the energetics for all possible collision processes are correct (e.g., heats of reaction, dissociation energies, etc.) and it is a minimum prerequisite for any reasonable potential.

2. It should represent the function accurately in those interaction regions for which nonempirical calculations or other kinds of information are available.

3. It should behave in a physically reasonable manner over the part of the interaction region where no potential information is available.

4. It should smoothly connect the asymptotic and interaction regions in a physically reasonable way.

5. It should reflect the symmetry properties of the system under study. For example, a potential function for the reaction $H + Cl_2 \rightarrow HCl$ ought to be symmetrical with respect to interchange of the two Cl atoms.

It is actually fairly difficult to obtain interpolation schemes (and especially model potentials) having all these properties, and many potential functions used in dynamical calculations fail in one or more of these points.

### 3.3.2. Local Interpolation

When many nonempirical points are available, it is advantageous to use a local interpolation scheme. An example is provided by McLaughlin and Thompson,[18] who use spline functions to represent the potential of $HeH^+ + H_2 \rightarrow He + H_3^+$ in $C_{2v}$ geometries. This is essentially a method of fitting the *ab initio* potential energies in a piecewise fashion and matching up the pieces at their end points so that the resulting function is smooth. The advantage is that the accuracy of the fit in each of the small regions is mainly determined by the shape of the function in that region itself, so that the representation is very flexible. Particular care must be taken with regard to the way the representation behaves at the boundary regions of the points being fitted (points 1, 3, 4, and 5 of Section 3.3.1).

### 3.3.3. Global Interpolation

For many collision systems, there is a lack of detailed information about the potential function. This situation favors the use of global fits to the potential energy function. Usually methods are chosen to meet requirements 1, 3, and 4 above, in the hope that an accurate fit to whatever information is available will be possible. In general, global fits are less accurate than local ones but, in the absence of much information, can be more convenient to use.

Global methods do not necessarily demand the use of an explicit functional expression to represent the potential. Any procedure or algorithm that can supply quick and accurate estimates of the potential could be used. For example, a semiempirical theory, such as the diatomics-in-molecules (DIM) method (see Section 3.4.3), would be suitable for those systems for which it possesses sufficient flexibility. Most global methods do use a functional expression, however, and we shall now consider these methods in greater detail.

## 3.4. Model Potentials

### 3.4.1. Introduction

Many dynamical studies are done without the benefit of any knowledge of the potential function in the interaction region, except perhaps for very limited information inferred from experiment (e.g., barrier heights). Such studies have as their goal either the discovery of the kinds of surfaces that are compatible with experimental observation[19-26] or the description of the effects of surface features on dynamics,[27-34] with which we are concerned here. Both types of study have prompted the use of model potential functions appropriate to the reactions under study.

The model potentials considered here are all relatively simple expressions for the potential function in terms of known analytical functions. They are intended to give a reasonable description of the potential in the interaction region, and they interpolate smoothly from this region to the asymptotic regions, thus satisfying criteria 1–4 of Section 3.3.1. They differ, however, in their flexibility, for some expressions are derived from semiempirical theories and are therefore necessarily restricted in form, whereas others are designed to be flexible enough to represent surfaces of almost arbitrary shape.

A nearly universal requirement of functions suitable for the description of rearrangement collisions is that the polyatomic interaction must not be written in terms of a sum of diatomic interactions. For example, suppose that a triatomic system had the potential

$$U(r_1, r_2, r_3) = U_1(r_1) + U_2(r_2) + U_3(r_3) \tag{7}$$

Each of the terms $U_i$ would have to represent the potential energy of the appropriate diatomic molecule if the asymptotic region were to be adequately represented. The additive expression would then have no remaining flexibility to describe the interaction region. Moreover, if AB, BC, and CA were all strongly bound molecules (as is often the case), then the pairwise potential would produce in general a very deep well in the corner of the surface, representing the *simultaneous* formation of two strong bonds. It is normally the case that the "old" bond breaks as the "new" bond forms, which means that the interaction between any two atoms is dependent on the positions of all the remaining atoms. This feature is of paramount importance, because it introduces coupling in the motions of the various coordinates, an effect that is called *potential coupling*.

### 3.4.2. Arbitrary Functions

In addition to the many different types of completely arbitrary functions used for fitting *ab initio* surfaces, there are some functional expressions designed to represent surfaces of specified shapes. The parameters in these expressions relate closely to the topological features, and the functions are often conceived in terms of fixed angle surfaces. A recent example of such a function is the hyperbolic map function of Bunker.[17] It is a descendant of a representation due to Wall and Porter[35] in which an FAS representation is generated by rotating a Morse function a quarter circle about some point in the upper right-hand corner (A+B+C region) of a FAS surface; the Morse parameters are functions of the rotation angle, and their angular variation directly controls the barrier height, width, and position as well as the characteristics of other features, such as potential wells. This type of representation has been applied in both systematic studies[35,36] and in cases where *ab initio* points need to be fitted.[37]

Another class of arbitrary functions offers a smooth interpolation from reagents to products, but without any close relationship between parameters and surface topology. An example is the earlier Blais–Bunker potential employing *switching functions*[17,38-41] This function represents the reaction A+BC → AB+C. The asymptotic regions AB+C and A+BC are each represented by Morse functions:

$$U_i(r_i) = D_i\{1 - \exp[-\beta_i(r_i - r_i^0)]\}^2 - D_i \tag{8}$$

where $i = 1, 2$. The parameters $D_i$, $\beta_i$, and $r_i^0$ are chosen to accurately represent the diatomic molecules AB and BC. The complete interaction potential is a sum of the two asymptotic terms plus two more terms: one that couples the $r_1$ and $r_2$ parts of the potential and another that provides a repulsion between atoms A and C.

$$U(r_1, r_2, r_3) = U_1(r_1) + U_2(r_2)$$
$$+ D_2[1 - \tanh(ar_1 + b)] \exp[-\beta_2(r_2 - r_2^0)]$$
$$+ D \exp[-\beta(r_3 - r_3^0)] \tag{9}$$

The function tanh $x$ goes smoothly from tanh $(-\infty)=-1$ through tanh $0 = 0$ to tanh $(+\infty)=1$. The values of the function in the asymptotic regions are A + B + C, $U_\infty = 0$; AB + C, $U_1(r_1^0) = -D_1$; and A + BC, $U_2(r_2^0) = -D_2$. The change in potential in going from reagents to products is $\Delta U = D_2 - D_1$ (<0 for exothermic reaction, $D_1 > D_2$). The entire interaction region is described by the third and fourth terms, and the parameters available for this are $a$, $b$, $D$, $\beta$, and $r_3^0$. This function and related versions of it have been used in several systematic trajectory studies.[38–41] Although it is a physically plausible representation, it is somewhat restrictive, and it cannot treat the region AC + B and therefore could not be applied in cases where BC was a homonuclear molecule.

### 3.4.3. Semiempirical Functions

*General Advantages and Disadvantages.* Semiempirical potential functions come from approximate solutions to the electronic Schrödinger equation, and therefore they often satisfy criteria 1, 3, 4, and 5 of a reasonable potential representation (see Section 3.3.1). Moreover, despite their crudeness, they do have a theoretical foundation and as such are more likely to embody the key features of the physical interaction than an arbitrary potential. On the other hand, they are somewhat restrictive and are not flexible enough to generate surfaces of arbitrary shape. Also, they are apt to be better representations for some families of reactions than for others. In short, their greatest virtue is their only fault.

*The Method of Diatomics-in-Molecules (DIM).* The DIM method[42,43] is treated here in some detail, and several examples explicitly given, because the majority of systematic studies on model potential surfaces have employed it. Basically, one starts with atomic states for each of the atoms making up the polyatomic collision system and couples these together to describe the states of interest of the diatomic, triatomic, ..., polyatomic groups. This coupling can be done in several ways (for given atomic states and a given order of coupling) and so leads to several basis functions for the polyatomic system. The trial wave function for the whole system is expressed as a linear combination of these basis functions, whose coefficients are determined by the usual variation principle. The matrix elements of the Hamiltonian operator in this basis set are evaluated in terms of the energies of the diatomic and atomic states that enter into the coupling scheme making up the particular polyatomic basis function. The method approximately expresses the energy of a polyatomic system in terms of atomic and diatomic energies, which are assumed known.

As an example, consider the triatomic system resulting from coupling together the three atomic states A $(^2P)$, B $(^2S)$, and C $(^2S)$, (e.g., FH$_2$). If A is first coupled to B, there results the set of AB diatomic states $^1\Sigma$, $^3\Sigma$, $^1\Pi$, $^3\Pi$. On coupling with C$(^2S)$ each of these leads to one basis function for the ground doublet state of ABC; hence, the variation problem involves a $4 \times 4$ secular equation. Linear combinations of these same four ABC basis functions are obtained if B is first coupled to C and the result to A, or if C is first coupled to A and the result to B. The matrix elements of the Hamiltonian operator in one of these basis sets can therefore be expressed in terms of the diatomic energies of the $^1\Sigma$, $^3\Sigma$, $^1\Pi$, $^3\Pi$ states of AB and AC plus the $^1\Sigma$ and $^3\Sigma$ energies of BC. These energies are evaluated at the distances of AB, BC, and CA appropriate to the geometry of ABC in question.

By its very nature the method yields the correct energies for those asymptotic regions that involve only atoms or diatomic molecules. As such, it provides a good interpolation method for connecting the reagent and product regions of triatomic systems. The method also takes full account of the symmetry of the polyatomic molecule.

*DIM Potential Functions.* In several instances the DIM energy expression can be written out explicitly, and in such cases the method has been extensively applied to dynamical calculations.[25–32] When the states of A, B, and C are all $^2S$, the London equation results:

$$U(r_1, r_2, r_3) = Q_1 + Q_2 + Q_3 - (J_1^2 + J_2^2 + J_3^2 - J_1J_2 - J_2J_3 - J_3J_1)^{1/2} \quad (10)$$

where $Q_i$ and $J_i$ (functions only of $r_i$) are defined in terms of the singlet and triplet diatomic energies:

$$Q_i(r_i) = (^1E_i + {}^3E_i)/2 \quad (11)$$

$$J_i(r_i) = (^1E_i - {}^3E_i)/2 \quad (12)$$

Equation (10) is appropriate to $H_3$ but has been used as a model potential for many triatomic systems, such as $HCl_2$ and $FH_2$; it has also been used in many systematic studies,[28–30] including those on which this chapter is based.

The London equation in the above form is also obtained for a four-atom system when all atoms are in $^2S$ states. The meaning of the terms is altered slightly: $Q_i$ and $J_i$ must be replaced by $Q_i + Q_{i+3}$ and $J_i + J_{i+3}$, respectively, where the indices 4, 5, and 6 refer to the diatomic pairs CD, DA, and BD, respectively.[31]

In the case for which $HeH_2^+$ is the prototype, a slightly simpler equation results[44]:

$$U(r_1, r_2, r_3) = Q_1 + Q_2 + Q_3 - [J_2^2 + (J_1 - J_3)^2]^{1/2} \quad (13)$$

where A is the He atom and $BC^+$ is $H_2^+$. Such an equation represents the potential function for $He + H_2^+ \rightarrow HeH^+ + H$. The quantities $Q_i$ and $J_i$ are defined just as in Eqs. (11) and (12), but in this case the $^1E_i$ and $^3E_i$ denote the interaction energies of the lowest $^2\Sigma_g$ and $^2\Sigma_u$ states of $H_2^+$, respectively, when

$i = 2$ and the $X^1\Sigma$ state of HeH$^+$ and $X^2\Sigma$ state of HeH, respectively, when $i = 1$ or 3.

*The LEPS Procedure* $(A + BC \to AB + C)$. In most applications of the DIM method, it is only the ground state potential surface that is required (see Section 7). In such cases the excited state diatomic curves are freely adjusted to provide a good representation of the surface in the interaction region, whereas the ground state diatomic curves are chosen to represent as accurately as possible the asymptotic regions. The London–Eyring–Polanyi–Sato (LEPS) procedure is a convenient recipe for obtaining a model DIM potential in this spirit.

As an example, consider the DIM equation for H$_3$. The ground state diatomic curves, $^1E_i$, are represented by Morse potentials:

$$^1E_i = {}^1D_i\{1 - \exp[-{}^1\beta_i(r_i - {}^1r_i^0)]\}^2 - {}^1D_i \tag{14}$$

The excited state diatomic curves are represented by "anti-Morse" potentials:

$$^3E_i = {}^3D_i\{1 + \exp[-{}^3\beta_i(r_i - {}^3r_i^0)]\}^2 - {}^3D_i \tag{15}$$

The nine parameters $^3D_i$, $^3\beta_i$, $^3r_i^0$ ($i = 1, 3$) could serve to adjust the shape of the potential in the interaction region, but in the LEPS procedure only the three $^3D_i$ are adjustable,* the other six being set equal to those for the ground state: $^3\beta_i = {}^1\beta_i$ and $^3r_i^0 = {}^1r_i^0$. This procedure allows one to specify the triatomic potential in terms of the spectroscopic parameters $^1D_i$, $^1\beta_i$, $^1r_i^0$ and the adjustable parameters $^3D_i$. The surfaces in Figs. 1–3 are examples of LEPS functions.

The above procedure is not restricted to H$_3$. The other model DIM equations for three- and four-atom systems can be manipulated in the same manner.

*Evaluation of DIM.* The DIM method provides a systematic, physically reasonable approach to the problem of selecting a potential energy surface for a many-body system. Moreover, for three- and four-body systems, the concise expressions it offers for the potential have proven well-nigh irresistible to many doing dynamical calculations, with the result that many of the conclusions and generalizations made about the effect of potential features on collisions have been based on studies of this one type of potential. It is important, therefore, to have some idea of the validity of these qualitative potentials. In most cases, no effort has been made to see exactly how well a DIM calculation can represent an *ab initio* calculation; however, there are some comparisons that are encouraging.

---

*Very often the LEPS potential is written in terms of "Sato parameters," $s_i$. The $Q_i$ and $J_i$ in Eq. (10) are replaced by $Q_i'/(1 + s_i)$ and $J_i'/(1 + s_i)$, and the $s_i$ also enter into the definition of the $Q_i'$ and $J_i'$ in terms of $^1E_i$ and $^3E_i$. The parameters $^3D_i$ are set equal to $^1D_i/2$, and the $s_i$ are used as adjustable parameters. This complicated procedure is entirely equivalent to the present formulation provided that $^3D_i = 0.5[(1 - s_i)/(1 + s_i)]^1D_i$.

The best *ab initio* calculation to date for the $H_3$ collinear surface[45] predicts a barrier height of $\simeq 9.8$ kcal/mole at a symmetrical geometry having $r_{HH} = 0.93$ Å $(= 1.757$ bohr). A DIM calculation of $H_3$ using unadjusted energies for the $^1\Sigma_g^+$ and $^3\Sigma_u^+$ curves of $H_2$ yields a barrier height of 12.5 kcal/mole at an $H_2$ distance of 0.95 Å (1.80 bohr).[46] Allowing the triplet curve free variation would give a definite improvement.

A nonempirical potential for collinear unsymmetric $HeH_2^+$ is also available[47] and is very well represented by Eq. (13). Unadjusted diatomic curves give a qualitatively good overall fit but fail to produce a small well in the surface. Slight adjustment of the excited state curves gives an excellent representation.[44]

The $H_3^+$ system has been studied very thoroughly by trajectory calculations using a DIM potential.[48] In this calculation two-potential surfaces enter into the dynamics, and so the need for a good potential representation is even more acute. A comparison of the DIM potentials with *ab initio* calculations[49] of the two lowest $^1A_1$ states of $H_3^+$ shows very good agreement, especially at the larger distances. In this case, the DIM diatomic curves needed are unadjusted, because they are all known theoretically.

An *ab initio* surface for the entrance channel of the four-atom system $HF + HF$ has also been computed.[50] Although the qualitative fit to an LEPS potential was good, the quantitative fit at small HF–HF distances was not too good. In this case, the proper DIM energies (as contrasted with the LEPS model) could provide a better representation, but this was not attempted. On the other hand, the LEPS equation for the system $FH_2$ gives a good qualitative representation of the features of the *ab initio* surface.[51]

These examples serve to illustrate that DIM potentials can provide reasonable representation of real potential surfaces. Their use in the study of the dynamical effects of potential energy surface features is therefore justified, and their restricted flexibility is not a great liability at all.

There are some practical difficulties involved with the model potentials when they are applied to a variety of systems. One is that they may be able to fit some geometries better than others. For example, Eq. (13) becomes oversimplified in $C_{2v}$ geometries, where the potential reduces to a sum of diatomic terms without the excited state curve of $H_2^+$. The omission of this curve (because of symmetry) restricts the equation's ability to handle $C_{2v}$ geometries. The London equation (10) for systems $AB_2$ also behaves oddly in $C_{2v}$ geometries, where $J_1 = J_3$ and $Q_1 = Q_3$. The potential reduces to

$$U = 2Q_1 + Q_2 - |J_1 - J_2| \qquad (16)$$

where 1 refers to AB and 2 to BB. Therefore, in some $C_{2v}$ geometry it is possible for the surface to have a cusp. The bump in the FMS representation for $FH_2$ (Fig. 1) along the line $x = 0$ is an example. Care must be taken in the integration of the equations of motion in this region.

# 4. Dynamical Principles of Reactive Collisions

In this section we shall present some guiding principles that form the background of later discussion. These principles are founded in the classical equations governing the motion, and as such they form the basis for all the generalizations to follow and a point of view for the interpretation of dynamical behavior.

## 4.1. The Problem of Rearrangement

### 4.1.1. Dynamical Requirements

In general, a reactive encounter involves the breaking of old bonds, the formation of new bonds, and an exchange or transfer of particles (i.e., mass) from one collision partner to the other, i.e., there is a change in energy (from potential to kinetic or vice versa) and a transfer of momentum. Accordingly, there are both energetic and dynamical requirements that must be met if reaction is to take place. The energetic requirements are obvious enough: There must be a path connecting the reagents to the products such that the kinetic energy everywhere along this path is positive, because regions of negative kinetic energy are the "classically forbidden" regions. The dynamical requirements are more subtle and, if they are not met, can cause a process to be "dynamically forbidden."[23,24]

In any collision system the available energy is divided among the various modes of motion. The amount of energy in each mode is continually changing as the collision proceeds, because of the influence of the potential function. For rearrangement to occur, momentum must be transferred from one mode to another in the time that it takes for the molecules to collide and separate. To accomplish this it is first necessary for the two modes to be *coupled* properly, that is, there must be some mechanism that can drain energy from one mode and pour it into the other. Second, the momentum changes corresponding to this coupling mechanism can be brought about only at a rate proportional to the forces that can be brought to bear upon the system through the potential surface. To effect the required changes in momentum, the forces must be able to act long enough. However, the maximum time that they can act is determined by the duration of the collision, and if this is too short, the rearrangement attempt will fail. Thus, a process can be dynamically forbidden even if it is energetically allowed and if a suitable coupling between modes is present. In fact it is quite possible for a reaction to fail even when the reagents have twice the necessary amount of energy.[29]

The coupling of the various modes of motion is always brought about through the interaction potential, but it is useful to distinguish several different types of mechanisms. Consider the reaction $A + BC \rightarrow AB + C$. For the products to separate, some motion must be developed in the B—C coordinate. If we consider the case where the reagents have only translational energy, this means that motion in the AB mode must develop into motion in the BC mode. One way this can occur is through a potential function that has a low-energy path from reagents to products only when the old bond gradually breaks as the new bond is gradually formed. In terms of a fixed angle surface, the trajectory turns the corner of the surface smoothly, and the collision is "direct." Another way for the coupling to occur is by a sequential mechanism involving several close collisions of the atom being transferred: Momentum is first transferred from A to B and then from B to C. This is typical of an "indirect" encounter involving the very repulsive parts of the diatomic potentials. The efficiency of both the direct and indirect coupling mechanisms is dependent on the masses of the atoms involved.

Even supposing that the shape of the surface and the collision conditions are suitable for reaction, there is still a great latitude in the type of reactive trajectory that can result. One kind of surface may have just the right forces acting to smoothly transform relative kinetic energy of the reagents into relative kinetic energy of the products. Such trajectories exhibit an internal energy *adiabaticity* and represent one extreme of reactive trajectory. At the other extreme are surfaces that are very poor at effecting this transformation and that tend to mix the modes of motion as the corner is rounded. Such surfaces tend to transpose reagent translational energy into product internal energy. How close trajectories lie to either of these extremes will depend on the distribution of reagents' energy at the beginning of the trajectory as well as on the surface features.

The general conclusion about rearrangement is that the properties of the reactive trajectories depend a great deal on the timing of momentum transfer between the various modes of motion. It should also be apparent that it is nearly impossible to discuss the effects of the potential surface in isolation from the other dynamical influences, in particular the masses of the atoms and the reagents' energy distribution.

### 4.1.2. Regions of Maximum Influence of Potential

*The Critical State.* We saw before that the potential energy function will have its greatest influence on a trajectory when two conditions are met: (1) The force in a given direction in coordinate space is large, and (2) the momentum in that same direction in coordinate space is small. This follows from Newton's second law: If a particular component of momentum is large compared with the perturbation caused by the force, the effect on that component is reduced in comparison to the effect the same force has on smaller components.

The potential function will exert its *maximum* influence in regions where the kinetic energy of the relevant mode of motion is small in comparison with the total energy available to that mode. Such regions include the turning points of the various modes and all points on the surface where the overall total energy equals the potential energy; the latter regions are possible candidates for "critical states." For example, for a total energy of $\approx 1.0\,\text{eV}$, the $F + H_2$ surface in Figs. 1–3 would exert its main influence near the 1.0 eV contour lines. There are two of these lines on both the FAS figures, and for $\theta = 90°$ they pass very near the crest of the barrier. Any of these regions could serve as a critical state of the system.

The critical state concept is quite useful because it is often possible to tell, on the basis of systematic trajectory studies, how the energy released in retreating from the critical state will apportion itself among the various degrees of freedom of the system. In the case that the critical state is the last one encountered by the trajectory, this allows inferences about the distribution of energy among the products in terms of the shape of the surface between the critical state and the exit valley.

*Direct and Indirect Collisions.* It is convenient to distinguish between direct collisions, which pass smoothly from reagents to products, encountering only one critical state, and indirect collisions, which take a more circuitous route across the potential surface, sampling more than one critical state. Direct trajectories require much simpler analysis than indirect ones, and most quantitative statements about the influence of surface properties on collisions are restricted to this type of collision.

Indirect collisions can come about either because the relative masses of the particles favor them or because of the nature of the potential. They can range from very simple collisions involving only two or three critical configurations to extremely complicated trajectories. So-called "complex" (or "snarled") trajectories have long collision times during which energy is constantly changed from one mode to another. Such trajectories are very sensitive to the initial conditions of the reagents, and it is very difficult to make simple statements about the relationship of the surface to the outcome of the collision; in these cases the outcome of trajectories is often statistical.

*Threshold Region Dynamics.* We shall consider here an exothermic reaction $A + BC \rightarrow AB + C$. In general there will be a barrier to reaction: Each possible path connecting reagents to products will have at some point in the interaction region a maximum potential energy, greater than the reagents' minimum potential energy, and the minimum of these maxima is the energy at the barrier, or "the barrier." The difference between this energy and the reagents' energy is the barrier height, $E_c$; the position of the barrier is denoted by $(r_1^{\ddagger}, r_2^{\ddagger}, r_3^{\ddagger})$, or just $(r_1^{\ddagger}, r_2^{\ddagger})$ for a collinear surface.

The existence of a barrier implies that there will be a certain threshold energy, $E_0 \geq E_c$, that the reagents must possess in order to form products. When the total energy of the system is in this threshold region, it is obvious that all reactive trajectories will encounter a critical state in the region around the crest of the barrier. Direct reaction, therefore, will be characterized by the shape of the potential between the barrier crest and the exit valley (see Section 4.3). Because many dynamical studies concern the threshold region, much of our later discussion will devolve upon this region.

At total energies greater than threshold, the barrier region gradually decreases in importance relative to possible critical states on the walls of the potential surface near the "corner." Dynamics in this energy region are correspondingly more difficult to predict, because it is harder to ascertain where on the surface the critical states will occur.

## 4.2. Mass Effects

### 4.2.1. Mass Ratios

The masses of the atoms involved in the collision process have an enormous influence on the dynamics. In the classical approximation, however, it is only the ratios of masses, and not the absolute masses, that are important. If the masses of all atoms in a process are changed by the same factor, then the trajectory will not change its path through coordinate space but will only traverse that path in a different amount of time. This is true provided that the kinetic energy of all atoms at the beginning of the trajectory is kept the same.

The unimportance of absolute mass follows from the very definition of mass in classical physics, and it is evident from the classical equations of motion. The position of each atom can be described by three Cartesian coordinates, and the equation for each of these is

$$m_i \frac{d^2 Q_i}{dt^2} = F_i = -\left(\frac{\partial U}{\partial Q_i}\right) \tag{17}$$

where $U$ is the potential energy function and $Q_i$ is one of the Cartesian coordinates. The force is independent of mass, and so the $Q_i$ coordinates will bear the same relationship to each other if all masses $m_i$ are multiplied by the same factor, $c$. This merely has the effect of scaling the time by an amount $c^{1/2}$; i.e., exactly the same trajectory is obtained if we use the masses $M_i = cm_i$ and the scaled time $T = c^{1/2} t$:

$$M_i \frac{d^2 Q_i}{dT^2} = F_i \tag{18}$$

The initial kinetic energy in these new variables is the same as in the old:

$$\frac{1}{2} \sum_i M_i \left(\frac{dQ_i}{dT}\right)^2 = \frac{1}{2} \sum_i m_i \left(\frac{dQ_i}{dt}\right)^2 \tag{19}$$

This fact can sometimes be of use in eliminating mass effects when comparing different experiments. For example, the systems $Ne + H_2(v = 0)$ and

$Ar + D_2(v = 1)$ are very nearly equivalent in this respect ($Ne = 20$ amu, $Ar = 40$ amu), there being only a slight difference in vibrational energy. Another example is $K + Cl_2(1 + 0.91 + 0.91)$ and $Rb + Br_2(1 + 0.94 + 0.94)$. Experiments carried out with the same collision energy for each member in the pair will therefore show up differences in the potential energy surfaces. These relations have recently been exploited by Grosser and Haberland[52] in experiments on the systems $H + IBr \rightarrow HI + Br$, $H + I_2 \rightarrow HI + I$, $H + Br_2 \rightarrow HBr + Br$, and $H + Cl_2 \rightarrow HCl + Cl$. They observe that the angular differential cross sections progressively shift toward more forward directions (see Section 6.3.1) along the series. Moreover, the identical distributions are obtained when one substitutes D for H. They argue that the shift in angular distribution cannot be a mass effect but must reflect some qualitative differences in the potential energy surfaces for the various systems, because the reactions ($H + Cl_2$, $D + Br_2$) have roughly the same mass ratios, as do to a lesser extent the pair ($H + Br_2$, $D + I_2$).

### 4.2.2. Inertial Coupling

We have already seen that, in general, rearrangement requires a transfer of energy in just the right way from one mode of motion to another during a collision. Coupling in the potential function is one way of bringing this about; inertial (or mass) coupling is another. This kind of coupling arises when the masses are such that motion in one degree of freedom implies motion in another, on purely inertial grounds.

As an example, consider the collinear triatomic systems **L—H—L** and **H—L—H**, where **H** and **L** denote heavy and light masses, respectively. In the **L—H—L** system, motion in one of the **L—H** coordinates has very little influence on the other, because the **H** particle need only move slightly to keep the center of mass fixed. (All motion is discussed in the center of mass system, because the motion of the center of mass itself is completely separable from the internal motion, the potential function involving only relative coordinates.) On the other hand, any motion in one of the **L—H** coordinates of **H—L—H** implies a great deal of motion in the other coordinate. This coupling can effect a flow of energy between the coordinates even if there is no coupling in the potential function. A more explicit treatment of mass effects follows later.

## 5. Collinear Dynamics

### 5.1. Equations of Motion

Collinear systems are very useful for describing the exchange reaction $A + BC \rightarrow AB + C$ in its simplest terms. Only two coordinates, $r_1$ (the A—B distance) and $r_2$ (B—C distance), are needed to specify the geometry; the third coordinate, $r_3$ (C—A distance), is a sum of the others: $r_3 = r_1 + r_2$. This substitution allows one to define the potential function in terms of $r_1$ and $r_2$ alone:

$$\tilde{U}(r_1, r_2) = U(r_1, r_2, r_1 + r_2) \tag{20}$$

The equations of motion in the variables $r_1$ and $r_2$ are

$$\ddot{r}_1 = -[(A+B)/(AB)](\partial\tilde{U}/\partial r_1)_{r_2} + (1/B)(\partial\tilde{U}/\partial r_2)_{r_1} \tag{21}$$

$$\ddot{r}_2 = (1/B)(\partial\tilde{U}/\partial r_1)_{r_2} - [(B+C)/(BC)](\partial\tilde{U}/\partial r_2)_{r_1} \tag{22}$$

The kinetic energy is

$$T(\dot{r}_1, \dot{r}_2) = \alpha_1^2 \dot{r}_1^2/2 + AC\dot{r}_1\dot{r}_2/M + \alpha_2^2\dot{r}_2^2/2 \tag{23}$$

where $\alpha_1^2 = A(B+C)/M$, $\alpha_2^2 = C(A+B)/M$, and $M = A+B+C$.

The role of potential coupling and inertial coupling is made explicit by these equations. The terms in $1/B$ in the equations of motion are responsible for the inertial coupling between the $r_1$ and $r_2$ coordinates. If these terms were absent, the only way that motion in one coordinate could influence that in the other would be via coupling in $\tilde{U}_1(r_1, r_2)$. On the other hand, if $\tilde{U}(r_1, r_2)$ were a sum of terms, $\tilde{U}_1(r_1) + \tilde{U}_2(r_2)$, and involved no coupling, the presence of the $1/B$ terms would serve to couple the motion.

## 5.2. Single-Particle Representation

### 5.2.1. Coordinates That Diagonalize the Kinetic Energy

It is possible to choose a set of coordinates $Q_i$ $(i = 1, N)$ in which the trajectory represents the motion of a single hypothetical particle moving in $N$ dimensions under the influence of the transformed potential, $U(Q_1, \ldots, Q_N)$.[53] Such a particle would have a kinetic energy consisting of a sum of the squared velocities, $\dot{Q}_i^2$, all multiplied by the same factor:

$$T(\dot{Q}_1, \ldots, \dot{Q}_N) = m\left(\sum_i \dot{Q}_i^2\right)/2 \tag{24}$$

Motion in such a coordinate system is then easier to visualize.

In the collinear system $A + BC \rightarrow AB + C$, $r_1$ and $r_2$ must be transformed into $Q_1$ and $Q_2$. A linear transformation that effects the required diagonalization of $T$ is[54]

$$Q_1 = r_1 + \beta r_2 \sin \phi$$
$$Q_2 = \beta r_2 \cos \phi \tag{25}$$

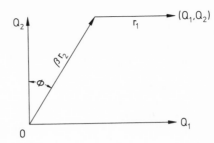

Fig. 5. Relationship between the coordinates $(Q_1, Q_2)$ and $(r_1, r_2)$ showing also the angle of skew, $\phi$, and the scaling factor $\beta$.

with the inverse transformation

$$r_1 = Q_1 - Q_2 \tan \phi$$
$$r_2 = Q_2 \beta^{-1} \sec \phi \qquad (26)$$

Substitution into Eq. (23) diagonalizes $T$ provided that

$$\sin \phi = \{AC/[(A+B)(B+C)]\}^{1/2} \qquad (27)$$

$$\beta = \alpha_2/\alpha_1 = \{C(A+B)/[A(B+C)]\}^{1/2} \qquad (28)$$

The effective mass is $m = A(B+C)/M = \alpha_1$, but its value is irrelevant to the dynamics, as was shown in Section 4.2.1.

The relationship between $(Q_1, Q_2)$ and $(r_1, r_2)$ is shown in Fig. 5, where it is seen that the axes $r_1$ and $r_2$ have been "skewed and scaled" relative to $Q_1, Q_2$. The potential surface is plotted in the $Q_1 Q_2$ plane and will always lie to the right of the skewed axis. The result of mapping $r_2$ onto the skewed axis $\beta r_2$ will be to extend the surface along the $\beta r_2$ axis if $\beta > 1$ (e.g., the points at $r_2 = 1.0$ will appear at the value $\beta r_2 = 2$ if $\beta = 2$) and to compress the surface along the $\beta r_2$ axis if $\beta < 1$ (the points at $r_2 = 1.0$ will now appear at $\beta r_2 = 0.5$ if $\beta = 0.5$). For later reference, some values of $\phi$ and $\beta$ are given in Table 1 for all possible combinations of heavy (**H**) and light (**L**) masses (**H/L** $= 80$) in the triatomic system $A + BC \rightarrow AB + C$.

Table 1. *Diagonalization Parameters for Various Masses*[a]

| A | B | C | $\phi(°)$ | $\beta$ |
|---|---|---|---|---|
| **L** | **L** | **L** | 30 | 1.00 |
| **L** | **L** | **H** | 44.65 | 1.41 |
| **L** | **H** | **L** | 0.71 | 1.00 |
| **L** | **H** | **H** | 4.51 | 6.36 |
| **H** | **L** | **L** | 44.65 | 0.71 |
| **H** | **L** | **H** | 80.99 | 1.00 |
| **H** | **H** | **L** | 4.51 | 0.16 |
| **H** | **H** | **H** | 30 | 1.00 |

[a] Reaction is $A + BC \rightarrow AB + C$; **L** $= 1$ amu, **H** $= 80$ amu.

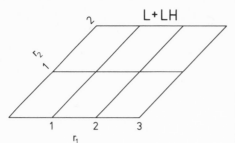

Fig. 6. Grid lines for the masses **L+LH** ($\beta > 1$) in the skewed and scaled coordinate system. The **L—L** distance is $r_1$; the **L—H** distance is $r_2$.

The usual way of drawing collinear surfaces in the skewed and scaled representation is then to draw the axis $r_1$ horizontally and to incline the $r_2$ axis an angle $\phi$ away from the vertical. Units for $r_1$ and $r_2$ are marked on each axis, but the units on the $r_2$ axis are scaled by the factor $\beta$. An example of the grid lines to use for the (complementary) systems **L+LH** and **H+LL** are shown in Figs. 6 and 7; in the former case the $r_2$ axis is extended relative to $r_1$, and in the latter case it is compressed. Both systems have the same angle of skewing.

### 5.2.2. Interpretation of Dynamics in the Skewed and Scaled System

For the same potential energy surface, systems with nearly the same skewing angle *and* scaling factor will exhibit nearly the same dynamics. In agreement with the previous comments concerning identical mass ratios, Table 1 shows that **L+LL** and **H+HH** have the same $\phi$ and $\beta$.

The parameter $\phi$ provides information about the extent of inertial coupling between the two coordinates. If $\phi$ is small, the axes $r_1$ and $\beta r_2$ are nearly perpendicular, and motion in one coordinate does not imply much motion in the other. This occurs if either A or C is much smaller than B. On the other hand, if $\phi$ is large, the motions in the two modes are strongly coupled.

The parameter $\beta$ determines the relative time scales of the motion in the two coordinates for a given skew angle. Forces in the relatively extended coordinate are diminished relative to those in the compressed coordinate. If $\beta > 1$, the exit coordinate $r_2$ is extended, so that motion in the entrance channel takes place more quickly for equal forces. As $\beta$ increases, the duration of the

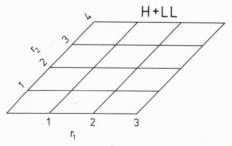

Fig. 7. Grid lines for **H+LL** ($\beta < 1$) in the skewed and scaled system. The **L—L** distance is $r_2$; the **H—L** distance is $r_1$.

collision decreases, leaving less time for the momentum transfer necessary for rearrangement; hence, reaction is favored by systems having small $\beta$.

The advantage of the skewed and scaled coordinate system in two dimensions is that it provides an analogy that allows one to guess more easily the outcome of a trajectory. One can imagine, for instance, that the hypothetical particle has an electric charge and is moving in a plane (the $Q_1 Q_2$ plane) under the influence of an electric field, whose potential lines are given by the (transformed) potential in the skewed and scaled system. The absence of the cross terms in the kinetic energy expression allows one to visualize the motion more easily. Another (less appropriate) analogy is that of a mass sliding over a three-dimensional surface. This analogy introduces motion in a direction perpendicular to the $Q_1 Q_2$ plane, a dimension that does not exist in the kinetic energy expression $T(\dot{Q}_1, \dot{Q}_2)$, so that it is only faithful when the potential surface is flat, or at any rate, not too steep. It is easier to think in terms of these analogies than in terms of the equations of motion, and we shall refer to them many times. It should be noted, however, that it is not always so simple to predict where a sliding mass will go when one has a real potential, and one must ultimately rely on the solution to the equations.

## 5.3. Collinear Rearrangement Dynamics

In Section 4 the problem of rearrangement was seen to be equivalent to being able to turn the corner of the potential surface. This aspect of a reactive trajectory can be discussed in more detail here with the help of the skewed and scaled coordinate system.

### 5.3.1. Mass Effects

Perhaps the easiest systems to discuss in terms of skewed and scaled axes are those for which A, B, and C are regarded as hard spheres moving in "square well" potentials, i.e., the potentials are simply lines forming two channels, one parallel to the $r_1$ axis and one parallel to the $r_2$ axis. The trajectories are then easily obtained as straight line segments that are reflected from the potential walls with the angle of reflection equal to the angle of incidence. A trajectory coming in parallel to the $r_1$ axis will behave in quite different ways depending on the angle of skew. If $\phi \approx 0°$, the trajectory will merely bounce out into the entrance channel again. If $\phi = 30°$, the trajectory will bounce twice, once in a diagonal direction, corresponding to B moving between stationary A and C, and once more in a direction parallel to the $r_2$ axis, corresponding to C leaving AB. If $\phi = 45°$, the trajectory bounces three times; the first collision of A with B converts all kinetic energy to vibration of BC, but after one complete vibration, this energy is again put back into translation of A relative to BC, and no reaction occurs. This illustrates that just the right amount of coupling is needed to turn the corner of the surface.

Inertial coupling can be demonstrated by considering a potential function that is devoid of coupling between the $r_1$ and $r_2$ degrees of freedom:

$$\tilde{U}(r_1, r_2) = \tilde{U}_1(r_1) + \tilde{U}_2(r_2) \tag{29}$$

The potential will have a minimum at the point $(r_1^0, r_2^0)$ where the entry and exit lines intersect, and around this point it will be approximately quadratic in the displacements from equilibrium:

$$\tilde{U}(r_1, r_2) \simeq \tilde{U}_1(r_1^0) + \tilde{U}_2(r_2^0) + \tfrac{1}{2}K(\Delta r_1^2 + \Delta r_2^2) \tag{30}$$

where we have chosen the same force constant $K$ for both modes for simplicity. The potential contours are circles in the $r_1 r_2$ plane, and the direction of steepest descent from a point in the entrance channel on the entry line lies along the entry line. In the $Q_1 Q_2$ plane, however, the potential has the form

$$U(Q_1, Q_2) = U_1(Q_1^0) + U_2(Q_2^0)$$
$$+ \tfrac{1}{2}K[\Delta Q_1^2 - 2\Delta Q_1 \Delta Q_2 \tan\phi + \Delta Q_2^2(\tan^2\phi + \beta^{-2}\sec^2\phi)] \tag{31}$$

The direction of steepest descent from points on the entry line is along the direction of the gradient:

$$\nabla U(Q_1, Q_2) = K[\Delta Q_1 \hat{Q}_1 - (\Delta Q_1 \tan\phi)\hat{Q}_2] \tag{32}$$

If inertial coupling is small, $\phi \simeq 0$, and there is no force in the $Q_2$ direction, but as inertial coupling increases, there develop strong forces in the $Q_2$ direction. Thus a sliding mass on such a surface will tend to be pulled away from the entry line, and an oscillation about the entry line will develop, corresponding to the transfer of energy into the relative motion of the BC bond.

### 5.3.2. Curvature and Minimum Path Characterization

When the angle $\phi = 0$ in the last example the path of steepest descent merely follows the entry line, but when $\phi > 0$ it departs from the entry line, before turning back again toward the minimum, i.e., it develops a curvature. Now, at any point on a collinear potential surface, the maximum force acts along the path of steepest descent, when this is defined. A curvature in this path implies a changing force (at least in direction). The hypothetical particle approaching along the entry line will experience this change in force and will tend to follow the path of steepest descent *if it is moving slowly enough.* At higher velocities, however, it will not follow the direction of maximum force, because its momentum in the original direction is too great. Hence, the behavior of the curvature in the path of steepest descent is a feature of a potential surface that will influence the collision dynamics, because a trajectory started off along the path of steepest descent will tend to follow that path if the *momentum* in the direction along the path is *small* in the region where the *curvature* is *large*.[55-57]

In the example above, the surface has a hollow, and the path of steepest descent depends on where one starts in the entrance channel. Most surfaces that we shall consider have no hollow, but rather a barrier, and they therefore can be categorized by the shape of the path of steepest descent leading from the top of the barrier into the entrance and exit channels, the so-called minimum reaction path (MRP). Of particular interest is the way in which the potential energy decreases along this path from the barrier crest into the exit channel (which we assume here is lower than the entrance channel), for this will determine the way in which the heat of reaction is distributed among the various degrees of freedom of the products for direct collisions in the threshold region. Three cases arise, depending on whether the energy is released before, during, or after the region of maximum curvature. If most of the energy is released before the maximum curvature region, the hypothetical sliding mass will have too much momentum to follow the minimum path, and it will tend to retain a large amount of vibrational motion in the AB bond, provided the products can separate smoothly. At the other extreme, if the particle encounters the large curvature region at low velocities, it will tend to have its motion more easily changed into relative motion along the exit channel, and the subsequent release of most of the energy will merely enhance this relative motion, producing AB with little vibration. If energy is released while the corner is being rounded, it is difficult to say what will happen in general, because the precise shape of the surface will play a more important role in this case.

The three categories of energy release are termed attractive, repulsive, and mixed. A method of classifying surfaces according to the energy released along the minimum path is shown in Fig. 8 for the case of a barrier in the entrance channel. The barrier is at the point $(r_1^\ddagger, r_2^\ddagger)$ marked by the X in the figure, and the energy here is an amount $E_c$ above that of the entrance channel; the exothermicity is $-\Delta U$. For the minimum path classification it is necessary to define *amplitude lines* $r_2 = r_2^a$ parallel to the entry line and $r_1 = r_1^a$ parallel to the exit line. The minimum path approaches both the entry and exit lines asymptotically, and it cuts the entry amplitude line at $r_1 = (\bar{r}_1)_M$ and the exit amplitude line at $r_2 = (\bar{r}_2)_M$. The attractive, mixed, and repulsive energies are defined in terms of the differences in energy between these various points, expressed as a percentage of the total energy available, $E_c - \Delta U$:

$$A_M = 100[\tilde{U}(r_1^\ddagger, r_2^\ddagger) - \tilde{U}(\bar{r}_1, r_2^a)]/(E_c - \Delta U)$$

$$= 0 \qquad \text{if } \bar{r}_1 > r_1^\ddagger \qquad\qquad (33)$$

$$M_M = 100[\tilde{U}(\bar{r}_1, r_2^a) - \tilde{U}(r_1^a, \bar{r}_2)]/(E_c - \Delta U)$$

$$= 0 \qquad \text{if } r_1^a > \bar{r}_1 \tag{34}$$

$$R_M = 100 - A_M + M_M \tag{35}$$

The amount of mixed energy release depends on $r_1^a$ and $r_2^a$. Obviously, one wants these quantities to reflect values of $r_1$ and $r_2$ that are significantly greater than $r_1^0$ and $r_2^0$, respectively, but it is clear that the exact amount is arbitrary. The convention[32] is to take distances greater than the vibrational amplitudes of AB and BC as significant; this amplitude depends on the vibrational energy, which is taken to be the larger of the zero-point energy and the thermal energy.

### 5.3.3. Dynamic Characterizations of Surfaces

The minimum path categorization is nice because it depends only on the potential surface itself, and trajectory studies show that for a given mass combination the sum of $A_M + M_M$ correlates with the products' vibrational excitation.[58] Nevertheless, it is better to use a classification scheme that also takes into account the inertial effects of the collision while still retaining some properties of the surface topology.

Properties of the surface itself are used in the *rectilinear method* of classification.[28,32] It ignores mixed energy release and attempts to divide the energy drop from the crest of the barrier to the bottom of the exit valley (at infinite separation of products) into attractive and repulsive components, $A_\perp$ and $R_\perp$. The portion $A_\perp$ is the drop in energy along a line parallel to the entry line that starts at the barrier crest and ends either at the intersection with the exit line or at the minimum of $\tilde{U}$ $(r_1, r_2^\ddagger)$, whichever comes first (the point marked by $\perp$ in Fig. 8); the value of $r_1$ at this point is denoted $\tilde{r}_1$. If the surface has no barrier, then the starting point is on the entry line at $r_1 = \infty$. Again, $A_\perp$ and $R_\perp$ are expressed as percentages of $E_c - \Delta U$:

$$A_\perp = 100[\tilde{U}(r_1^\ddagger, r_2^\ddagger) - \tilde{U}(\tilde{r}_1, r_2^\ddagger)]/(E_c - \Delta U) \tag{36}$$

$$R_\perp = 100 - A_\perp \tag{37}$$

Dynamical effects are taken into account by the *trajectory classification method*, in which a collinear trajectory for the appropriate mass combination is computed. The initial translational energy is equal to the average thermal energy, or the threshold energy if there is a barrier; there is no vibrational energy in BC. The trajectory cuts the entry amplitude line at $r_1 = (\bar{r}_1)_T$ and the *exit line* at $(\bar{r}_2)_T$. The attractive, mixed, and repulsive components are therefore

$$A_T = 100[\tilde{U}(r_1^\ddagger, r_2^\ddagger) - \tilde{U}(\bar{r}_1, r_2^a)]/(E_c - \Delta U) \tag{38}$$

$$M_T = 100[\tilde{U}(\bar{r}_1, r_2^a) - \tilde{U}(r_1^0, \bar{r}_2)]/(E_c - \Delta U)$$

$$= 0 \qquad \text{if } \bar{r}_2 < r_2^a \tag{39}$$

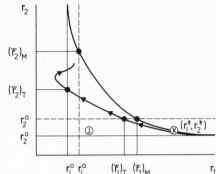

Fig. 8. Points on a collinear surface needed for the minimum path, rectilinear, and trajectory classification methods. The minimum reaction path is shown as a solid line and the schematic trajectory by the solid line with arrows.

$$R_T = 100 - A_T - M_T \tag{40}$$

A schematic classification trajectory is shown by the solid line with arrows in Fig. 8.

The dynamic classification scheme presented here is quite useful for surfaces with barriers in the entrance valley.[28] Three of these surfaces are shown in Figs. 9 and 10; all have the same exoergicity, but they differ in the amount of their energy release, $A_\perp$ ranging from 3% to 47% to 72% for surfaces 1, 8, and 4, respectively. Values of $A_T$, $M_T$, and $A_\perp$ are shown in Table 2 for two different mass combinations, **L+HH** and **L+LL**.

### 5.3.4. Dynamical Role of Mixed Energy Release

It is convenient to use the rectilinear classification scheme when *comparing surfaces*, whereas discussion of *dynamics* is best done in terms of the quantities $A_T$, $M_T$, and $R_T$. The idea behind the classification schemes is based on the qualitative notion that attractive energy release, $A_T$, tends to become vibration of the new bond, whereas repulsive release, $R_T$, will become product translational energy. Mixed energy release can go into vibration or translation, depending on circumstances.[36]

On some surfaces, the potential energy rises steeply as A approaches B (i.e., after the barrier is climbed), which means that $A_\perp \simeq 0$. In such a case, the quantity $M_T$ can be quite high, and the barrier crest is not far from the region of maximum curvature of the minimum reaction path. As previously pointed out, the influence of the surface in guiding the sliding mass smoothly around the corner is very efficacious, so that a substantial proportion of $M_T$ tends to become translational energy of the products.

More often, a surface has a significant attractive component of energy release, $A_\perp$, and the trajectory has considerable latitude (on energetic grounds) as to where it enters the exit valley. If it enters at the valley's head, $M_T$ will be small, and the trajectory will experience a repulsion almost equal to $R_\perp$, which

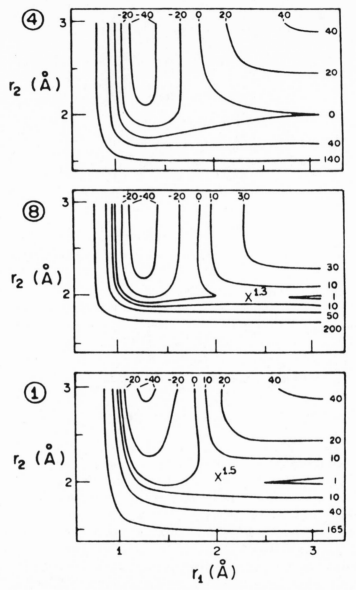

Fig. 9. Collinear surfaces for an exothermic reaction with varying amounts of attractive energy release $A_\perp$: surface 1, 4%; surface 8, 47%; surface 4, 72%. Contours are in kilocalories per mole. (After Kuntz et al.[28])

favors product translation. On the other hand, if it enters at the side of the valley, it can *avoid* the large repulsion; $M_T$ is increased at the expense of $R_T$, and the products tend to receive more vibrational excitation. This is summarized by the hypothesis that $A_T + M_T$ correlates well with the vibrational excitation of the products. Also, if trajectories can avail themselves of the

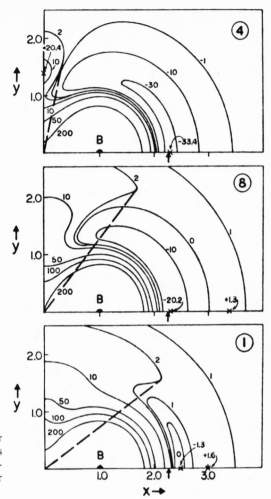

Fig. 10. Fixed molecule surfaces for potentials 1, 8, and 4 for BC at its internuclear distance. Contour energies are in kilocalories per mole. (After Kuntz et al.[28])

mixed energy release mechanism of avoiding repulsion, then surfaces with large $R_\perp$ can still yield highly vibrationally excited products.

An example of this effect of mixed energy release is shown in Fig. 11 for a surface with $R_\perp \approx 43\%$. The trajectories start from the lower right-hand corner, which can be considered as the barrier. For the mass combination **L+HH** a typical trajectory enters the valley much closer to the head than for **L+LL** and thereby encounters more repulsion. The average product vibrational excitation for the two cases is $\approx 75\%$ and $\approx 85\%$, respectively.

It is apparent that the amount of repulsion, $R_\perp$, that can be avoided via "mixed energy release" is quite dependent on the masses for a given potential energy surface. For example, for the mass group **L+HH** the **LH** molecule being formed can easily reach its equilibrium distance before the **HH** distance has a chance to increase. The bulk of the repulsion must then be released at

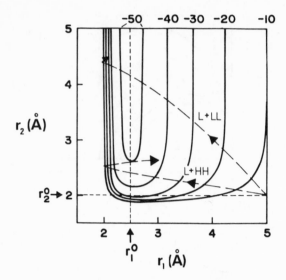

Fig. 11. Trajectories illustrating that mixed energy release is greater for **L+LL** than for **L+HH**.

small **LH** distances, a situation unfavorable to vibrational excitation of the **LH** bond. On the other hand, the "reverse" mass group **H+HL** is just the opposite and allows almost all repulsion to be released while the **HH** bond is still greatly extended. Table 1 shows that both of these masses have the same skew angle $\phi$, but that the *exit* valley is stretched for **L+HH**, whereas the *entrance* valley is stretched for **H+HL**; hence, for a given skew angle, mixed energy release is favored by masses that stretch the entrance coordinate relative to the exit coordinate (i.e., $\beta < 1$). For example, on surfaces 1, 8, and 4 in Fig. 9, the values of $M_T$ are 0, 14, and 0% for **L+HH** compared with 99, 97, and 80% for **H+HL**; also, they are 46, 85, and 80% for **L+LH** compared with 69, 92, and 84% for **H+LL**. The differences are not so pronounced in the latter case, because the scaling factor $\beta$ is closer to unity.

The values of $M_T$ are especially low for **L+HH** compared with *any* other mass group in Table 1. This means that $A_T + M_T \simeq A_\perp$ (Table 2), and therefore the amount of vibrational excitation in the product molecule will correlate quite well with $A_\perp$ itself. In other words, this mass combination is especially suited to probing the shape of the potential surface, at least as regards energy distributions. A correlation of the average vibrational energy with $A_\perp$ (for a 2D trajectory calculation[28]) is shown in Fig. 12 for the surfaces in Fig. 9. The

*Table 2. Some LEPS Surface Properties*

| Surface | Masses | % $A_\perp$ | % $A_T$ | % $M_T$ | % $\langle V' \rangle$ |
|---|---|---|---|---|---|
| 1 | **L+HH** | 3 | 4 | 0 | 13.5 |
|   | **L+LL** | 3 | -2 | 53 | 44 |
| 8 | **L+HH** | 47 | 33 | 14 | 58.5 |
|   | **L+LL** | 47 | 1 | 86 | 84 |
| 4 | **L+HH** | 72 | 70 | 0 | 71 |
|   | **L+LL** | 72 | 14 | 76 | 84 |

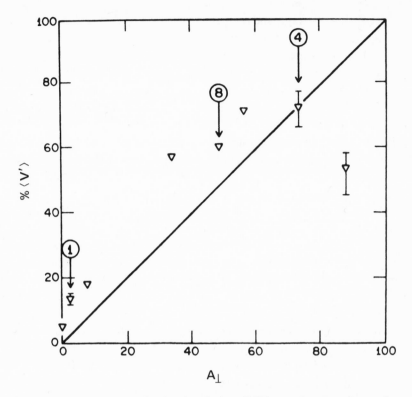

Fig. 12. Mean product vibrational excitation, $\%\langle V'\rangle$, as a function of attractive energy release, $A_\perp$, for a series of surfaces of the same exothermicity, including the surfaces 1, 8, and 4. The reaction is $\mathbf{L+HH \rightarrow LH+H}$. (After Kuntz et al.[28])

behavior of $\mathbf{L+HH}$ in this regard has come to be known as the "light atom anomaly."[16]

## 5.4. Utility of Collinear Calculations

### 5.4.1. An Illustration

The chief advantage of collinear calculations is their simplicity compared with three-dimensional ones. The complete potential surface is present in a single contour diagram, and only two kinds of motion, translational and vibrational, need be considered. Also, the visualization of the rearrangement in terms of the sliding mass analogy is an aid to understanding.

As an illustration, consider using the equations of motion (21) and (22) in a trajectory study. The collinear trajectories are defined by these equations once the initial values of $r_1$, $r_2$, $\dot{r}_1$, and $\dot{r}_2$ are given. These, in turn, are related to the initial separation of A and BC, $R_{A,BC}$; the initial relative kinetic energy, $T$; the initial vibrational energy, $V$; and the phase angle of the BC vibration, $\Phi$. The value of $R_{A,BC}$ is a constant, and once the energies $T$ and $V$ are chosen, there remains only the variable $\Phi$ over which to average, so that the vibrational excitation of the products, $V'$, is readily obtained as a function of $\Phi$. The total energy of the products is $T' + V'$, and so the product energy distribution can be characterized by the fraction of available energy that enters the vibrational mode, $f_{V'} = V'/(T' + V')$. A typical plot of $f_{V'}$ against phase angle is shown in Fig. 13 for a potential surface appropriate to the reaction $He + H_2^+ \rightarrow HeH^+ + H$.[44] The trajectories have a total energy of 25 kcal/mole distributed as $T = 22$, $V = 3$ kcal/mole. The different regions of phase angle lead to trajectories with different outcomes: The largest region is for unreactive trajectories $(A+BC)$, the next largest for reactive ones $(AB+C)$, and the smallest two for complex encounters for which the trajectories are incomplete $(ABC)$.

Figure 13 is a fairly complete description of the dynamics of rearrangment at the chosen values of $T$ and $V$. It was obtained with only 65 trajectories, whereas reliable three-dimensional results would require many more trajectories and could not be described in a single diagram. The results indicate that the requirements for rearrangement are adequately met by reagents with sufficient translational energy to surmount the barrier (the reaction is endothermic by $\sim 17$ kcal/mole). Consideration of Fig. 14 shows that "rounding the corner" is accomplished by two relatively hard indirect collisions: A hits B, B hits C, and AB separates from C. The last critical state before the products separate lies high up the wall of the surface below the entry line, which

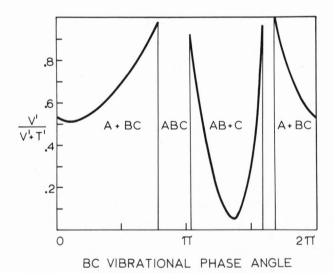

Fig. 13. Fraction of available energy present as vibrational excitation of the products plotted against vibrational phase angle of the reagent molecule for various rearrangement channels. The ABC region is for complex collisions.

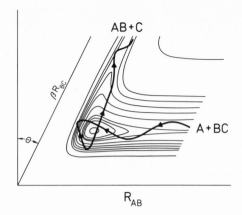

Fig. 14. Collinear trajectory illustrating the indirect collision mechanism that effects the momentum transfer required for rearrangement.

corresponds to repulsion in the BC bond. This is released as the products separate, tending to give large amounts of translational energy to the products (after the barrier has been surmounted). This is the predominant mechanism at this particular value of $T$ and $V$, because the curve of $f_{V'}$ against $\Phi$ in Fig. 13 has its minimum at low values of $V'$ (therefore at high $T'$), meaning that the distribution of $f_{V'}$ would be peaked at low $V'$. The indirect collision coupling mechanism seems to prevail in both the reactive and nonreactive regions of BC phase angle, as is shown by the trajectories in Fig. 15.

### 5.4.2. Application to 3D Dynamics

The whole point in developing concepts from (simple) collinear dynamics is that they will aid in the understanding and interpretation of the more complicated three-dimensional results. Accordingly, a three-dimensional potential function is often characterized by the shape of its collinear cut, which is used in a diagnostic fashion. It is not expected, therefore, that the one-dimensional and three-dimensional dynamics will be exactly the same but only that dynamical changes in going from one diagnostic surface to another in one dimension will be indicative of similar changes in the three-dimensional dynamics.

The fact that surface characterizations based on collinear dynamics are often useful rests on the coincidence that potential energy functions for many systems actually favor a collinear approach of the reagents (see Fig. 4). This means that the barrier to reaction is lowest in the collinear approach, so that in the threshold energy region most collisions that lead to reaction must pass through a nearly linear geometry. As the collision energy increases, this happy state of affairs does not apply. Even then, however, the collinear surface may embody many of the features that are characteristic of the remainder of the surface and would then remain a useful diagnostic device.

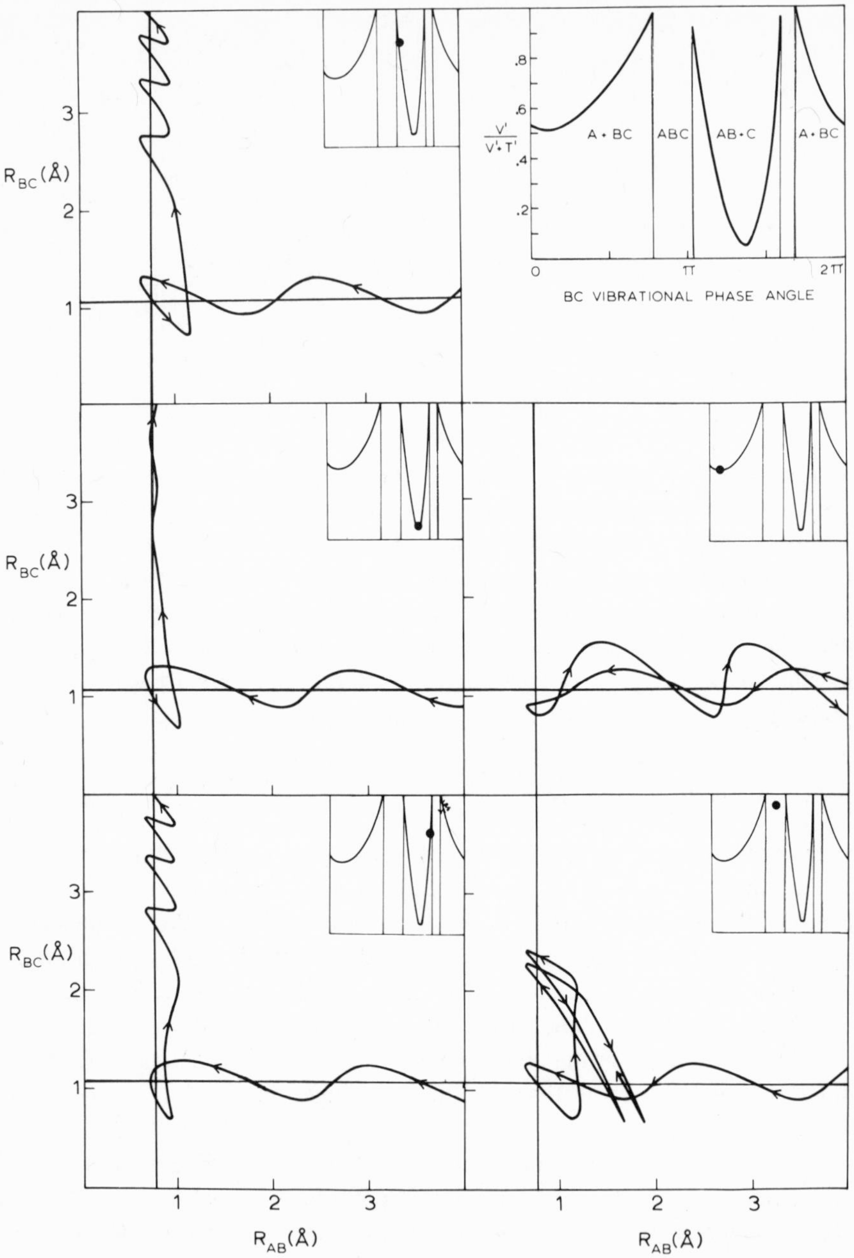

Fig. 15. Series of collinear trajectories at various initial vibrational phases of the reagent molecule corresponding to the different regions in Fig. 13, which is reproduced here as an inset, the dot indicating the phase angle corresponding to each trajectory.

One three-dimensional dynamical variable that collinear dynamics cannot mimic is rotation. Often, repulsive energy release can lead to rotational excitation, especially when the last critical state is in a bent geometry. Consequently, one-dimensional conclusions about product translational energy can prove erroneous when extended to three dimensions, unless translation and rotation $(T' + R')$ are considered together.

### 5.4.3. Misleading Aspects of Collinear Calculations

Collinear calculations err in two obvious ways: They exaggerate mechanisms that are peculiar to one dimension, and they ignore mechanisms peculiar to three dimensions.

The mass combination **H + LH** provides a good example of the first error. The skew angle $\phi$ is very large, meaning that inertial coupling is large. The hypothetical mass on a collinear surface can spend a lot of time in the corner, corresponding to atom B bouncing back and forth between A and C. In three dimensions, however, the first few collisions usually cause B to rotate as well, taking it away from the region between A and C, thereby altering the dynamical picture completely.

The second error could occur when the dynamics at large impact parameters are significantly different from those at small impact parameters. An example is again the reaction $He + H_2^+ \rightarrow HeH^+ + H$. Collinear trajectory results for different values of $T$ and $V$ (similar to those shown in Figs. 13 and 15) indicate that translational energy is slightly more effective than vibrational energy in bringing about reaction. A similar result is obtained for the same surface with collinear quantum dynamics.[59] In three dimensions, however, the result is quite different; translation is again effective to some extent, but vibrational energy is of enormously greater utility in promoting reaction.[11,60] Moreover, close examination of three-dimensional trajectories reveals that it is the high impact parameter trajectories that lead to the enhanced cross sections, which occur at low translational energy. This suggests, but does not prove, that noncollinear geometries play a large role in the dynamics, something suggested by the DIM representation of the potential surface itself, which does not strongly favor a linear approach.

Despite the fact that collinear dynamics have obvious deficiencies, some of the notions developed with their aid are transferable to three dimensions. Of particular use are the importance of the last critical state to energy distributions, the classification of surfaces for threshold energies, and the mechanism of mixed energy release. All these will be referred to later in connection with three-dimensional work.

## 6. Triatomic Systems

### 6.1. Introduction

Here we shall consider transfer reactions $A + BC \rightarrow AB + C$. We shall discuss these reactions in some detail, because much experimental and most of the theoretical work has been done on these systems. The connection between the collision phenomena and the potential features that are found for these systems can be extended to other reactions as well.

Most of the generalizations about product energy and angular distributions hold only for direct collisions for which the critical state is known. It would be possible to apply these generalizations to product distributions for indirect collisions as well, provided the appropriate critical state before product separation could be identified. Sometimes this can be done, as in the collinear treatment of $He + H_2^+$ (Section 5.4). Sometimes, too, the indirect collisions complicate the product distributions but in a way that is predictable, even though the individual trajectories defy simple classification. In such cases, one must be satisfied with more qualitative generalizations.

In the discussion below, the following conventions hold: An exothermic reaction is denoted as $A + BC \rightarrow AB + C$ but an endothermic reaction as $AB + C \rightarrow A + BC$. Vibrational and rotational energies are denoted $V, R$ for the BC molecule and $V', R'$ for the AB molecule; relative translational energy of A relative to BC is $T$, and of AB relative to C, $T'$. The internuclear distances for AB, BC, and CA are still denoted by $r_1, r_2$, and $r_3$, as before.

## 6.2. Energy Distributions

### 6.2.1. Threshold Region

Early trajectory studies[34,35,38–41] on a range of potential surfaces having different amounts of attractive and repulsive energy release are consistent with the notion that large attractive $(A_\perp)$ energy release leads to vibrational excitation, and large repulsive $(R_\perp)$ energy release leads to translation. Many of these studies used very attractive potential surfaces, which are suitable for reactions of the alkali metal atoms with halogen compounds, and on some of these surfaces upwards of 80% of the available energy could become product internal excitation. The surfaces were not of the LEPS variety but rather of the Blais–Bunker type (Section 3.4.2).

This early work led to other work on LEPS surfaces that was aimed at providing the more quantitative classifications discussed in Section 5.3.3.[28] Eight potential surfaces (three of which are in Figs. 9 and 10) with $A_\perp$ ranging from ~3% to ~87% produced a wide variation in the mean product vibrational excitation, and for the mass combination **L + HH**, the correlation between %$\langle V' \rangle$ and $A_\perp$ shown in Fig. 12 was obtained. Other mass combinations (eight in all were used) did not show this correlation, and the mechanism of mixed energy release was recognized as being responsible for the different behavior of various masses. It now appears as if the quantity $A_T + M_T$ is the proper index of vibrational excitation. This quantity is compared with %$\langle V' \rangle$ in

Table 2 for two mass combinations on each of the three potential surfaces in Fig. 9. One mass combination, **L + HH**, does not give rise to mixed energy release, as we have seen, whereas the other, **L + LL**, does; in each case, the quantity $A_T + M_T$ is not a bad indicator for $\%\langle V' \rangle$. Similar linear relationships have since been found in a study on the reaction $F + H_2 \rightarrow FH + H$,[58] and since then further studies on this system and other similar systems have been consistent with the relationship.[21,61–65] All the surfaces used are of the type where the barrier is in the entrance channel, and the region of maximum curvature of the minimum reaction path appears to come significantly after the barrier, so that the mixed energy release mechanism can operate to avoid repulsion.

The previous generalizations can be invalidated for surfaces where there are a substantial number of secondary encounters. Such encounters occur either because the potential function is too attractive (see Fig. 12 for $A_\perp$ very

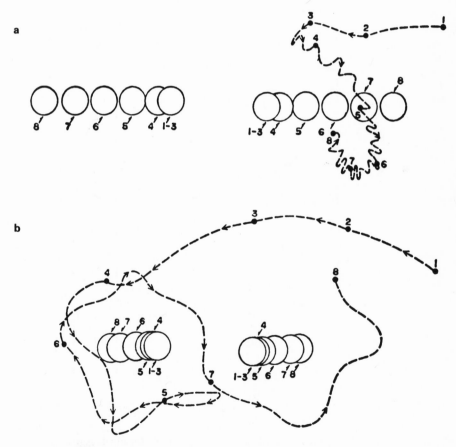

Fig. 16. Trajectories for the masses **L + HH** (a) on the repulsive surface 1 and (b) on a very attractive surface. The dots 1–8 show the positions after successive equal time intervals. (After Kuntz et al.[28])

large) or when the skew angle is very large. In the first instance, not enough energy is invested in the BC coordinate, and the products cannot separate before the atom A or B has collided again with C. These extra collisions can be of either an attractive or a repulsive character, and they complicate the simple pictures put forth before. Usually the collinear surfaces are not of too much help in analyzing the resulting motion, because the secondary encounters occur at all sorts of geometries. An example of a trajectory on a surface with $A_\perp = 87\%$ is given in Fig. 16(b).[28] Here, the mass combination used is **L + HH**, one of the most conducive to secondary encounters in this case, because the light atom can travel an enormous distance in the time taken for the two heavy atoms to separate. For comparison, a "normal" trajectory [Fig. 16(a)] is also shown for the same mass combination on a surface where $A_\perp = 3\%$.

An example of the second case is the mass combination **H + LH** for which $\phi = 81°$. The collinear surface indicates that the trajectory bounces back and

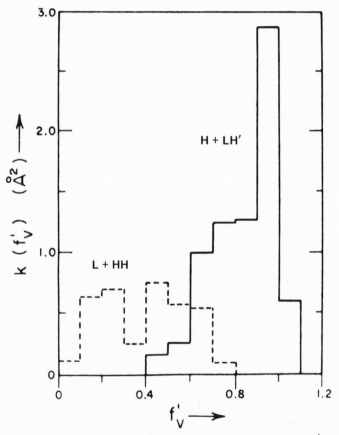

Fig. 17. Detailed rate constants against fraction of available energy in vibration of products for an exothermic reaction for two sets of masses: solid line, **H + LH′**; dashed line, **L + HH** (**H** = 35, **H′** = 127, **L** = 1 amu). (After Parr et al.[25])

forth from the one channel to the other many times before reaction is over. This picture is reflected in three-dimensional trajectory calculations as well but is not so exaggerated.[25] The light atom can again travel rings around the two heavy atoms while they separate. Here, the scattering can almost be described as a two-body interaction between the two heavy particles, with the light atom darting around about them.

We have seen that the mass combination **L+HH** tends to allow little mixed energy release, whereas the combination **H+LH** favors mixed energy release, because the two heavy atoms take some time to approach each other compared with the time taken for the new bond to contract. This mass effect is clearly demonstrated by a comparison of two three-dimensional trajectory calculations on a potential surface of the LEPS type, with parameters chosen to represent the exothermic reaction $Cl+HI \rightarrow HCl+I$, $\Delta U = -31.7$ kcal/mole.[25] With a mass combination corresponding to $Cl+HI$ (which we symbolize as **H+LH'**), a collinear trajectory categorization yields $A_T = 2\%$, $M_T = 89\%$, $R_T = 9\%$; the rectilinear classification gives $A_\perp = 24\%$. On this basis, one expects the masses **H+LH'** to have products whose vibrational excitation is much higher than those of **L+HH** (compare $A_T + M_T = 91\%$ and $\sim 24\%$, respectively). The trajectories in each case, however, are complicated by secondary encounters (expecially for **H+LH'**), which can markedly alter the expected distributions. Nevertheless, the distributions of the fraction of the available energy in vibration, $f_{v'}$, in Fig. 17 distinctly differ from each other in the expected manner. This is a clear demonstration of a large mass effect, and it underlines the importance of the mechanism of mixed energy release for surfaces of this type. It also confirms the applicability of the collinear classification scheme to three-dimensional collisions.

### 6.2.2. Excess Reagent Energy Region

The previous discussions concerned trajectories for which the reagents' energy, $V+T+R$, was just sufficient to promote reaction (i.e., threshold region). When the reagents are given energy in excess of this one expects the product distribution to become more dependent on the history of the trajectory and on the potentials in the corner of the surface and somewhat less dependent on the surface categories based on a collinear trajectory with a threshold energy. The excess energy can be supplied either as translational energy or as internal energy of the reagents. For the reactions $Cl+HI \rightarrow HCl+I$, $F+HCl \rightarrow HF+Cl$, $F+D_2 \rightarrow DF+D$, $H+Cl_2 \rightarrow HCl+Cl$, and $K+ICH_3 \rightarrow KI+CH_3$, there is some experimental evidence that a modest increase in reagent translational energy, $\Delta T$, is reflected by an increase of translation ($\Delta T'$) and rotation ($\Delta R'$) in the products,[12,66,67] For ion–molecule reactions, more substantial amounts of excess translational energy in the reagents often appear to be partially converted to translational energy according to the relationship

$\Delta T' = \sin^2 \phi \, \Delta T$, where $\phi$ is the angle of skew (Table 1).[68,69] In these experiments, internal energy of the products cannot be measured, and so the hypothesis $\Delta T \to \Delta T' + \Delta R'$ cannot be tested; however, the chemilumines-cence experiments on $F + D_2 \to DF + D$ show a substantial amount of rotational excitation, $\Delta R'$, coming from $\Delta T$,[67] which is consistent with the ion–molecule results for similar mass combinations **H + LL** (e.g., $Ar^+ + D_2$).

Trajectory calculations [25,67] on the systems $Cl + HI$, $F + HCl$, $H + CL_2$, and $F + D_2$ show these same effects of excess reagent energy. The fact that excess translational energy becomes translation or rotation of the products is interpreted in terms of a critical state that corresponds to compression of the BC bond and hence large repulsive energy release. The trajectories are less influenced by the potential immediately past the barrier and make their way into the corner of the potential surface, from where they experience a greater repulsion than they would at lower collision energies. Moreover, this increased repulsion cannot be converted easily into vibration, because the AB molecule is already contracted, so that it tends to go mainly into translational energy or, if

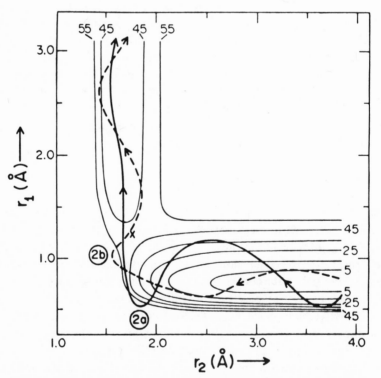

Fig. 18. Three-dimensional trajectories with excess reagent translational energy: solid line, $T = V = 35$ kcal/mole.; dashed line, $T = 64$, $V = 6$ kcal/mole. The positions $2a$ and $2b$ mark the regions of low kinetic energy. Contours (kilocalories per mole.) represent an FAS surface for $\angle ABC = 100°$, the geometry near the low kinetic energy region. (After Perry et al.[32])

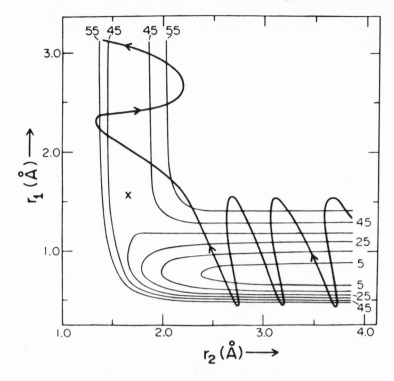

Fig. 19. Three-dimensional trajectory with a large excess of reagent vibrational energy ($V = 70$, $T = 3$ kcal/mole.) illustrating induced attractive energy release. The potential contours represent an FAS surface for $\angle ABC = 130°$, the geometry of the molecule near the region of least kinetic energy. (After Perry et al.[32])

the ABC intermediate is bent, into translation and rotation of the products. Trajectories with such critical states are said to exhibit *induced repulsive energy release*.

Examples of trajectories having critical states high on the repulsive wall of the surface are shown in Fig. 18 for the endothermic direction.[32] The critical states are marked by the numbers 2*a* and 2*b*. It is clear that in the excess energy region the position of the critical state will depend very much on the initial conditions of the trajectory.

An increase in reagent vibrational energy, on the other hand, can lead to extreme cutting of the corner of the surface. This tends to produce trajectories with critical states on the attractive wall of the exit channel (extended AB bonds), a situation that gives products with large vibrational excitation, i.e., $\Delta V \rightarrow \Delta V'$. This result is expected when the BC molecule is extended when the strong interaction region is entered, and trajectories with such critical states exhibit *induced attractive energy release*. A diagram illustrating this is shown in Fig. 19, where this time the critical state corresponds to an extension in both bonds. The trajectories in Figs. 18 and 19 are three-dimensional ones, and the surfaces are fixed angle surfaces drawn to correspond to the geometry of the critical state and only represent the potential along the trajectory in this region. For some values of the initial vibrational phase angle, there might be trajectories for which BC is greatly compressed as the trajectory reaches the head of the exit valley. These trajectories could lead to enhanced translational excitation of the products. If the two types of trajectories reflect two distinct modes of reaction, the vibrational distribution of the product AB molecules could have two peaks,[70] each coming from critical states from different regions of the potential surface.

The channeling of excess reagent translational energy into product translation and rotation seems to remain operative for higher translational energies as well. For example, in a trajectory study[71] on potential surfaces designed to represent ion–molecule reactions of the type $Ar^+ + D_2 \rightarrow ArD^+ + D$, the amounts of vibrational energy at collision energies of 5 and 50 kcal/mole were 9.5 and 14.0, respectively, i.e., of the extra 45 kcal/mole of translational energy, only 10% went into increasing vibration, whereas 90% went into translation and rotation.

When the reagents have a great deal of excess energy, the product distributions may change because of a falloff in the reactive cross section. This could occur, for example, because many of the internally excited product molecules that would have been stable at lower total energies can now go on to dissociate. This is, in fact, the preferred mode for collision-induced dissociation in many cases: Molecules are "torn" apart rather than "smashed" apart. The process $A + BC \rightarrow AB^\dagger + C \rightarrow A + B + C$ is more likely than $A + BC \rightarrow A + BC^\dagger \rightarrow A + B + C$, where the dagger represents highly internally excited molecules.[72] Because a lot of vibration in AB stems more readily from excess vibrational energy in BC than relative energy between A and BC, it follows that vibrational excitation of BC can enhance collision-induced dissociation cross sections. This effect has been observed both experimentally[11,73] and in trajectory calculations [60] for the system $He + H_2^+ \rightarrow HeH^+ + H$.

## 6.3. Angular Distributions

### 6.3.1. Forward Direction

In the collision between A and BC, the direction of the initial velocity vector of atom A in the center of mass system is taken as a reference for the purpose of defining the scattering angle of an atom or molecule, $\theta$. This direction is often called the forward direction. In our discussions of reactive collisions we shall reserve the term *forward scattering* to mean scattering of the product *molecule* in the forward direction, or at least in the angular range about the forward direction. Forward scattering of the molecule AB implies backward scattering of the atom C, from the conservation of linear momentum. Unless otherwise stated, a scattering angle of $\theta = 180°$ means forward scattering of the product molecule. In many of the diagrams taken from trajectory calculations, this angle is denoted $\theta_{at}$.

For surfaces on which direct trajectories take place, the angular distributions are not symmetric in $\theta$ about $\theta = 90°$, but there is usually a bias toward the forward or backward direction. Such a bias means that the products of the collision retain some information about the reagents, because all angles are referred to the velocity of the incoming atom. There is a general dynamical bias that tends to favor the forward direction, namely the momentum of the initial atom in that direction. At low collision energies the forces arising from the interaction potential can counteract this effect of initial momentum, but at higher energies this is more difficult, and the angular scattering tends to shift to the forward direction. This general observation is often seen in trajectory calculations. However, the limit of the scattering angle at very high energies is not necessarily $180°$.[74]

## 6.3.2. Influence of Potential Surface

Generalizations about the influence of potential on angular scattering are much less certain than those about energy distributions, because the effect on energy distributions is due to the shape of the surface from the last critical state onward, whereas the angle of scattering depends in a cumulative way on the complete past history of the trajectory. For example, the scattering angle may be quite sensitive to the long-range part of the potential, which can cause very large deflections. All generalizations about angular distributions must therefore be used with caution.

The effect of the potential energy surface on the scattering angle distribution can, however, be rationalized in terms of the dynamical bias for the forward direction that accompanies large reagent momentum. Attractive surfaces tend to accelerate the reagents toward one another. They expend part of the exoergicity reinforcing an already existing bias due to the initial relative momentum, and consequently have even less (repulsive) energy available to counteract this bias. The general shift from backward scattering on repulsive surfaces to forward scattering on attractive surfaces is seen in Fig. 20 for the various mass combinations.

## 6.3.3. Special Influences

We have seen previously that the **L + HH** combination was much more sensitive to properties of the potential surface than other masses; this holds true with regard to angular scattering as well. Backward scattering is more prevalent for this mass combination, because there is not much mixed energy

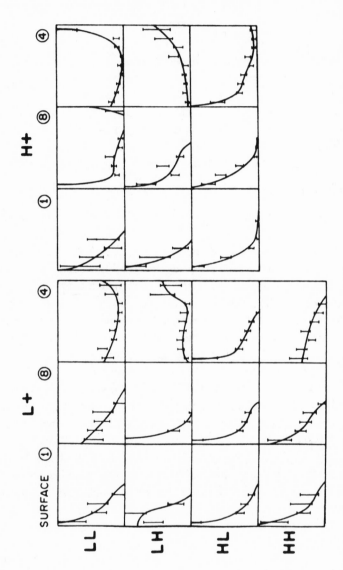

Fig. 20. Center of mass angular distributions for all masses on surfaces 1, 8, and 4. Abscissa runs from backward direction (left) to forward scattering (right). (After Polyani and Schreiber.[16])

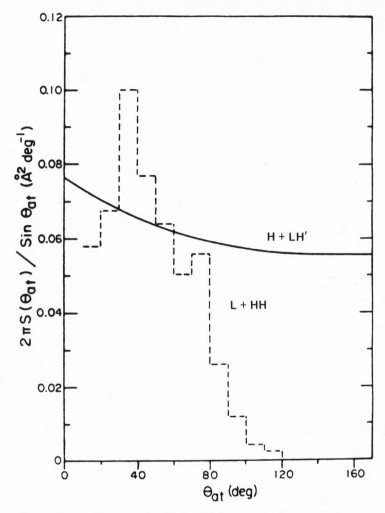

Fig. 21. Center of mass angular distributions for two sets of masses on the same exothermic surface. See caption of Figure 17. (After Parr et al.[25])

release, and consequently the effect of product repulsion is felt all the more strongly. The mass combination **H + LH**, on the other hand, is fairly insensitive to the details of the surface, and the angular scattering is fairly well described in terms of elastic scattering between the two heavy masses interacting through an average potential[25] (Fig. 21).

The angular distributions from a reaction are often complicated by the simultaneous presence of different *mechanisms* of reactions. For example, in the reaction $A + BB \rightarrow AB + B$, the A atom can react with either end of the BB molecule. In some instances, the nearer B atom turns out to be the unreactive one but only after there is a strong encounter between it and A, which then attaches itself to the second B atom. Such a mechanism can provide forward

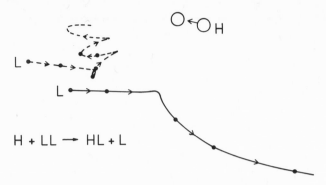

Fig. 22. Planar trajectory illustrating an abortive reactive encounter with the (initially) nearer atom of the BC molecule. [After P. J. Kuntz, *in*: *The Physics of Electronic and Atomic Collisions* (T. R. Govers and F. J. de Heer, eds.), Vol. VII, ICPEAC, North-Holland Publishing Company, Amsterdam, (1972).]

scattering at quite low collision energies, because it allows A to "hop" over the first B and carry on with the second in the forward direction, despite possible repulsion between products. Trajectory calculations[71] show that such a mechanism can profoundly influence the angular scattering at intermediate reagent translational energies. A typical trajectory is shown in Fig. 22 for the mass combination **H** + **LL**. Other mechanisms can occur at quite high collision energies, when the atoms can be approximated by hard spheres; in these cases the angular scattering is determined largely by considerations of energy and momentum conservation, and detailed features of the surface are no longer of importance.[74]

There are some instances where the angular scattering may be computed by two-body models. We have seen one instance of this already: **H** + **LH** on a repulsive surface. Another example occurs with repulsive surfaces for cases in which the departing atom C is effectively pushed along the BC bond direction in a time short compared to the rotational period of BC. This situation allows the calculation of the scattering angle purely in terms of the angle $\varepsilon$ between the relative velocity vector of A and BC and the BC bond, i.e. it is determined geometrically. Information about the distribution of $\varepsilon$ at the onset of the repulsion is then transformed into distributions of scattering angle $\theta$. In fact, this can be done analytically.[75] For the mass combination **H** + **HL** (actually $K + ICH_3$), it can be shown that the angular distributions are dependent mainly on steric effects, i.e., on the range of geometries for which reaction is allowed (by the potential function).[76] Moreover, this effect can be independent of the way in which the energy release of the various FAS surfaces varies with angle, something which has a marked effect on energy distributions. This means that for some mass combinations the angular distributions and energy distributions are *uncoupled*; they provide different kinds of information about the potential function. The energy–angle contour maps obtainable from molecular beam experiments for $K + ICH_3 \rightarrow KI + CH_3$ do in fact exhibit this type of uncoupling.[77] The mass combination **H + HL** is perhaps most amenable to such a two-body treatment, because the light atom can leave very quickly, and the $r_2$ motion is not mass-coupled to the $r_1$ motion.

Another mass combination for which the same two-body treatment has been applied is **L + HH**. On a repulsive surface, the **H** atom can leave before **H—H** has rotated very much. Experimental angular distributions for HX molecules formed from reactions of hydrogen atons with $X_2$ molecules are peaked between the forward and backward directions, and the distributions are different for X = Cl, Br, and I.[52] The geometric model interpretation of this is that reaction takes place mainly through nonlinear geometries; computed angular distributions made under these assumptions are in fact not in bad agreement with experiment. Trajectory calculations on surfaces where a nonlinear approach is favored also reflect the behavior indicated by the model in that they give angular distributions with sideways peaking.[27]

## 6.4. Cross Sections

### 6.4.1. Introduction

In trajectory calculations, the cross section $\sigma_P$ for a process is defined in terms of some characteristic impact parameter, $b_{max}$, that is large enough to include all collisions that lead to the process:

$$\sigma_P = N_P \pi b_{max}^2 / N \tag{41}$$

where $N$ is the number of collisions with $b \leq b_{max}$ and $N_P$ is the number of these collisions that lead to the process. The absolute value of the cross section is determined in part, therefore, by the "size" of the colliding molecules. Reactive cross sections are usually many times smaller than this, however, because not all collisions are successful in leading to reaction.

The value of $b_{max}$ that one must use in the above formula is determined by the effective range of interaction. In general, if there is a long-range attraction between the reagents, $b_{max}$ will be larger than if there is not, and the chance that reaction will occur might be correspondingly increased.

For surfaces without a barrier one can calculate a so-called capture cross section, which is sometimes a good indication of $b_{max}$. One can assume (for the sake of argument) that the long-range potential between reagents is of a particular form, for example,

$$V(R) = -C/R^N, \quad N > 2 \tag{42}$$

The effective potential for a collision of orbital angular momentum $L = \mu v b$ and kinetic energy $E = \mu v^2/2$ is

$$V_{\text{eff}}(R) = -C/R^N + Eb^2/R^2 \qquad (43)$$

This potential energy will have a centrifugal barrier that is dependent on the impact parameter, $\hat{V}_{\text{eff}}(b)$. This is obtained by solving $V'_{\text{eff}}(\hat{R}) = 0$ for the position of the barrier, $\hat{R}$, and substituting into Eq. (43) to obtain $\hat{V}_{\text{eff}}(b)$. As long as the collision energy is greater than this barrier height, the reagents will be able to approach closely together. As $b$ increases from 0, $\hat{V}_{\text{eff}}$ also increases, so one can define $b_c$ as that impact parameter for which $\hat{V}_{\text{eff}}(b) = E$. The capture cross section is then $\pi b_c^2$, where

$$b_c^2 = [2E/(NC)]^{-(2/N)}[1 - (2/N)]^{-(N-2)/N} \qquad (44)$$

For example, when $N = 4$, the capture cross section is

$$\sigma_c = 2\pi(C/E)^{1/2} \qquad (45)$$

This equation shows how the capture cross section becomes larger as the collision energy decreases and as the long-range forces increase. Equation (44) can in *some* cases provide a rough indication about the magnitude of reactive cross sections, but its greatest value should be in providing an estimate of the maximum impact parameter to use in dynamical calculations [e.g., in estimating $b_{\text{max}}$ in Eq. (41)]. In *most* cases, however, it says nothing about the reactive cross section because it does not take into account the requirements for rearrangement; it is not enough for reagents to closely approach one another in order for reaction to occur.

There has been very little work directed at the problem of systematically studying the effect of surface features on reaction cross sections. Clearly, a potential barrier is one of the most important features. If two processes are possible, the one with the lowest barrier is favored near the threshold region of that barrier. At higher energies, the other process also begins to occur, but whether it does so at the expense of the lower-energy process or at the expense of the nonreactive cross section will depend a great deal on the surface. If the new process is a reaction with a different site on the molecule, the nonreactive component may be depleted, but if the new process is dissociation into fragments, it usually grows at the expense of the first process, expecially if the surface is one that favors high internal excitation of the molecule. However, it is very difficult to ascertain the particular property of the potential surface that makes a particular trajectory undergo one process and a second trajectory another. Especially at higher energies, the outcome is the result of a delicate balance between strong forces that vary considerably along the trajectory.

At lower energies, the prediction of isotope effects is also particularly difficult. There are several trajectory studies that compute cross sections for different isotopic reactions, but the general features of the potential function which determine the branching ratios are not systematically known.[61–65] In fact, dynamical effects, such as the rotation of the reagent molecule, may in the end prove to have a very important influence, alongside the specific features of the potential surface. We shall now consider some systematic studies of these dual influences on cross section.

### 6.4.2. Selectivity

*Substantial Barriers.* Reactions that are not exoergic usually have a fairly large energy barrier compared with those we have so far discussed. For example, the thermoneutral hydrogen exchange reaction $H + H_2 \rightarrow H_2 + H$ has a (symmetric) barrier of $\sim 9.8$ kcal/mole. All endothermic reactions have barriers at least as high as their endothermicity. The position of a barrier on a potential surface is one of the most significant features of a potential surface that can affect the dynamics.

The most obvious effect of a barrier is that it prevents reaction to all those trajectories having energies below the threshold energy, which is often approximately equal to the barrier height. For collisions in the threshold region, the amount of repulsive, mixed, and attractive energy release in the region *after* the barrier will depend critically on the location of the barrier, and so the location of the barrier will determine the energy distribution of the products. Even more important, however, the *location* of the barrier is found to determine *what kind* of energy is needed by the reagents in order to react at all.

First consider the different requirements of a surface with a barrier deep in the entrance channel as against one with a barrier deep in the exit channel.[78] In the former case, energy in the form of relative translation will be directed along the entrance channel and so will be effective in surmounting the barrier; vibrational energy, however, is in the wrong coordinate, and unless there exists a mechanism for converting vibration into translation while the barrier is surmounted (unlikely in the entrance channel), the trajectory will reach a turning point in the $r_1$ coordinate long before it reaches the barrier crest, and reaction will be forestalled. In the latter case, whatever motion is present in the reagents must be transformed into motion along the direction of the exit valley. If the reagents have translational energy, the trajectory may not be able to "turn the corner" of the potential surface and may never reach the region of the barrier. Vibrational energy, on the other hand, might be effective in allowing the trajectory to turn the corner because there is already some motion in the required coordinate. It is interesting to note that there are two conditions here: surmounting the barrier and turning the corner of the surface. One could conceive of a surface where the energy requirements of one are not those of the other, and in this case it would be possible for the threshold energy to lie far above the barrier height, or perhaps the reagents would require a very special mixture of translational and internal energy to react.

A more realistic situation occurs when the *crest* of the barrier lies in the entrance or exit channel but the *rest* of the barrier penetrates the interaction region. Accordingly, we shall consider two surfaces whose barriers can be considered as coming from shifts of a symmetrically placed barrier slightly into the entrance channel (surface I) and slightly into the exit channel (surface II), respectively. Such thermoneutral surfaces have been used in a trajectory study that examined the effect of supplying the energy required to surmount the barrier (~7 kcal/mole) in the forms of vibration or translation.[29] Collinear cuts through the two surfaces are shown in Fig. 23. The line dividing the entrance valley from the exit valley is arbitrarily taken as a line of unit slope passing through the intersection of the entry and exit lines: $r_1 - r_1^0 = r_2 - r_2^0$. On each surface, trajectories are computed for a total reagent energy only slightly in excess of the barrier height but with different amounts of vibrational and translational energy. The effect of the disposition of energy among the reagents on the reactive cross section is enormous, even though the crests of the barriers are only slightly shifted (by 0.3 Å) from the symmetric position. For surface I, supplying the necessary energy in the form of translation allows reaction, whereas vibrational energy equal to double the barrier height produces no reaction. Conversely, on surface II reagent vibration is required, and transla-tion is ineffective.

The considerations of the dynamics on surfaces I and II indicate that surfaces with symmetrically placed barriers ought to partake of some of the properties of both I and II. An example of a reaction with such a surface is $H + H_2 \rightarrow H_2 + H$. This has been studied with the aid of trajectory calculations, and, indeed, it turns out that both reagent vibration and translation are effective in promoting reaction.[79]

The qualitative results are for **L + LL** but hold as well for other masses. To interpret the effect of a change in the various masses, the surfaces must be plotted in skewed and scaled coordinates, such as those for the mass combina-tions **L + HH** (Fig. 24) and **H + HL** (Fig. 25); these two cases have the same angle of skew but are scaled differently. The effect of scaling is to shift the barrier further into the exit valley for **L + HH** and further into the entrance valley for **H + HL**. Figures 24 and 25 show two trajectories each, one with almost all translation and one with almost all vibration. On surface I, it is easier for a sliding mass with translation to surmount the barrier, whereas on surface II, vibrational energy is more effective. The trajectories on skewed and scaled collinear surfaces very nicely illustrate this qualitative difference in the dynamics, which persists in three-dimensional trajectories too. The qualitative behavior is the same as for the **L + LL** cases, but the overall cross sections for **H + HL** are larger than those for **L + HH**, a fact that is interpreted in terms of the differences in the widths of the exit valleys: The sliding mass on the surface for **H + HL** can more easily make its way into the broad exit channel than one on the surface for **L + HH** (see also Section 5.2.2).

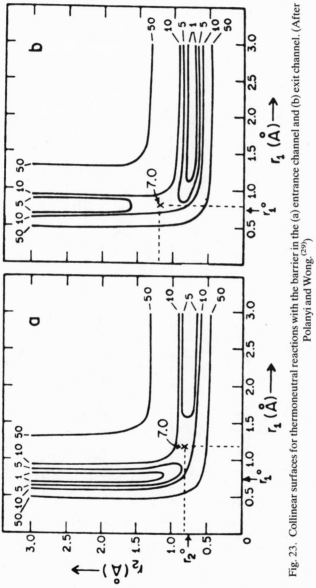

Fig. 23.  Collinear surfaces for thermoneutral reactions with the barrier in the (a) entrance channel and (b) exit channel. (After Polanyi and Wong.[29])

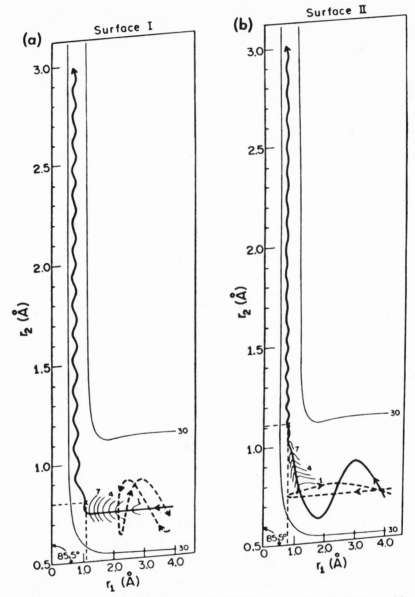

Fig. 24. Collinear trajectories for masses **L + HH**. Reagent energy distribution is (a) solid line, $T = 9$, $V = 0$; dashed line, $T = 1.5$, $V = 7.5$; (b) solid line, $T = 1.5$, $V = 7.5$; dashed line, $T = 9$, $V = 0$. All energies are in kilocalories per mole. (After Hodgson and Polanyi.[30])

Another observation is the effect of different reagent energy distributions on the product energy distributions. For surfaces with barriers located in the entrance valley, the motion required to surmount the barrier is the same as that for vibration in the new bond. It is found, therefore, that such surfaces tend to effect a transformation of reagent translational energy into product vibrational energy. On the other hand, the motion required to surmount the exit channel barrier corresponds to relative translation of the products, and it is found that these surfaces convert reagent vibrational energy into product translational energy. In summary, in the *threshold* region, *reaction* is accompanied by the *transformation* of energy: $T \rightarrow V'$, $V \rightarrow T'$. The partitioning of the energy released in sliding down the barrier depends on the shape of the potential surface in the region following the barrier crest and is governed by the same considerations that were applied to the exoergic reactions. The energy released can be categorized into attractive, mixed, and repulsive components for the various mass combinations, and, as before, the correlation of the dynamical results with these surface features holds true.

Angular scattering on these surfaces tends to be backward. Whereas the product energy distributions are determined by the shape of the surface after the barrier, the scattering angle is sensitive to the interactions before *and* after the ascent to the crest of the barrier. Because these interactions are on balance repulsive in character, the scattering is backward. The mass combination **L + HH** has a more pronounced sharply backward-peaked angular distribution, because it cannot avail itself of the mixed energy release mechanism so easily and thereby takes the full brunt of product repulsion.

When *excess* energy is used in collisions on surfaces of types I and II, it does *not* undergo a transformation as noted above for energy in the threshold region. Instead, translational energy of the reagents tends to remain translational energy of the products and reagent vibration becomes product vibration. This effect is entirely the same as that for exoergic reactions and can again be understood in terms of induced repulsive or attractive energy release.

*Hollows.* The previous discussion concerned surfaces where the substantial barrier was the only dominant feature. The question arises as to how the dynamics alter when there are other influences present. Limited trajectory studies have been done for surfaces with hollows in conjunction with barriers, and we shall see that some aspects of the dynamics can alter significantly.[16] First, however, it is useful to consider the effect of a hollow by itself.

We shall consider two surfaces that are virtually upside-down versions of the previous two. Surface −I has a 7 kcal/mole hollow in the entrance valley, whereas surface −II has a similar hollow in the exit valley; the surfaces are thermoneutral. The collisions on these surfaces are, on the whole, indirect but not so much so as to be long-lived complex collisions. Consequently, a dynamical effect is able to manifest itself: Hollows in the entrance channel require vibration; those in the exit channel require translation. The interpretation of these effects is not so straightforward, because of the indirect nature of the collisions. It is also difficult to fit these results in with the calculations by Borne and Bunker.[33] which indicate that dynamics are insensitive to the shape, depth, and position of the well. The two studies do *not* strongly contradict one another because they were carried out with completely different ends in view, and the variables studied were different; nevertheless, there

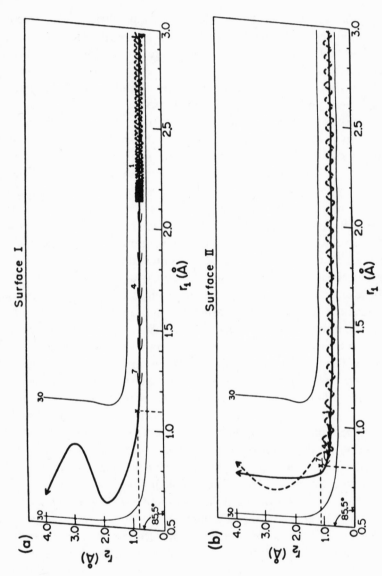

Fig. 25. Collinear trajectories for masses **H** + **HL**. Reagent energy distribution is (a) solid line, $T = 9$, $V = 0$; dashed line, $T = 1.5$, $V = 7.5$; (b) solid line, $T = 9$, $V = 0$; dashed line, $T = 1.5$, $V = 7.5$ kcal/mole. (After Hodgson and Polanyi.[30])

seems to be little *reinforcement* between the two. It may very well be that some essential features about surfaces with hollows are still not grasped very well or that the effects observed here are intimately related to the indirectness of the collisions.

The angular distributions on surfaces −I and −II are rather broad, which is consistent with the indirect nature of the collisions. Also, strong, long-range attraction between the reagents (present on both surfaces) allows them to spiral inward about each other, and attraction between the products allows them to spiral outwards, so that the scattering angle is very sensitive to initial conditions and tends to be broad on this account alone.

*Barriers and Hollows.* When the hollow (7 kcal/mole) is combined with a barrier (4 kcal/mole), the energy requirements tend to reinforce one another. A barrier in the entrance valley requires translational energy, as does a hollow in the exit channel, so that a surface with a hollow in the exit channel and a barrier in the entrance channel (type +I −II) requires reagent translation. On the other hand, a hollow in the entrance channel together with a barrier in the exit channel (type −I +II) require reagent vibration. The surface with the hollow after the barrier (+I − II) is able to produce a great degree of forward scattering. Other instances of attraction between products giving rise to forward scattering are seen in trajectory studies on the systems $K + Br_2 \rightarrow KB_r + Br^{(80,81)}$ and $Ar^+ + D_2 \rightarrow ArD^+ + D.^{(71)}$

## 6.4.3. Endothermic Reactions

Endothermic reactions have barriers at least equal to the endothermicity. The barrier could really occur anywhere on the surface; however, studies of families of surfaces indicate that this is not the case.[82] Instead, it is found that in related families of exothermic reactions the barrier is in the entry valley, which implies that for substantially endothermic reactions it is in the exit valley. Furthermore, for both exothermic and endothermic families of reactions in which there is a qualitative correlation of reaction energy (exothermicity or endothermicity) with barrier height, the barrier moves to successively later positions along the reaction coordinate with increasing barrier height. Because substantially endothermic reactions have their barriers in the exit valley, it follows from the previous work that vibrational excitation of the reagents is required for reaction. This conclusion receives support from the laser excitation experiments[4] on $K + HCl \rightarrow KCl + H$ and the chemiluminescence experiments [6] on $HCl + Br \rightarrow Cl + HBr$. Both reactions are endothermic, and both are promoted markedly by vibrational excitation of the HCl.

Extensive trajectory calculations have been done on a model LEPS potential surface having an endothermicity of 36 kcal/mole and parameters suitable for the reaction $H_2 + I \rightarrow H + HI.^{(32)}$ The barrier crest on this surface lies well into the exit channel, and the amount of attractive energy release in the

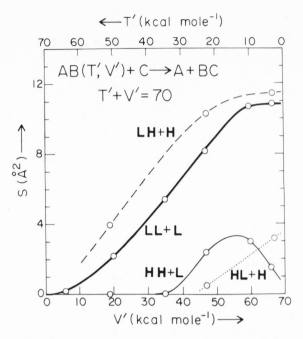

Fig. 26. Reactive cross sections for various masses for an endothermic reaction as a function of the distribution of reagent energy between translation ($T'$) and vibration ($V'$). (After Perry et al.[32])

exothermic direction is $A_\perp = 11\%$. The dynamics on this surface resemble in most respects those on surface II for thermoneutral reaction, including energy requirements for reaction, transposition of reagent translation (vibration) into product vibration (translation) in the threshold region, retention of energy type in the excess region, and behavior of cross section for the various mass combinations. The mass combinations used were **LL + L, LH + H, HH + L**, and **HL + H**. Figure 26 shows the main result, the variation of reactive cross section as a function of reagent vibrational excitation at a fixed *total* energy equal to approximately twice the endothermicity. Here the **LH + H** case has the highest cross sections because the exit channel is much broader than the entrance channel, and the barrier is not shifted so far down the exit channel.

## 6.5. Complex Collisions

If a trajectory spends a long time (i.e., many rotational periods of the complex molecule) in the interaction region, it is generally difficult to correlate the outcome of the collision with the shape of the surface. Instead, angular and energy distributions tend to be statistical in nature, and their shapes are determined largely by considerations of total energy and angular momentum conservation. Angular distributions have a symmetry about the sideways direction, one of the criteria by which complex collisions are recognized experimentally.[83] For example, a potential energy surface that gives rise to this kind of behavior is the one calculated by Roach and Child[84] for the reaction $K + NaCl \rightarrow KCl + Na$. This surface has no activation barriers and has a well (13.5 kcal/mole deep with respect to dissociation of the products) corresponding to stable KClNa in a bent geometry; the reaction itself is only ~4 kcal/mole exothermic. Trajectories computed on this surface over a range of collision energies from 0.5 to 42.4 kcal/mole exhibit complex behavior and at the lower energies give angular distributions that are roughly symmetric about 90°.[85]

It is not completely certain what features of a surface are necessary to produce long-lived complexes, although one favorable feature is a potential well corresponding to a stable intermediate. Thus $H_3^+$ has a very deep potential well,[86] and a trajectory study on this surface shows long lifetime trajectories, with accompanying symmetric angular distributions, even at a collision energy of 4.5 eV.[37]

The mere presence of a potential energy well, however, is not a sufficient condition for such behavior.[33] The shape of the potential at large internuclear distances in the entrance and exit channels is also very important, because the complex must essentially be trapped by *centrifugal barriers*. The original total angular momentum of the complex is the sum of the orbital angular momentum, $L = \mu v b$, and the rotational angular momentum of the reagent molecule, $J$. Here $\mu$ is the reduced mass of the colliding system, $v$ is the relative collision speed, and $b$ is the impact parameter. The orbital angular momentum contributes a repulsive term to the effective potential in the asymptotic region

$$V_{\text{eff}}(R) = V(R) + L^2/(2\mu R^2) \tag{46}$$

where $R$ is the distance from the atom to the center of mass of the molecule. Even when $V(R)$ is attractive (as it usually is for these cases), the term in $L$ gives rise to a barrier. The collision of A and BC reapportions the energy among the various degrees of freedom of the complex, and subsequent attempts at separation fail because there is not enough energy to mount the barrier.

For exoergic reactions, such as the $K + NaCl$ reaction, it is very helpful to complex formation if the exit channel is "blocked" by a high centrifugal barrier that compensates for the possible gain in relative kinetic energy provided by the exoergicity. Such a barrier is favored when the reduced mass of the products is less than that of the reactants [see Eq. (46)], i.e., when the scaling factor $\beta < 1$. Table 1 shows that masses **H + HL** fulfill this requirement. For the $K + NaCl$ reaction, $\beta \approx 0.87$.

That the existence of a potential well is not sufficient for the formation of complexes is seen in two trajectory studies. The first is for the mass combination $Ar^+ + D_2 \rightarrow ArD^+ + D$ on a potential surface exothermic by 20 kcal/mole, and possessing a well 20 kcal/mole deep for the geometry $H—Ar—H$.[71] The well in this case was an artifact of the function used to represent the potential. Trajectories at $T = 5$ kcal/mole showed no evidence of significant complex formation, even though some of the trajectories were known to sample the geometries in the region of the well. The mass combination is favorable to complex formation as well, but in this case the energy gained from exothermicity is too great, and the long-range part of the potential not attractive enough. The second trajectory study was an investigation of the mass combination $Br + I_2 \rightarrow BrI + I$ ($\beta = 1.14$) on a series of 13 potential surfaces all having an exothermicity of 6.3 kcal/mole but differing in the position, depth, and shape of a small well ($\sim 10$ kcal/mole deep). Initial relative kinetic energy was about 2.8 kcal/mole in most cases. None of the calculations on *any* of these surfaces produced long-lived complexes, and, if anything, the presence of the well tended to speed up the trajectories.[33] These surfaces therefore possess the incorrect type of behavior at long range to hold the complex together.

## 7. Polyatomic Systems

### 7.1. Four-Atom Systems

Polyatomic systems are more difficult to treat than triatomic ones, both dynamically and conceptually. The potential function depends on more variables and so is more difficult to represent, graphically and functionally. Two-dimensional contour maps are less representative of the full hypersurface than in the triatomic case, which means that "diagnostic surfaces" must be taken with even greater reserve. It is all the more encouraging, therefore, that many of the generalizations made for triatomic systems can be extended quite well to four-atom ones.

A greater variety of four-atom rearrangements is possible. One of these, $AB + CD \rightarrow A + BCD$, allows the whole reactive event to take place in a $C_{2v}$ geometry. This allows some simplification in the calculations, but the surface is still dependent on three variables and cannot be represented by a single contour diagram. Isosceles trajectories have been computed for the reaction $HeH^+ + H_2 \rightarrow He + H_3^+$ ($\Delta U = -2.6$ eV) using an *ab initio* potential function.[18] For reagent translational energies of 0.001 and 0.13 eV and two vibrational energies, most of the available energy becomes vibration of $H_3^+$. The surface is fairly attractive because of the interaction between the ion and the molecule, and one therefore expects large product vibration. In addition to this, the $H_2$ molecule is initially compressed relative to the $H_3^+$ equilibrium bond distance, so that the relaxation of this bond during the formation of $H_3^+$ also contributes to product vibration.

Four-atom collinear calculations can also be done on this system and its reverse reaction $A + BCD \rightarrow AB + CD$. Simulation of the reaction $O(^3P) +$

SCS → SO + CS ($\Delta U = -26.6$ kcal/mole) on surfaces related to the LEPS variety with different barrier heights (between 0 and 12 kcal/mole) shows that the amount of vibrational energy in the *new* bond (SO) is correlated with the sum $A_T + M_T$, as for triatomic systems.[87] Here, however, the repulsive energy release can be divided between relative translation energy of the products and vibration of the *old* bond. The partitioning of this energy is sensitive to the slope of the potential in the exit channel; a sharp force, rather than a gradual one, is more efficient in exciting the old bond. Three-dimensional calculations on attractive surfaces simulating the reaction K + I(CH$_2$) (CH$_3$) by a four-body system also indicate that most of the exothermicity becomes internal energy of the new (KI) bond.[88]

The rearrangement AB + CD → AC + BD cannot be represented by collinear or isosceles calculations. Planar calculations are possible, but the planar potential surface requires four coordinates for its representation. Rectangular diagnostic surfaces have been proposed[31] in which the four particles are kept in the shape of a rectangle, and the surface variables are taken as the AB ( = CD) and AC ( = BD) distances. These surfaces cannot represent the potential along real trajectories (because the equations of motion do not preserve the rectangular geometry), but they do serve as useful indicators of the topology of the full hypersurface.

The effect of barrier location on the energy requirements for the reactions AB + CD → AC + BD parallels the results for triatomic reactions. Reagent translation promotes reaction for entrance channel barriers, vibration for exit channel barriers, and both translation and vibration for symmetric barriers; also $T \rightarrow V'$ and $V \rightarrow T'$ near threshold. An additional variable in four-atom systems is the distribution of reagent internal energy between AB and CD; symmetric distribution is more conducive to reaction than unsymmetric.

## 7.2. Six-Atom Systems

The complexity of six-atom systems is many times greater than three- and four-atom ones, especially when all the atoms can take part in bond breakage and formation. Such a possibility arises in the tritium-methane system. At energies below the CH$_3$—H dissociation level, two reactive processes are possible: T + CH$_4$ → TH + CH$_3$ (abstraction) and T + CH$_4$ → CH$_3$T + H (displacement). The abstraction reaction has a threshold around 12 kcal/mole, and the displacement at about 35 kcal/mole. The history of trajectory calculations on this system illustrates the difficulties involved in trying to identify the surface features responsible for experimental observations.

An early study simulated these reactions by a three-atom model, with the CH$_3$ group treated as a single point of mass 15 amu.[89] A suitable potential function was found only after 23 others were rejected because they incorrectly favored displacement over abstraction at $T = 2$ eV. Results of any three-atom model must, of course, be suspect, because it ignores the symmetry of the $CH_4$ molecule.

Bunker and Pattengill[90] performed a calculation in which all six atoms were represented dynamically. Unfortunately, the potential energy function did not handle the molecular symmetry fully, and they found that all their surfaces could not explain the substitution reaction but had no trouble in obtaining abstraction. They concluded from this negative result that substitution was a four-or-more-center process that might proceed through inversion of the $CH_3$ group.

In a later paper, Valencich and Bunker[91] reported another six-atom calculation, this time with equivalent H atoms and inversion allowed. They were able to obtain substitution and found that a substantial part of it went via the inversion mechanism.

Recently, a new potential representation of this system has been developed by Raff,[92] who used thermodynamic and spectroscopic data to fix the potential representation for the reagents and products and semiempirical plus *ab initio* quantum chemistry results to fix various interaction regions. The surface was not adjusted to fit any kinetic data, but managed, via a six-atom trajectory calculation, to account quite well for many experimental observations on this system. Contrary to the previous six-atom work, displacement did not occur via the inversion mechanism. This work emphasizes the need to obtain a good potential representation, and its success in describing the dynamic properties of a complicated system is very encouraging.


## 8. Multipotential Systems

Previous discussion has dealt exclusively with dynamics on a single potential surface. Sometimes, however, more than one surface is necessary, even for the simplest treatment. Because Chapter 5 treats this subject in detail, we shall mention it here only briefly.

Many reactions occur by way of a sudden rearrangement of electrons, i.e., an electronic configuration that adequately describes the ground state of the reagents does not correlate at all with the configuration for the product state of interest. An example is $K + Br_2 \rightarrow KBr + Br$, which goes from a *covalent* $Br_2$ molecule to an *ionic* KBr molecule. Somewhere along the reaction path, the potential surfaces approach each other closely, and the change of configuration occurs. In this region, it is possible for the system to undergo a transition from one electronic state to another, and such transitions will have a profound effect on the dynamics and cross section of the rearrangement process.

If the transition region is very localized and lies far in the entrance or exit channel, the rearrangement dynamics can often be treated separately from the passage through the transition region. For example, at large internuclear distances, the potential for $K + Br_2$ is nearly flat, and there is a fairly well-defined point at which the potential abruptly changes to a strongly attractive surface. A single-surface treatment can therefore say nothing about the magnitude of the total reaction cross section, but it can say that those trajectories that *do* pass the transition region will produce KBr molecules with high vibration.[93]

In other systems, the transitions can occur all along the entrance channel, because they involve motion in the $r_2$ coordinate.[94] An example is the reaction $Ar^+ + H_2 \rightarrow ArH^+ + H$. Far out in the entrance channel, the $H_2$ molecule vibrates continually back and forth through the transition region. At each vibration it has the chance of losing an electron to $Ar^+$ to form the configuration $Ar + H_2^+$ (which correlates with the ground state of $ArH^+ + H$), but this is unlikely until the $Ar-H_2$ distance is small (but still in the entrance valley). After the electron "jump" the $H_2^+$ begins to vibrate, because it is formed in a compressed condition, and the trajectory continues on the lower surface. In this case, a multisurface treatment is necessary because it is impossible to pin down the region where the jump will occur[95]

## 9. Summary

Of the various collision phenomena that concern one in reactive scattering, it would appear that the distribution of the energy among the products is the best understood in terms of features of the potential surface. Particularly when the geometry of the critical state can be identified in a direct (or not *too* indirect) collision, a semiquantitative index to energy distributions can probably be obtained in terms of features of the potential surface. At present, one is available for the case value the critical state occurs near the barrier, and this is particularly successful when the barrier is in the entrance channel.

Closely related to energy disposal in the products is the energy selectivity of the reagents. This appears to be understandable in terms of the position of the barrier on the surface. It is encouraging that this correlation between energy requirement and surface features is robust enough to apply to systems other than three-atom ones. The role of excess reagent energy and its fate in reactive collisions is qualitatively understood, but there is more room for development here because of the greater number of critical states through which the trajectories can pass. In particular, the conditions under which multiply peaked product distributions can appear could be studied in terms of surface features.

*Quantitative* correlations between angular distributions and surface features are not yet available. These, if possible, will probably have to take the form of quantities calculated along a diagnostic trajectory, because the scattering angle is subject to the cumulative influence of the potential.

The features of a potential surface that lead to complex collisions seem to be only half understood. The arguments invoking angular momentum and centrifugal barriers are certainly true enough, but they are somewhat tautological and do not explain things in terms of the surface itself. A more quantitative understanding of the prerequisite surface properties would be desirable.

The situation for many-body systems will hopefully soon be improved. The trial and error procedures of selecting surfaces to fit various bits of experimental data are simply not good enough, because the degree of uniqueness of fit that one can obtain in this way is not high and, in most cases, cannot even be guessed at. There are really too many variables at play, and "uniqueness of fit" usually means that "imagination has run out." Fortunately, the work of Raff on the $T + CH_4$ system[92] gives hope that surfaces fitted as well as possible to many kinds of nonkinetic data can actually provide a satisfactory basis for good dynamical calculations. Interpretation of dynamics in terms of such surfaces, which, like the LEPS surface for three-atom systems, have a physical basis, will hopefully lead to meaningful correlations and stimulate new discoveries.

ACKNOWLEDGMENTS

I am indebted to Dr. W. N. Whitton, Mrs. H. Gadewoltz, Mrs. L. Lembke, Mrs. C. Kiwi, and Mrs. K. Gfrörer for their help in preparing the manuscript.

## References

1. R. D. Levine and R. B. Bernstein, Energy disposal and energy requirements for elementary chemical reactions, *Faraday Discuss. Chem. Soc.* **55**, 100–112 (1973).
2. K. T. Gillen, A. M. Rulis, and R. B. Bernstein, Molecular beam study of the $K + I_2$ reaction, *J. Chem. Phys.* **54**, 2831–2851 (1971).
3. G. Hancock, C. Morley, and I. W. M. Smith, Vibrational excitation of CO in the reaction $O + CS \rightarrow CO + S$, *Chem. Phys. Lett.* **12**, 193–196 (1971).
4. T. J. Odiorne, P. R. Brooks, and J. V. V. Kaspar, Molecular beam reaction of K with HCl: Effect of vibrational excitation of HCl, *J. Chem. Phys.* **55**, 1980–1983 (1971).
5. J. G. Pruett, F. R. Grabiner, and P. R. Brooks, Molecular beam reaction of K with HCl: Effect of translational excitation of reagents, *J. Chem. Phys.* **60**, 3335–3336 (1974).
6. D. J. Douglas, J. C. Polanyi, and J. J. Sloan, Effect of reagent vibrational excitation on the rate of a substantially endothermic reaction: HCl ($v' = 1$–4) + Br → Cl + HBr, *J. Chem. Phys.* **59**, 6679–6680 (1973).
7. K. G. Anlauf, P. J. Kuntz, D. H. Maylotte, P. D. Pacey, and J. C. Polanyi, Energy distribution among reaction products, *Discuss. Faraday Soc.* **44**, 183–193 (1967).
8. J. L. Kinsey, *in: MTP International Review of Science* (J. C. Polanyi, ed.), Vol. 9, "Chemical Kinetics," p. 173, Butterworth & Company (Publishers) Ltd., London (1972).

9. T. Carrington and J. C. Polanyi, *in: MTP International Review of Science* (J. C. Polanyi, ed.), Vol. 9, "Chemical Kinetics," p. 135, Butterworth & Company (Publishers) Ltd., London (1972).

10. P. R. Brooks, Scattering of K atoms from oriented $CF_3I$, *Faraday Discuss. Chem. Soc.* **55**, 299–306 (1973).

11. W. A. Chupka, *in: Ion–Molecule Reactions* (J. L. Franklin, ed.), Vol. 1, Chap. 3, Plenum Press, New York (1972).

12. R. B. Bernstein and A. M. Rulis, Translational energy dependence of product energy and angular distribution for the $K + CH_3I \rightarrow KI + CH_3$ reaction, *Faraday Discuss. Chem. Soc.* **55**, 293–298 (1973).

13. H. Eyring and S. H. Lin, *in: Physical Chemistry, An Advanced Treatise* (W. Jost, ed.), Vol. VIA, Chap. 3, Academic Press, Inc., New York (1974).

14. J. C. Slater, *Quantum Theory of Molecules and Solids*, Vol. 1, "Electronic Structure of Molecules," McGraw-Hill Book Company, New York (1963).

15. H. Eyring, J. Walter, and G. E. Kimball, *Quantum Chemistry*, John Wiley & Sons, Inc., New York (1944).

16. J. C. Polanyi and J. L. Schreiber, *in: Physical Chemistry, An Advanced Treatise* (W. Jost. ed.), Vol. VIA, Chap. 6, Academic Press, Inc., New York (1974).

17. D. L. Bunker, Classical trajectory methods, *Methods Comput. Phys.* **10**, 287–325 (1971).

18. D. R. McLaughlin and D. L. Thompson, *Ab initio* dynamics: $HeH^+ + H_2 \rightarrow He + H_3^+(C_{2v})$ classical trajectories using a quantum mechanical potential energy surface, *J. Chem. Phys.* **59**, 4393–4405 (1973).

19. N. C. Blais and J. B. Cross, Molecular beam kinetics: The differential cross section of the reaction $Cl + Br_2$, *J. Chem. Phys.* **52**, 3580–3586 (1970).

20. G. R. North and J. J. Leventhal, Classical superposition phenomena in $H_2^+$ $(v = 0) + He$ reactive collisions, *Chem. Phys. Lett.* **23**, 600–602 (1973).

21. N. C. Blais and D. G. Truhlar, Monte Carlo trajectories: Dynamics of the reaction $F + D_2$ on a semi-empirical valence-bond potential energy surface, *J. Chem. Phys.* **58**, 1080–1108 (1973).

22. J. M. White and D. L. Thompson, Monte Carlo quasiclassical trajectory study of $Br + HBr$ and $H + HBr$: Effect of reagent vibration and rotation on reaction rates and energy transfer, *J. Chem. Phys.* **61**, 719–732 (1974).

23. J. M. White, Trajectory study of reactions in $HBr–Br_2$ systems, *J. Chem. Phys.* **58**, 4482–4495 (1973).

24. R. N. Porter, L. B. Sims, D. L. Thompson, and L. M. Raff, Classical dynamical investigations of reaction mechanism in three-body hydrogen–halogen systems, *J. Chem. Phys.* **58**, 2855–2869 (1973).

25. C. A. Parr, J. C. Polanyi, and W. H. Wong, Distribution of reaction products (theory). VIII. $Cl + HI$, $Cl + DI$, *J. Chem. Phys.* **58**, 5–20 (1973).

26. J. B. Anderson and R. T. V. Kung, Vibrational population inversion in hydrogen iodide from $H + I_2 \rightarrow HI + I$, *J. Chem. Phys.* **58**, 2477–2479 (1973).

27. J. D. McDonald, Classical trajectory studies of angular distributions of reactions of deuterium atoms with iodine molecules, *J. Chem. Phys.* **60**, 2040–2046 (1974).

28. P. J. Kuntz, E. M. Nemeth, J. C. Polanyi, S. D. Rosner, and C. E. Young, Energy distribution among products of exothermic reactions. II. Repulsive, mixed, and attractive energy release, *J. Chem. Phys.* **44**, 1168–1184 (1966).

29. J. C. Polanyi and W. H. Wong, Location of energy barriers. I. Effect on the dynamics of reactions A + BC, *J. Chem. Phys.* **51**, 1439–1460 (1969).

30. B. A. Hodgson and J. C. Polanyi, Location of energy barriers. IV. Effect of rotation and mass on the dynamics of reactions A + BC, *J. Chem. Phys.* **55**, 4745–4757 (1971).

31. M. H. Mok and J. C. Polanyi, Location of energy barriers. III. Effect on the dynamics of $AB + CD \rightarrow AC + BD$, *J. Chem. Phys.* **53**, 4588–4604 (1970).

32. D. S. Perry, J. C. Polanyi, and C. W. Wilson, Jr., Location of energy barriers. VI, The dynamics of endothermic reactions AB + C, *Chem. Phys.* **3**, 317–331 (1974).

33. T. B. Borne and D. L. Bunker, Trajectory studies of halogen atom–molecule exchange reactions, *J. Chem. Phys.* **55**, 4861–4866 (1971).

34. D. L. Bunker and N. C. Blais, Monte Carlo calculations. V. Three-dimensional study of a general bimolecular interaction potential, *J. Chem. Phys.* **41**, 2377–2386 (1964).
35. F. T. Wall and R. N. Porter, Sensitivity of exchange-reaction probabilities to the potential energy surface, *J. Chem. Phys.* **39**, 3112–3117 (1963).
36. J. W. Duff and Donald G. Truhlar, Effect of curvature of the reaction path on dynamic effects in endothermic chemical reactions and product energies in exothermic reactions, *J. Chem. Phys.* **62**, 2477–2491 (1975).
37. I. G. Csizmadia, J. C. Polanyi, A. C. Roach, and W. H. Wong, Distribution of reaction products (theory). VII. $D^+ + H_2 \rightarrow DH + H^+$ using an ab initio potential energy surface, *Can. J. Chem.* **47**, 4097–4099 (1969).
38. N. C. Blais and D. L. Bunker, Monte Carlo calculations. II. Reactions of alkali atoms with methyl iodide, *J. Chem. Phys.* **37**, 2713–2720 (1962).
39. N. C. Blais and D. L. Bunker, Monte Carlo calculations. III. A general study of bimolecular exchange reactions, *J. Chem. Phys.* **39**, 315–323 (1963).
40. M. Karplus and L. M. Raff, Theoretical investigations of reactive collisions in molecular beams: $K + CH_3I$, *J. Chem. Phys.* **41**, 1267–1277 (1964).
41. L. M. Raff and M. Karplus, Theoretical investigations of reactive collisions in molecular beams: $K + CH_3I$ and related systems, *J. Chem. Phys.* **44**, 1212–1229 (1966).
42. F. O. Ellison, A method of diatomics-in-molecules. I. General theory and application to $H_2O$, *J. Am. Chem. Soc.* **85**, 3540–3544 (1963).
43. J. C. Tully, Diatomics-in-molecules potential energy surfaces. I. First-row triatomic hydrides, *J. Chem. Phys.* **58**, 1396–1410 (1973).
44. P. J. Kuntz, Use of the method of diatomics-in-molecules in fitting *ab initio* potential surfaces: The system $HeH_2^+$, *Chem. Phys. Lett.* **16**, 581–583 (1972).
45. B. Liu, *Ab initio* potential energy surface for linear $H_3$, *J. Chem. Phys.* **58**, 1925–1937 (1973).
46. E. Steiner, P. R. Certain, and P. J. Kuntz, Extended diatomics-in-molecules calculations, *J. Chem. Phys.* **59**, 47–55 (1973).
47. P. J. Brown and E. F. Hayes, Non-empirical LCAO–MO–SCF study of the energy surface for linear $HeH_2^+$, *J. Chem. Phys.* **55**, 922–926 (1971).
48. J. R. Krenos, R. K. Preston, R. Wolfgang, and J. C. Tully, Molecular beam and trajectory studies of reactions of $H^+$ with $H_2$, *J. Chem. Phys.* **60**, 1634–1659 (1974).
49. C. W. Bauschlicher, Jr., S. V. O'Neil, R. K. Preston, H. F. Schaefer III, and C. F. Bender, Avoided intersection of potential energy surfaces: The $(H^+ + H_2, H + H_2^+)$ system, *J. Chem. Phys.* **59**, 1286–1292 (1973).
50. D. R. Yarkony, S. V. O'Neil, H. F. Schaefer III, C. P. Baskin, and C. F. Bender, Interaction potential between two rigid HF molecules, *J. Chem. Phys.* **60**, 855–865 (1974).
51. C. F. Bender, P. K. Pearson, S. V. O'Neil, and H. F. Schaefer III, Potential energy surface including electron correlation for the chemical reaction $F + H_2 \rightarrow FH + H$. I. Preliminary surface, *J. Chem. Phys.* **56**, 4626–4631 (1972).
52. J. Grosser and H. Haberland, Reactive scattering of hydrogen and deuterium atoms from halogen molecules, *Chem. Phys.* **2**, 342–351 (1973).
53. J. O. Hirschfelder, Coordinates which diagonalize the kinetic energy of relative motion, *Int. J. Quantum Chem.* **III S**, 17–31 (1969).
54. S. Glasstone, K. J. Laidler, and H. Eyring, *The Theory of Rate Processes*, McGraw-Hill Book Company, New York (1941).
55. G. L. Hofacker and R. D. Levine, A non-abiabatic model for population inversion in molecular collisions, *Chem. Phys. Lett.* **9**, 617–620 (1971).
56. R. A. Marcus, On the analytical mechanics of chemical reactions: Classical mechanics of linear collisions, *J. Chem. Phys.* **45**, 4500–4504 (1966).
57. G. Miller and J. C. Light, Quantum calculations of collinear reactive triatomic systems. III. $H + Cl_2 \rightarrow HCl + Cl$, *J. Chem. Phys.* **54**, 1643–1651 (1971).
58. J. T. Muckerman, Classical dynamics of the reaction of fluorine atoms with hydrogen molecules. II. Dependence on the potential energy surface, *J. Chem. Phys.* **56**, 2997–3006 (1972).

59. D. J. Kouri and M. Baer, Collinear quantum mechanical calculations of the $He + H_2^+$ proton transfer reaction, *Chem. Phys. Lett.* **24**, 37–40 (1974).

60. P. J. Kuntz and W. N. Whitton, Trajectory calculations for the reactions $H_2^+ + He \rightarrow H + HeH^+$, *Chem. Phys. Lett.* **34**, 340–342 (1975).

61. J. T. Muckerman, Monte Carlo calculations of energy partitioning and isotope effects in reaction of fluorine atoms with $H_2$, HD, and $D_2$, *J. Chem. Phys.* **54**, 1155–1164 (1971).

62. J. T. Muckerman, Classical dynamics of the reaction of fluorine atoms with hydrogen molecules. III. The hot-atom reactions of $^{18}F$ with HD, *J. Chem. Phys.* **57**, 3388–3396 (1972).

63. R. L. Wilkins, Monte Carlo calculation of reaction rates and energy distributions among reaction products. I. $F + H_2 \rightarrow HF + H$, *J. Chem. Phys.* **57**, 912–917 (1972).

64. R. L. Wilkins, Monte Carlo calculation of reaction rates and energy distributions among reaction products. II. $H + HF(v) \rightarrow H_2(v') + F$ and $H + HF(v) \rightarrow HF(v') + H$, *J. Chem. Phys.* **58**, 3038–3046 (1973).

65. R. L. Wilkins, Monte Carlo calculation of reaction rates and energy distributions among reaction products. III. $H + F_2 \rightarrow HF + F$ and $D + F_2 \rightarrow DF + F$, *J. Chem. Phys.* **58**, 2326–2332 (1973).

66. L. T. Cowley, D. S. Horne, and J. C. Polanyi, Infrared chemiluminescence study of the reaction $Cl + HI \rightarrow HCl + I$ at enhanced collision energies, *Chem. Phys. Lett.* **12**, 144–149 (1971).

67. A. M. G. Ding, L. J. Kirsch, D. S. Perry, J. C. Polanyi, and J. L. Schreiber, Effect of changing reagent energy on reaction probability and product energy distribution, *Faraday Discuss. Chem. Soc.* **55**, 252–276 (1973).

68. P. M. Hierl, Z. Herman, and R. Wolfgang, Chemical accelerator studies of isotope effects on collision dynamics of ion–molecule reactions: Elaboration of a model for direct reactions, *J. Chem. Phys.* **53**, 660–673(1970).

69. M. Chiang, E. A. Gislason, B. H. Mahan, C. W. Tsao, and S. W. Werner, Dynamics of the reaction $Ar^+$ with $D_2$, *J. Chem. Phys.* **52**, 2698–2708 (1970).

70. C. A. Parr, J. C. Polanyi, W. H. Wong, and D. C. Tardy, General discussion, *Faraday Discuss. Chem. Soc.* **55**, 308–309 (1973).

71. P. J. Kuntz and A. C. Roach, Classical trajectory study of exothermic ion–molecule reactions, *J. Chem. Phys.* **59**, 6299–6311 (1973).

72. R. A. LaBudde, P. J. Kuntz, R. B. Bernstein, and R. D. Levine, Classical trajectory study of the $K + CH_3I$ reaction, *J. Chem. Phys.* **59**, 6286–6298 (1973).

73. R. B. Bernstein, Potential practical applications of basic research in molecular dynamics, *Isr. J. Chem.* **9**, 615–635 (1971).

74. K. T. Gillen, B. H. Mahan, and J. S. Winn, Dynamics of the $O^+ + H_2$ reaction, *J. Chem. Phys.* **59**, 6380–6396 (1973).

75. P. J. Kuntz, Analytical properties of a direct interaction model for gas-phase chemical reactions $A + BC \rightarrow AB + C$, *Trans. Faraday Soc.* **66**, 2980–2996 (1970).

76. P. J. Kuntz, The $K + CH_3I \rightarrow KI + CH_3$ reaction: Interpretation of the product angular and energy distributions in terms of a direct interaction model, *Mol. Phys.* **23**, 1035–1050 (1972).

77. A. M. Rulis and R. B. Bernstein, Molecular beam study of the $K + CH_3I$ reaction: Energy dependence of the detailed differential reactive cross section, *J. Chem. Phys.* **57**, 5497–5515 (1972).

78. N. N. Hijazi and K. J. Laidler, Dynamics of collinear $A + BC$ systems, *J. Chem. Phys.* **58**, 349–353 (1973).

79. M. Karplus, R. N. Porter, and R. D. Sharma, Exchange reactions with activation energy. I. Simple barrier potential for $(H, H_2)$, *J. Chem. Phys.* **43**, 3259–3287 (1965).

80. N. C. Blais, Monte Carlo trajectories: The dynamics of harpooning in alkali–halogen reactions, *J. Chem. Phys.* **49**, 9–14 (1968).

81. M. Godfrey and M. Karplus, Theoretical investigations of reactive collisions in molecular beams: $K + Br_2$, *J. Chem. Phys.* **49**, 3602–3609 (1968).

82. M. H. Mok and J. C. Polanyi, Location of energy barriers. II. Correlation with barrier beight, *J. Chem. Phys.* **51**, 1451–1469 (1969).

83. W. B. Miller, S. A. Safron, and D. R. Herschbach, Exchange reactions of alkali atoms with alkali halides: A collision complex mechansm, *Discuss. Faraday Soc.* **44**, 108–122 (1967).

84. A. C. Roach and M. S. Child, Electronic potential energy surfaces for the reaction K+NaCl→KCl+Na, *Mol. Phys.* **14**, 1–15 (1968).

85. G. H. Kwei, B. P. Boffardi, and S. F. Sun, Classical trajectory studies of long-lived collision complexes. I. Reaction of K atoms with NaCl molecules, *J. Chem. Phys.* **58**, 1722–1734 (1973).

86. I. G. Csizmadia, R. E. Kari, J. C. Polanyi, A. C. Roach, and M. A. Robb, *Ab initio* SCF–MO–CI calculations for $H^-$, $H_2$, and $H_3^+$ using Gaussian basis sets, *J. Chem. Phys.* **52**, 6205–6211 (1970).

87. I. W. M. Smith, Experimental and computer studies of the kinetics and distribution of vibrational energy in both products of the reaction $O(^3P)+CS_2 \rightarrow SO+CS$, *Discuss. Faraday Soc.* **44**, 1964–1974 (1967).

88. L. M. Raff, Classical Monte Carlo analysis of four-body reactions: $K + C_2H_5I$ system, *J. Chem. Phys.* **44**, 1202–1211 (1966).

89. P. J. Kuntz, E. M. Nemeth, J. C. Polanyi, and W. H. Wong, Distribution of reaction products. VI. Hot-atom reactions, T + HR, *J. Chem. Phys.* **52**, 4654–4674 (1970).

90. D. L. Bunker and M. D. Pattengill, Trajectory studies of hot-atom reactions. I. Tritium and methane, *J. Chem. Phys.* **53**, 3041–3049 (1970).

91. T. Valencich and D. L. Bunker, Energy-dependent cross sections for the tritium–methane hot-atom reactions, *Chem. Phys. Lett.* **20**, 50–52 (1973).

92. L. M. Raff, Theoretical investigations of the reaction dynamics of polyatomic systems: Chemistry of the hot-atom systems ($T + CH_4$) and ($T + CD_4$), *J. Chem. Phys.* **60**, 2220–2244 (1974).

93. P. J. Kuntz, M. H. Mok, and J. C. Polanyi, Distribution or reaction products. V. Reactions forming an ionic bond, M + XC (3D), *J. Chem. Phys.* **50**, 4623–4652 (1969).

94. P. J. Kuntz and A. C. Roach, Ion–Molecule reactions of the rare gases with hydrogen, *J. Chem. Soc. Faraday Trans.* II **68**, 259–280 (1972).

95. S. Chapman and R. K. Preston, Nonadiabatic molecular collisions: Charge exchange and chemical reaction in the $Ar^+ + H_2$ system, *J. Chem. Phys.* **60**, 650–659 (1974).

# 3

# Dynamics of Unimolecular Reactions

*William L. Hase*

## 1. Introduction

The subject of unimolecular dynamics deals with the intermolecular and intramolecular microscopic details of unimolecular reactions. Theories of unimolecular dynamics are concerned with molecular motion over potential energy surfaces and the behavior of molecular coordinates as a function of time. Most studies of unimolecular reactions have involved measurements and predictions of the rate at which an energized molecule will undergo a unimolecular reaction. The basic postulate of all unimolecular theories is the rapidity of intramolecular vibrational energy relaxation. Experimentalists were awarded a rare opportunity to test two conflicting assumptions regarding this postulate by the simultaneous advent of the Slater[1] and Rice–Ramsperger–Kassel–Marcus (RRKM)[2] theories in the 1950s. Slater's theory, which is dynamical, pictures a molecule as an assembly of harmonic oscillators. Within this framework vibrational energy relaxation between the normal modes is forbidden, and reaction occurs only when some coordinate, the reaction coordinate, reaches a critical extension by superposition of the various normal modes. In contrast, the RRKM theory, which is an extension by R. A. Marcus of the statistical theory developed by O. K. Rice, H. C. Ramsperger, and L. S. Kassel, assumes rapid relaxation of vibrational energy. The experimental tests overwhelmingly endorsed the RRKM theory, and the controversy involving intramolecular vibrational energy relaxation was seemingly laid to rest. It also appeared as though dynamical treatments of unimolecular reactions were unnecessary.

*William L. Hase* • Department of Chemistry, Wayne State University, Detroit, Michigan

Within the last few years there has been an increasing interest in understanding the dynamics of unimolecular reactions, including the efficiency of intramolecular vibrational energy relaxation. This has been brought about in part by new experimental techniques such as molecular beams[3-9] and infrared chemiluminescence,[10-14] which have provided information about the unimolecular dynamics. In addition, both theoretical[15] and experimental studies[16,17] have indicated that the RRKM theory does not have universal applicability and that for some situations dynamical theories are necessary to describe the unimolecular reaction. We may expect that some general criteria will be developed for employing the RRKM theory. The relationships between the dynamical properties of unimolecular reactions and potential energy surfaces are just beginning to be unfolded. The next decade should be exciting times for unimolecular dynamicists.

The remainder of this chapter is divided into four parts. In Section 2 a brief derivation of the RRKM theory is given along with its inherent assumptions. In Section 3 we shall describe the application of the RRKM theory to thermal and chemical activation experiments. A comparison is also made between experimental results and the RRKM prediction. In Section 4 a discussion is given of measurements and interpretations of unimolecular dynamics. Finally, in Section 5, new experimental and theoretical directions of study are described.

## 2. RRKM Theory

The RRKM theory was developed by R. A. Marcus using the RRK model and extending it to consider explicitly individual vibrational and rotational modes and include zero-point energies. Several revisions of the theory have been made since its conception, primarily as a result of improved treatments of external rotational degrees of freedom.[18-21] Discussions of the RRKM theory are given in several kinetics books[22-24] and in two recent books devoted entirely to unimolecular reactions.[25,26] We shall give below a brief presentation of the RRKM theory. In most cases we have followed Marcus' notation; however, adherence is not complete.

### 2.1. Assumptions

The following mechanism is used in the RRKM theory to describe the isomerization or decomposition of an energized molecule $A^*$ that contains a total internal energy of $E^*$:

$$A^* \xrightarrow{k_a(E^*)} A^+ \to \text{products} \tag{1}$$

The critical configuration $A^+$ represents a unique molecular configuration (or molecular phase point) intermediate between the reactants $A^*$ and products so that the flux from $A^+$ toward products is equal to the flux from $A^+$ toward $A^*$. We may then express $k_a(E^*)$ as

$$k_a(E^*) = \frac{\nu}{2} \frac{[A^+]}{[A^*]} \tag{2}$$

where $[A^+]/[A^*]$ is the ratio of critical configurations to energized molecules and $\nu$ is the frequency of passage from $A^+$. The factor of 2 selects only those critical configurations moving toward products. According to Eq. (2), we see that $k_a(E^*)$ is the intrinsic rate constant for decomposition of $A^*$. It is equal to the frequency at which $A^+$ passes to products times the fraction of the energized molecules that have the critical configuration structure.

In writing Eqs. (1) and (2) we have made the following two very important assumptions:

1. All internal molecular states of energy $E^*$ are accessible and will ultimately lead to products.
2. Vibrational energy redistribution within the energized molecule is much faster than unimolecular reaction.

Assumption 1 means that the phase space of $A^*$ cannot be subdivided into smaller regions so that a representative phase point cannot move freely between the regions, i.e., vibrational energy may be transferred between all modes. As a result of assumption 2, all vibrational quantum states of $A^*$ have equal probabilities of proceeding to products per unit time. These assumptions allow us to relate $k_a(E^*)$ to the statistical ratio of critical configurations to energized molecules, $[A^+]/[A^*]$.

The above assumptions imply that $A^*$ will have a random lifetime distribution[1,22] given by

$$P(\tau) = k_a(E^*) \exp[-k_a(E^*)\tau] \tag{3}$$

where $P(\tau)$ is the probability that $A^*$, which contains energy $E^*$, has a lifetime $\tau$. Equation (3) is like the probability of radioactive decay, which is what we expect if passage of $A^*$ to $A^+$ is purely statistical. It should be recognized that acceptance of the random lifetime assumption frees the RRKM theory from the "equilibrium" hypothesis and the assumption of a "separable" reaction coordinate, which are inherent in the activated complex theory.[27]

## 2.2. Microscopic Unimolecular Rate Constant

The energized molecule may contain both vibrational, $E_v^*$, and external rotational, $E_r^*$, energies. The sum $E_v^* + E_r^*$ is $E^*$. Molecular motion from $A^*$ to

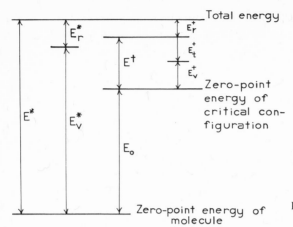

Fig. 1.  Diagram of the energies for the RRKM theory.

$A^+$ involves a change in potential energy. As a result, the total energy of the critical configuration is

$$E^+ + E_r^+ = E^* - E_0 \tag{4}$$

where $E_0$ is the potential energy difference between $A^*$ and $A^+$. Part of the energy $E^+$ will be associated with vibration $E_v^+$. A portion of $E^+$ also consists of translational motion in the critical configuration $E_t^+$. The sum $E_v^+ + E_t^+$ equals $E^+$. These energies are illustrated in Fig. 1.

The internal degrees of freedom are designated either active or adiabatic. Active modes are assumed to exchange energy freely. An adiabatic mode is one that remains in the same quantum state during the reaction. Conservation of angular momentum usually requires the external rotational modes to be treated as a adiabatic. In contrast, the vibrational and internal rotational modes are considered to be active. In our discussion we shall classify the internal rotations with vibrations.

Applying statistical mechanics, the ratio $[A^+]/[A^*]$ may be equated to the ratio of the density of quantum states in the critical configurations and molecules, $N(E^+)/N(E_v^*)$. We then have

$$k_a(E^*) = \frac{\nu}{2} \frac{N(E^+)}{N(E_v^*)} \tag{5}$$

The density $N(E^+)$ may be expressed as

$$N(E^+) = \sum_{E_v^+=0}^{E^+} P(E_v^+) N(E^+ - E_v^+) \tag{6}$$

where $N(E^+ - E_v^+) = N(E_t^+)$. The quantity $P(E_v^+)$ is the sum of vibrational quantum states of $A^+$ at energy $E_v^+$, and $N(E^+ - E_v^+)$ is the density of translational quantum state of $A^+$ with energy in the range $E_t^+ \to E_t^+ + dE_t^+$.

Treating the transformation of the critical configuration into products as the translation of a particle of mass $\mu$ in a one-dimensional box of length $x$, the energy $E_t^+$ of the $n$th quantum level is

$$E_t^+ = n^2 h^2 / 8\mu x^2 \tag{7}$$

where $h$ is Planck's constant. The density $N(E_t^+)$ may then be written as

$$N(E_t^+) = dn/dE_t^+ = (2\mu x^2 / h^2 E_t^+)^{1/2} \tag{8}$$

Substituting Eq. (8) into Eq. (6), we have

$$N(E^+) = \sum_{E_v^+=0}^{E^+} P(E_v^+) \left[ \frac{2\mu x^2}{h^2(E^+ - E_v^+)} \right]^{1/2} \tag{9}$$

The frequency of decomposition or isomerization of $A^+$, $\nu$, is found by dividing the velocity of the particle of mass $\mu$, in a one-dimensional box, by the length of the box to give

$$\nu = (2E_t^+ / \mu)^{1/2} / x \tag{10}$$

Substituting of Eqs. (10) and (9) into Eq. (5) and cancelling terms, we obtain

$$k_a(E^*) = \frac{1}{h} \frac{\sum\limits_{E_v^+=0}^{E^+} P(E_v^+)}{N(E_v^*)} \tag{11}$$

which is the complete RRKM expression for the microscopic unimolecular rate constant. To use it we need to evaluate the sum and density terms in the numerator and denominator. This requires information about the structural details (frequencies and moments of inertia) of both the molecule and the critical configuration. We also need to know the energies $E^*$ and $E^+$ and how the internal energy of the molecule is distrubuted between $E_v^*$ and $E_r^*$. The requirement of conservation of angular momentum will usually allow us to calculate $E_r^+$ from $E_r^*$ and the moments of inertia. It should be noted that the critical configuration has one less vibrational degree of freedom than the molecule, because it has an internal translational motion.

For small values of $E^*$ and $E^+$ the sum of quantum states in the critical configuration and the density of quantum states in the energized molecule have to be evaluated by exact counting procedures. However, for larger values of $E^*$ and $E^+$ approximate counting procedures, developed by Laplace transform methods[28] and by empirically fitting a function to results of direct counts,[28,29] may be used. The latter treatment has received the most use, and we have selected it for illustration. By empirical reasoning it was found that the sum of states in the critical configuration could be expressed as

$$\sum_{E_v^+=0}^{E^+} P(E_v^+) = \frac{Q_{ir}^+(E^+ + aE_z^+)^{s-1+d/2}}{(kT)^{d/2}\Gamma(s+d/2)\pi h \nu_i^+} \tag{12}$$

where $s-1$ and $d$ are the number of vibrational and internal rotational modes in the critical configuration, $E_z^+$ is the zero-point energy, and $Q_{ir}^+$ is the internal rotational partition function. The empirical factor $a$ equals $1 - \beta\omega$, where both $\beta$ and $\omega$ are also empirical factors; $\beta$ is related to the frequency dispersion, and $\omega$ is an analytical function of the reduced energy $E^+/E_z$. Replacing $E^+$ by $E^*$ in Eq. (12) and differentiating by $E_v^*$, we obtain an analytical expression for $N(E_v^*)$. A harmonicity correction to the sum and density may also be included if necessary.[28]

We have nearly completed our formulation of the RRKM expression for $k_a(E^*)$. The remaining item is the relationship between $E_r^*$ and $E_r^+$. In terms of the moment of inertia $I$ and angular momentum $L$, $E_r$ equals $L^2/2I$. The rotational energies of $A^*$ and $A^+$ will then be related by $E_r^*/E_r^+ = I^+/I^*$. Extension of this treatment to three dimensions gives

$$\begin{aligned} E_r^+ &= E_{rx}^*(I_x^*/I_y^+) + E_{ry}^*(I_y^*/I_y^+) + E_{rz}^*(I_z^*/I_z^+) \\ &= E_{rx}^+ + E_{ry}^+ + E_{rz}^+ \end{aligned} \tag{13}$$

## 2.3. Critical Configuration

Choosing the critical configuration poses the main problem in calculating $k_a(E^*)$, because its physical properties are hardly ever known. As discussed above, the critical configuration should be chosen so that the rates of outward and inward passage through $A^+$ are equal. Bunker and Pattengill[19] suggested that this be done by requiring the critical configuration to have a value of the reaction coordinate $r^+$ such that the number of accessible internal states of the molecule is minimized,

$$\partial N(E^+)/\partial r^+ = 0 \tag{14}$$

We shall designate the value of the reaction coordinate at the critical configuration as $R^+$.

The reaction coordinate need not be only one vibrational degree of freedom. It may represent the change in several internal coordinates as the potential energy changes in transforming the reactant molecule into product. For small values of $E^+$ Eq. (14) must be solved by numerical integration of Eq. (9). An approximation for Eq. (14) involves using Eq. (12) to represent the sum of quantum states in the critical configuration. We then have

$$N(E^+) \propto \frac{Q_{ir}^+}{\Gamma(s+d/2)\pi h \nu i^+} \int_{E_v^+=0}^{E^+} \frac{(E_v^+ + aE_z^+)^{s-1+d/2}}{(E^+ - E_v^+)^{1/2}} \, dE_v^+ \tag{15}$$

where

$$E^+ = E_v^* - V(r^+) + (E_r^* - E_r^+) \tag{16}$$

Integration of Eq. (15), assuming that $a$ is constant, gives

$$N(E^+) \propto \frac{Q_{ir}^+ [E_v^* - V(r^+) + E_r^* - E_r^+ + aE_z^+]^{s-3/2+d/2}}{\Gamma(s-1+d/2)\pi h \nu_i^+} \tag{17}$$

Here $V(r^+)$ is the potential energy associated with the reaction coordinate. We should note that in addition to $V(r^+)$ the $\nu_i^+$'s and $I^+$'s are also functions of $r^+$.

We expect the criterion of minimum state density to be most important for unimolecular reactions that do not have a well-defined maximum in the potential energy barrier, e.g., dissociation reactions. For unimolecular reactions where the barrier is well defined the critical configuration will be located at or very near the top of the potential energy barrier. Let us consider a dissociation reaction where there is considerable molecular loosening, i.e., a decrease in force constants, as $r^+$ is extended. As $r^+$ is initially increased from its equilibrium value there will be a decrease in $N(E^+)$, due to an increase in $V(r^+)$. Then as the top of the potential energy barrier is being approached there will be an increase in $N(E^+)$, and the minimum in the molecular state density or "bottleneck" in the molecular phase space will be located. This increase in $N(E^+)$ is due to lower molecular frequencies and larger moments of inertia as the molecule dissociates. These effects have been referred to as structural[30] or entropic[31] effects.

In practice Eq. (14) is difficult to apply because the potential energy surfaces for unimolecular reactions are seldom known in enough detail. However, it has been applied to several bond dissociation reactions by using the bond energy–bond order method[32] to account for the dependence of force constants on bond lengths and making a local normal mode analysis at each value of the reaction coordinate.[30,33] Such studies have been very instructive, and for display we shall show some results for $C_2H_6$ decomposition.[30] In this study the C—C stretch was assumed to be the reaction coordinate, and $V(r^+)$ was represented by a Morse function with appropriate parameters. Loosening in the molecule resulted in lower $CH_3$-rocking frequencies as $r^+$ was extended. Plots of $N(E^+)$ vs. $r^+$ assuming varying degrees of loosening are shown in Fig. 2. Model I represents the case where there is no decrease in the $CH_3$-rocking frequencies as $r^+$ is increased, and model III represents the maximum amount of loosening. Model II is the one that fits the experimental results. We see that increasing the loosening in the molecule results in smaller values of $R^+$, the value of $r^+$ at the critical configuration. In addition, we see that for models II and III the effective dissociation energy $E_0$ [the value of $V(r^+)$ for the critical configuration] is less than the C—C bond dissociation energy. Similar effects were found by Bunker and Pattengill[19] in their study of triatomic dissociations. It has previously been assumed that for dissociation reactions the critical

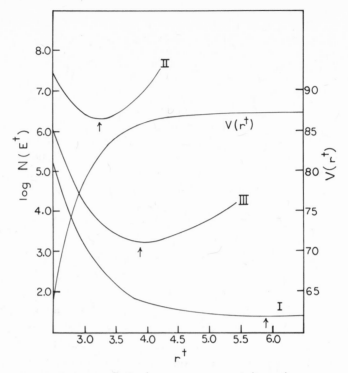

Fig. 2. Plots of $V(r^+)$ vs. $r^+$ and plots of log $N(E^+)$ vs. $r^+$ for three different potential surface models for ethane.[30] The arrows locate the minimums in the plots.

configuration should be located at the rotational barrier.[20,27] For unimolecular dissociation reactions the criterion of minimum state density predicts a significantly smaller value of $R^+$ than the rotational barrier.[19,30] According to Eq. (14)–(17), we see that the location of the critical configuration is not constant but is a function of both $E_v^*$ and $E_r^*$. In Figs. 3 and 4 the value of $R^+$ vs. $E_v^*$ and $E_r^*$ are shown for the $C_2H_6$ model II described above.

Because the criterion of minimum state density is hard to apply, critical configurations have usually been chosen by deduction from the thermal high-pressure frequency factor, $A_\infty$. Absolute rate theory[27] gives

$$A_\infty = kT/h \exp(\Delta S^+/R + 1) \tag{18}$$

where $\Delta S^+$ is the entropy difference between the molecule and critical configuration. Frequencies in the critical configuration are chosen so that the correct $\Delta S^+$ is obtained. This does not define a unique critical configuration but only a set of vibrational frequencies. However, it has been found that the RRKM results are not critically dependent on the particular choice of vibrational frequencies so long as $\Delta S^+$ is matched.[25,26] The quantity $E_0$ is then given by

$$E_0 = E_\infty^a - [\bar{E}^+(T) - \bar{E}(T)] \tag{19}$$

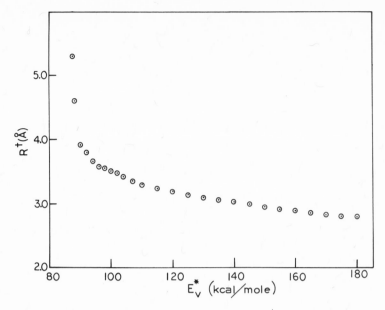

Fig. 3. Relationship between the critical configuration $r^+$ and molecular vibrational energy for ethane model II in Fig. 2.[30] The molecule has no angular momentum.

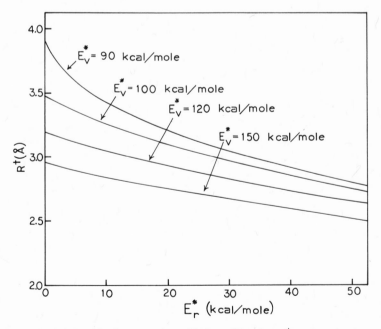

Fig. 4. Relationship between the critical configuration $r^+$ and molecular rotational energy for ethane model II in Fig. 2.[30]

where $E_\infty^a$ is the high-pressure thermal activation energy and $\bar{E}^+(T) - \bar{E}(T)$ is the difference in the thermal energies of the critical configuration and molecule.

What is often done is to use Eqs. (18) and (19) to define the critical configuration and then use this same model for calculating values of $k_a(E^*)$ for a wide assortment of $E_v^*$s and $E_r^*$s.[25,26] This prescription may be useful for reactions with well-defined barriers so that the criterion of minimum state density is not important. However, there is some question of using this procedure for dissociation reactions where structural effects are important and the critical configuration structure is not fixed but is dependent on the specific values of $E_v^*$ and $E_r^*$. This point will be discussed in more detail in Section 3. We should expect to find a more general application of the criterion of minimum state density in the near future as *ab initio* computer calculations of potential energy surfaces become available.

### 2.4. Non-RRKM Behavior

The assumptions of the RRKM theory infer that isolated molecules behave as if all their accessible states were occupied with equal probabilities. This means that the probability of A* transformation to $A^+$ is due entirely to statistical considerations and that there is no bias for *all* of A* to become products soon after formation or to exist for a longer length of time until decomposing. Figure 5(a) illustrates this schematically and shows random transitions among states at some energy high enough for eventual reaction (toward the right). In reality, transitions between states are not equally probable; some are more likely than others. Therefore, we have to rely on the molecular motion to be disorderly enough for the RRKM assumptions to be mimicked, as crudely illustrated in Fig. 5(b).

The lifetime distribution will depend in part on the manner in which the energy needed for reaction is supplied. As we shall see in Section 3, in the usual understanding of thermal reactions it is assumed that the activating collisions randomly prepare the internal vibrational states of A*. However, it has been pointed out previously[34] that thermal collisions surely obey at least an approximate Franck–Condon principle. Direct collisional preparation of a state with a large extension of some bond, in the direction of product formation, is suggested to be an implausible event.[34] We might therefore expect some shortage of lifetimes less than one fourth of a vibration period, as illustrated in Fig. 5(c).

The above is an example of apparent non-RRKM behavior arising from state selection. More interesting examples are those where the molecular vibrational states are intentionally excited nonrandomly. This is done in a large number of experiments, among which the most common are those based on

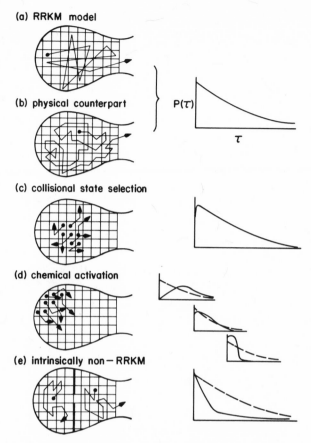

Fig. 5. Relation of state occupation (schematically shown at constant energy) to lifetime distribution for the RRKM theory and for various actual situations.[15] Dashed lines in lifetime distributions indicate RRKM behavior.

chemical activation. Regardless of the pattern of initial energization, the RRKM model in Fig. 5(a) would require that it become a random one in a negligibly short time. How well will the actual molecular motion approach this ideal? Three different possibilities are represented by Fig. 5(d). The probability of a short lifetime with respect to reaction may be enhanced or reduced, depending on the location of the initial excitation within the molecule. If there were strong internal coupling, $P(\tau)$ would not be very different from a random one.

A situation completely distinct from apparent non-RRKM behavior is intrinsic non-RRKM behavior. By this we mean that $A^*$ has a nonrandom $P(\tau)$ even if the internal vibrational states of $A^*$ are prepared randomly. This situation arises when transitions between two or more individual molecular vibrational states are slower than transitions leading to products. These slow transitions will arise when there is at least one bottleneck in the molecular phase space other than the one defining the critical configuration. An example of such a bottleneck, with a possible resultant $P(\tau)$, is shown in Fig. 5(e). It should be noted that if the molecular vibrational states are prepared randomly, a molecule that displays intrinsic non-RRKM behavior will still have a zero-time value of $P(\tau)$ equal to $k_a(E^*)$.

## 3. Applications of the RRKM Theory

Historically, the development of the theories of unimolecular reactions has been closely tied to thermal experiments. This is the case for the RRKM theory. Only in recent years have other experimental techniques been developed. One of the first was chemical activation. It provided an important additional method for studying unimolecular reactions. The ability of the RRKM theory to interpret the thermal and chemical activation experiments has contributed greatly to our understanding of unimolecular kinetics. In this section we shall describe the application of the RRKM theory to thermal and chemical activation unimolecular studies. Comparisons are made with experimental results.

### 3.1. Thermal Unimolecular Reactions

#### 3.1.1. Derivation of $k_{uni}$

The Lindemann theory forms the basis for the RRKM theory of thermal unimolecular reactions. The important concepts proposed by Lindemann are (1) a time lag between the moment a molecule becomes activated and the actual unimolecular dissociation or isomerization step and (2) bimolecular collisions for activating and deactivating the molecules. According to these concepts, the mechanism for the unimolecular reaction may be written as

$$A + M \xrightarrow{k_1} A^* + M \tag{20}$$

$$A^* + M \xrightarrow{k_2} A + M \tag{21}$$

$$A^* \xrightarrow{k_3} \text{product(s)} \tag{22}$$

Here $A^*$ represents a molecule with sufficient energy to react and $M$ is some collision partner. If we apply the steady state hypothesis to the concentration of

A*, the overall rate of reaction is given by

$$\text{rate} = k_{\text{uni}}[A] = k_3[A^*] = \frac{k_1 k_3 [A][M]}{k_2[M] + k_3} \tag{23}$$

The significance of the Lindemann mechanism is its ability to explain the change in reaction order, from first to second, as the pressure is decreased, i.e., $k_3 \gg k_2[M]$. However, Lindemann failed to recognize that $k_3$ was dependent on $E^*$, and he did not describe correctly the equilibrium probability that A* will have energy $E^* > E_0$. This resulted in several important differences between the theory and experiments.[25]

The essential modifications to the Lindemann theory were made by Rice, Ramsperger, and Kassel and ended with the work of Marcus. Two additional postulates were made. The first is called the strong collision assumption. According to this assumption, large amounts of energy are transferred in the molecular collisions so that deactivation and activation may be viewed as single-step processes, in contrast to ladder-climbing processes. This assumption also requires the collision to be so violent that the distribution of energy resulting from energizing or deenergizing collisions is random. As a result, the distribution of activated molecules, A*, may be determined directly from statistical considerations. We may therefore express

$$k_1/k_2 = (A^*/A)_{\text{eq}} \tag{24}$$

The second postulate is that of rapid intramolecular vibrational energy relaxation. This assumption was implicit in our derivation of the RRKM expression for $k_a(E^*)$.

We are now prepared to write the RRKM expression for the overall thermal unimolecular rate constant, $k_{\text{uni}}$. It is

$$k_{\text{uni}} = \int_0^\infty \frac{k_a(E^*) P(E^*) \, dE^*}{1 + k_a(E^*)/\omega} \tag{25}$$

where $P(E^*)$ is the probability that the activated molecule has energy $E^*$ and $\omega$ is the collision frequency, $k_2[M]$. If integrations are performed over the distributions for $E_v^*$ and $E_r^*$, Eq. (25) becomes

$$k_{\text{uni}} = \int_0^\infty \int_0^\infty \frac{k_a(E_v^*, E_r^*) P(E_v^*) P(E_r^*) \, dE_v^* \, dE_r^*}{1 + k_a(E_v^*, E_r^*)/\omega} \tag{26}$$

According to Eq. (24), $P(E_r^*)$ and $P(E_v^*)$ are Boltzmann distributions,

$$P(E_v^*) = N(E_v^*) \exp(-E_v^*/kT)/Q_v$$

$$P(E_r^*) = N(E_r^*) \exp(-E_r^*/kT)/Q_r \tag{27}$$

where $N(E_v^*)$ and $N(E_r^*)$ are the densities of vibrational and rotational quantum states and $Q_v$ and $Q_r$ are the molecular vibrational and external rotational partition functions.

An exact calculation of $k_{uni}$ from Eq. (26) is a very difficult task because the critical configuration is a function of both $E_v^*$ and $E_r^*$. However, for unimolecular reactions with well-defined potential barriers the critical configuration is only weakly dependent on the molecular vibrational and rotational energies. A good approximation then is to assume a constant critical configuration structure. Substituting Eqs. (11) and (27) into Eq. (26), we have

$$k_{uni} = \frac{1}{hQ_rQ_v} \int_0^\infty \int_0^\infty \frac{\sum\limits_{E_v^+=0}^{E^+} P(E_v^+)N(E_r^*) \exp[-(E_v^*+E_r^*)/kT]\, dE_v^*\, dE_r^*}{1+k_a(E_v^*,E_r^*)/\omega} \quad (28)$$

It should be noted that only because we have assumed a fixed critical configuration is $E_0$ identical for all values of $E_v^*$ and $E_r^*$. If the integration is performed over the energies of the critical configuration instead of those for the molecule, Eq. (26) becomes

$$k_{uni} = \frac{\exp(-E_0/kT)}{hQ_rQ_v} \int_0^\infty \int_0^\infty \frac{\sum\limits_{E_v^+=0}^{E^+} P(E_v^+)N(E_r^+) \exp[-(E^++E_r^+)/kT]\, dE^+\, dE_r^+}{1+k_a(E^+,E_r^+)/\omega}$$

$$(29)$$

At high pressures ($\omega \to \infty$)

$$k_\infty = \frac{\exp(-E_0/kT)}{hQ_rQ_v} \int_0^\infty N(E_r^+) \exp(-E_r^+/kT)\, dE_r^+$$

$$(30)$$

$$\times \int_0^\infty \sum\limits_{E_v^+=0}^{E^+} P(E_v^+) \exp(-E^+/kT)\, dE^+$$

The first integral in Eq. (30) is the external rotational partition function, and the second is the vibrational partition function times $kT$. As a result, we have

$$k_\infty = \frac{kT}{h} \frac{Q_r^+ Q_v^+}{Q_r Q_v} \exp(-E_0/kT) \quad (31)$$

in agreement with absolute rate theory.[27]

The quantity $k_a(E^+,E_r^+)$ in the denominator of Eq. (29) is

$$k_a(E^+,E_r^+) = \frac{\sum\limits_{E_v^+=0}^{E^+} P(E_v^+)}{hN(E^++E_0+E_r^+-E_r^*)} \quad (32)$$

To extend Eq. (29) to the falloff region and low-pressure limit ($\omega \to 0$) and avoid the laborious double integration, Marcus[18] suggested that $E_r^* - E_r^+$ in Eq. (32) be replaced by the average difference of rotational energies of reacting molecules $\langle E_r^* \rangle$ and critical configurations $\langle E_r^+ \rangle$. The quantity $\langle E_r^+ \rangle$ may be found by averaging $E_r^+$ over $\exp(-E_r^+/kT)$; $\langle E_r^+ \rangle = kT$. Because $E_r^*/E_r^+ = I^+/I^*$, the result for two dimensions is

$$\Delta E_r = \langle E_r^* \rangle - \langle E_r^+ \rangle = (I^+/I^* - 1)kT \tag{33}$$

Let us define

$$k_a(E^+) = \frac{\sum_{E_v^+=0}^{E^+} P(E_v^+)}{hN(E^+ + E_0)} \tag{34}$$

and introduce the factor $F$, where

$$F = N(E_0 + E^+ - \Delta E_r)/N(E_0 + E^+) \tag{35}$$

Combining Eqs. (32)–(35), inserting them into Eq. (29), and separating the terms including $E_r^+$, we obtain

$$k_{\mathrm{uni}} = \frac{Q_r^+ \exp(-E_0/kT)}{hQ_rQ_v} \int_0^\infty \frac{\sum_{E_v^+=0}^{E^+} P(E_v^+) \exp(-E^+/kT)\, dE^+}{1 + k_a(E^+)/F\omega} \tag{36}$$

In the low-pressure limit ($\omega \to 0$)

$$k_{\mathrm{uni}} = \frac{F\omega Q_r^+ \exp(-E_0/kT)}{hQ_rQ_v} \int_0^\infty N(E^+ + E_0) \exp(-E^+/kT)\, dE^+ \tag{37}$$

By performing exact numerical integrations of Eq. (29), Waage and Rabinovitch[20,35] found that a very good approximation for $F$ is

$$F = [1 + (s + d/2 - 1)(I^+/I^* - 1)kT/(E_0 + aE_z^*)]^{-1} \tag{38}$$

Other approximations have been made for $F$ that are similar to but not so accurate as that of Waage and Rabinovitch.[20,35] Because Eq. (38) applies to the two-dimensional case, where the moments of inertia are identical and change equally in going from the molecule to critical configuration, it should be used only for approximate calculations in situations where all three external rotational moments of inertia change.

Let us now consider the case when there is not a well-defined barrier separating reactants and products. Here the critical configuration is a function of $E_v^*$ and $E_r^*$, and as a result $E_0$, the $I^+$s, and the $\nu^+$s are not fixed. An approximate solution to Eq. (26), one that avoids the laborious double integration, still remains to be derived. Waage and Rabinovitch[20] have given a solution to Eq. (26) for the Gorin model.[36] In this case, the dissociation fragments are assumed to have the same geometry and vibrational frequencies as the free radicals, and each fragment has two tumbling motions about its own axes. Therefore, the frequencies of the critical configuration are not dependent on $r^+$ and the value of $R^+$, for the critical configuration, is defined by the rotational barrier. Although the Gorin model simplifies the solution to Eq. (26), it does not appear to be compatible with either the criterion of minimum state density as described above or experimental results. Rate constants measured for the dissociation of alcohols,[37] alkanes,[38,39] and alkylsilanes[40] are consistent with much tighter critical configurations than the Gorin model. The assumption of freely rotating fragments does not appear to be valid.

### 3.1.2. Experimental Measurements of $k_{uni}$

One of the most significant achievements of the RRKM theory is its ability to match measurements of $k_{uni}$ with pressure. The isomerization of cyclopropane and methyl isocyanide are probably two of the most thoroughly studied thermal unimolecular reactions.[25,26] In Fig. 6 we compare the experimental results of Pritchard et al.[41] and of Chambers and Kistiakowsky[42] with the RRKM calculation of Lin and Laidler[43] for cyclopropane isomerization. The information about the critical configuration used in the RRKM prediction was derived from the high-pressure Arrhenius parameters as discussed earlier, and $\omega$ was determined from the molecular collision diameter. We see that there is good agreement between theory and experiment. In addition, recent measurements of cyclopropane isomerization at much higher temperatures using shock tubes are also in agreement with the RRKM theory.[44]

The unimolecular falloff measurements of Rabinovitch and co-workers of isocyanide isomerization have provided some of the most thorough comparisons with the RRKM theory. The first reaction studied was[45]

$$CH_3NC \rightarrow CH_3CN \tag{39}$$

A cyclic structure was chosen for the critical configuration. Its frequencies were

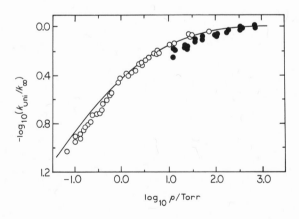

Fig. 6. Comparison of calculated and experimental falloff curves for the isomerization of cyclopropane at 492°C. Open circles, experimental results of Pritchard et al.[41]; filled circles, experimental results of Chambers and Kistiakowsky.[42] Calculated curve by Lin and Laidler.[43]

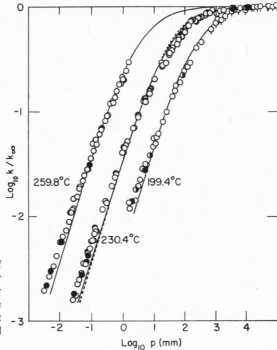

Fig. 7. Pressure dependence of unimolecular rate constant for $CH_3NC$ isomerization.[45] For clarity the 260° and 200° curves are displaced one log unit to the left and right, respectively. The solid lines are RRKM calculations.

estimated by the bond energy–bond order method[32] and by analogy with cyclopropane. Unknown frequencies were chosen by matching $A_\infty$, and $E_0$ was deduced from the activation energy, Eqs. (18) and (19). The figure axis rotation was also treated as being active. A comparison of the RRKM calculations and falloff measurements is shown in Fig. 7. In addition to $CH_3NC$, Rabinovitch and co-workers have also studied the isomerization of $CD_3NC$,[46] $CH_2DNC$,[47] $C_2H_5NC$,[48] and $C_2D_5NC$.[49] They have found that the same critical configuration that fits $CH_3NC$ isomerization also gives a good fit to the falloff behavior of these molecules. This agreement provides some assurance in the choice made for the critical configuration.

    The above are examples of unimolecular reactions with well-defined potential energy barriers, and Eq. (36) is applicable in calculating $k_{uni}$ vs. pressure. For dissociation reactions there is no simplified expression for calculating $k_{uni}$. Approximate calculations have been performed using Eq. (36) and thereby assuming a fixed critical configuration structure for all $E_v^*$ and $E_r^*$. Two examples of such calculations are those for $C_2H_5 \rightarrow C_2H_4 + H$[50] and $C_2H_6 \rightarrow 2CH_3$.[39] Some initial attempts have been made to perform exact RRKM calculations for dissociation reactions in the high-pressure regime.[30,33] For a thermal unimolecular reaction at high pressure the average critical configuration may be deduced by a Boltzmann average of the minimum state density criterion,[31]

$$\frac{\partial}{\partial r^+} \int_0^\infty \frac{dN(E^+)}{dE^*} \exp(-E^*/kT) \, dE^* = 0 \tag{40}$$

which is identical to the maximum free energy criterion.[23,51] An approximation to Eq. (40) has been applied to ethane dissociation at 870°K.[30] The potential energy surface was empirically deduced by assuming that the loosening in the critical configuration was due to lower H—C—C bending forces and by representing the C—C stretching mode, the reacting coordinate, by a Morse function. It was found that the average critical configuration has a C—C bond extension of 3.6 Å in contrast to 5.2 Å for $r^+$ at the rotational barrier.[52] This is expected because only as $E_v^* \to E_0$ will the critical configuration be located at the top of the potential energy barrier (see Fig. 3). In addition, it was found that $E_0$ is 2.3 kcal/mole less than the C—C bond dissociation energy. As a result, a negative activation energy for $CH_3$ recombination is predicted.

In concluding this section, we should ask if the ability of the RRKM theory to fit a thermal falloff measurement is a complete and sufficient test of the postulates of the RRKM theory. If not, what other tests are necessary? Also, what is the meaning of a critical configuration that fits a falloff measurement if it is found that the RRKM postulates are not valid for the molecule? Attempts will be made to answer these questions in Section 4.

### 3.2. Chemical Activation

Chemical activation is the technique of forming molecules with sufficient energy to undergo unimolecular processes by the energy changes of chemical reactions. Some of the more important chemical activation processes are H addition to olefins, $CH_2$ insertion and addition, and radical recombination. Chemical activation forms energized molecules with non-Boltzmann energy distributions and often with much higher energies than those attained in thermal experiments. The molecule $A^*$ formed by chemical activation may undergo either a unimolecular reaction or be collisionally stabilized. If the strong collision assumption is made and $A^*$ is assumed to be monoenergetically excited, we have

$$A^* \xrightarrow{k_a(E^*)} \text{decomposition products } (D)$$

$$\xrightarrow{\omega} A(S) \tag{41}$$

where $\omega$ is the collision frequency. From Eq. (41) we see that the rate constant $k_a(E^*)$ equals $\omega D/S$. When $A^*$ is formed with a distribution of energies given by $f(E^*)$ the average rate constant $\langle k_a(E^*)\rangle$ equals $\omega D/S$. The ratio $D/S$ may be found by averaging over $f(E^*)$, and the following expression is obtained for $\langle k_a(E^*)\rangle$:

$$\langle k_a(E^*)\rangle = \omega \frac{\int \{k_a(E^*)/[k_a(E^*)+\omega]\}f(E^*)\,dE^*}{\int \{\omega/[k_a(E^*)+\omega]\}f(E^*)\,dE^*} \tag{42}$$

At high pressures, $\omega \gg k_a(E^*)$, we have

$$\langle k_a(E^*)\rangle_\infty = \frac{\int k_a(E^*)f(E^*)\,dE^*}{\int f(E^*)\,dE^*} \tag{43}$$

Similarly, at low pressures, $\omega \ll k_a(E^*)$, Eq. (42) yields

$$\langle k_a(E^*)\rangle_0 = \frac{\int f(E^*)\,dE^*}{\int [f(E^*)/k_a(E^*)]\,dE^*} \tag{44}$$

For some chemical activation processes it is possible to assume an equilibrium between the two reactants and $A^*$ in order to determine the form for the distribution function $f(E^*)$.[25,26]

The logical choice of molecules to study by chemical activation would be those for which there are reliable measurements of $k_{uni}$ vs. pressure. Comparisons with the thermal studies would be quite meaningful, because chemical and thermal activation may produce molecules with much different ranges of energies. A simple test would be to see if the same critical configuration or potential energy surface can accommodate both sets of data. However, due to experimental complexities and uncertainties of the energetics of chemical activation experiments, such tests have rarely been performed. The tests that have been made have generally yielded results in agreement with the RRKM theory. An example is given in Fig. 8 where a critical configuration that fits $C_2H_5$ decomposition by chemical activation is also capable of fitting the thermal dissociation of the $C_2H_5$ radical.[50] A less comprehensive test involves comparisons where the high-pressure thermal rate constants are measured but the unimolecular falloff is not. Here, activated complex theory is used to calculate the A factor and activation energy from a set of $\nu^+$s that fit the chemical activation rate constant to compare with the thermal Arrhenius parameters.[38–40]

The most comprehensive applications of RRKM theory to chemical activation measurements have been for various homologous series. Agreement with the RRKM theory results when one critical configuration is suitable for the complete series. Rabinovitch and co-workers have made the major contributions to the studies involving H atom addition to olefins. Their most comprehensive studies have been H+ olefins to yield excited alkyl radicals that decompose by loss of $CH_3$, $C_2H_5$, $C_3H_7$, $C_4H_9$, and $C_5H_{11}$ radicals.[53] In all cases the results were found to be in satisfactory agreement with the RRKM theory.

Another series of chemical activation systems that have been studied is the decomposition of alkanes formed by singlet methylene insertion into C—H bonds, e.g., $^1CH_2 + CH_4 \rightarrow C_2H_6^*$. This technique has been used to study ethane, propane, isobutane, and neopentane decomposition by C—C bond rupture.[38,39] Critical configurations were chosen to match the measured decomposition rates and are shown in Table 1. We see that all the molecules have critical configurations with similar degrees of loosening (lower frequencies). The calculations were approximate in that the criterion of minimum state density was not used and $E_0$ was assumed to equal the C—C bond dissociation energy at 0°K. Of particular interest is the finding of much "tighter" critical configurations than those predicted by the Gorin model. If the criterion of minimum state density were applied, the critical configurations would be even tighter. The quantity $E_0$ would be lower than the bond dissociation energy at 0°K,[30] and there would have to be an increase in the frequencies in the critical configuration relative to those given in Table 1 in order to fit the experimental rate constants.

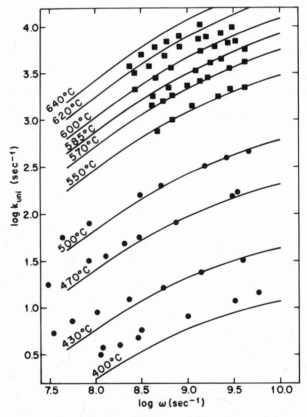

Fig. 8. Plot of $\log k_{uni}$ vs. $\log \omega$ for ethyl radical dissociation. Curves were calculated using the RRKM theory.[50]

*Table 1.  Critical Configurations for C—C Dissociation*[a]

| Mode | Normal mode frequencies,[b] $cm^{-1}$ | |
| --- | --- | --- |
| | Molecule | Critical configuration |

| $C_2H_6$ decomposition, $E_0 = 85.8$ kcal/mole | | |
| --- | --- | --- |
| C—C stretch | 955 | Reaction coordinate |
| $CH_3$ rock | 1206(2) | 252(2) |
| $CH_3$ rock | 822(2) | 172(2) |

| $C_3H_8$ decomposition, $E_0 = 82.6$ | | |
| --- | --- | --- |
| C—C stretch | 924 | Reaction coordinate |
| $CH_3$ rock | 1151 | 228 |
| $CH_3$ rock | 903 | 181 |
| C—C—C bend | 382 | 77 |
| $CH_2$ rock | 747 | 149 |

| Iso-$C_4H_{10}$ decomposition, $E_0 = 81.4$ | | |
| --- | --- | --- |
| C—C stretch | 791 | Reaction coordinate |
| $CH_3$ rock | 1168 | 299 |
| $CH_3$ rock | 904 | 229 |
| C—C—C bend | 418 | 106 |
| C—C—C bend | 381 | 94 |

| Neo-$C_5H_{12}$ decomposition, $E_0 = 78.4$ | | |
| --- | --- | --- |
| C—C stretch | 724 | Reaction coordinate |
| $CH_3$ rock | 1035 | 215 |
| $CH_3$ rock | 981 | 203 |
| C—C—C bend | 403 | 87 |
| C—C—C bend | 355 | 76 |

[a] The alkanes were excited by singlet methylene insertion reactions.[38,39]
[b] The degeneracy is given in parentheses.

The molecular elimination of hydrogen halides from chemically activated alkyl halide molecules has been studied extensively by Setser and co-workers.[24,54,55] Chemical activation results from radical recombination, e.g., $CH_3 + CH_2Cl \rightarrow CH_3CH_2Cl^*$. This technique has been used to study HF, HCl, and HBr elimination from F-, Cl-, and Br-alkanes with varying degrees of halo substitution. A four-centered critical configuration with the structure

$$\begin{array}{c} H \cdots X \\ \diagdown C - C \diagup \\ \diagup \quad \diagdown \end{array} \qquad X = F, Cl, or Br$$

was chosen, and a similar model fits all the chemical activation results as well as the high-pressure Arrhenius parameters. A comparison between the experimental results and the RRKM calculations for chloro- and dichloroalkanes is given in Fig. 9. There is good agreement.[54] These alkyl halide elimination reactions have also provided some insight into the dynamics of unimolecular reactions, and we shall refer to them again later.

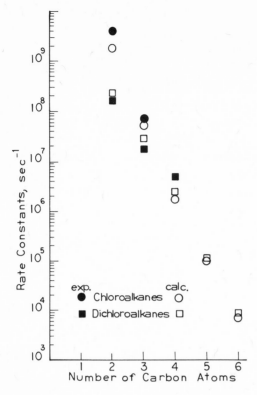

Fig. 9. Comparison of calculated (RRKM) and experimental rate constants for the series of chemically activated mono- and dichloroalkanes at a common energy of 89 kcal/mol.[54]

Another important chemical activation technique is hot atom processes.[56,57] In the hot atom experiments, translationally excited atoms, often formed by nuclear reactions, form chemically activated molecules or radicals by a substitution reaction, e.g., $T + CH_4 \rightarrow CH_3T + H$, or an addition reaction, e.g., $F + C_2H_4 \rightarrow CH_2CH_2F$. An interesting aspect of the hot atom technique is that the chemically activated molecules and radicals may often contain large amounts of rotational energy.[21] This "rotational effect" may strongly influence the unimolecular reaction if there are large moments of inertia changes between the molecule and critical configuration. RRKM calculations that included the effect of rotational energy have been compared with experimental results for $CH_2TNC$ isomerization, $CH_3CF_2{}^{18}F$ and $CH_2{}^{18}FCF_3$ decomposition, $CH_3T$ decomposition, and $c\text{-}C_4H_7T$ and $c\text{-}C_4D_7T$ decomposition.[21,58]

## 4. Unimolecular Dynamics

Unimolecular dynamics refers to the microscopic events involved in the production of $A^*$ and its ensuing isomerization or decomposition. Many of the first experimental studies of unimolecular dynamics involved tests of the random lifetime[59] and strong collision[41,60] assumptions of the RRKM theory. However, more recently it has become increasingly clear that unimolecular dynamics is a field in itself, distinct from the RRKM theory. It is obvious that if A* is formed with a nonrandom energy distribution, the initial rate of dissociation of A* will be governed by purely dynamical considerations instead of statistical ones. Experimental procedures are required for making measurements on A* at short times. The most meaningful measurements are those under isolated molecule conditions so that collisional disturbances of A* need not be considered. Recent experimental advancements in molecular beam techniques[3-9] and molecular spectroscopy[10-14] are beginning to provide the necessary dynamical information.

In this section we shall discuss experimental and theoretical investigations of unimolecular dynamics. At the conclusion a summary of the status of the RRKM theory is presented.

## 4.1. Intermolecular Energy Transfer

The strong collision assumption can be tested experimentally for thermal unimolecular reactions by measuring $k_{uni}$ in the second-order regime ($\omega \to 0$). The low-pressure unimolecular rate is simply the rate at which the molecules are excited to vibrational levels having energy greater than $E_0$, and the rate constant is one for bimolecular reaction, $k_{bi}$. If $k_{bi}$ is measured for a bath gas M, the pressure-for-pressure activation efficiency $\beta_p$ of M relative to reactant A is found from

$$\beta_p = (k_{bi})_M / (k_{bi})_A \qquad (45)$$

where $(k_{bi})_M$ is the slope of $k_{uni}$ vs. pressure of M with reactant pressure kept constant and $(k_{bi})_A$ is the slope of $k_{uni}$ vs. reactant pressure. We may convert this to relative efficiency per collision by using the kinetic theory of gases to remove the M—A and A—A collision frequencies. The resulting expression for the relative activation efficiency of A is

$$\beta_c = \beta_p (\sigma_{AA}/\sigma_{AM})^2 (\mu_{AM}/\mu_{AA})^{1/2} \qquad (46)$$

where the $\mu$s are reduced masses and the $\sigma$s are the collision diameters. The most extensive measurements of $\beta_p$ and thus $\beta_c$ are those of Rabinovitch and co-workers for methyl isocyanide isomerization.[61] They measured the relative efficiencies of over 100 bath gases, and some of their results are shown in Table 2.

For the noble gases and smaller molecules we see that $\beta_c$ relative to $\beta_c$ of $CH_3NC$ is much less than 1.00. As the molecular complexity increases the collisional efficiencies approach a limiting value. It seems reasonable that the limiting efficiency is unity and that the strong collision assumption of the Lindemann mechanism is valid for large molecules. However, for smaller molecules with $\beta_c$ less than 1, a multistep activation and deactivation

Table 2. *Energy Transfer Efficiencies for Methyl Isocyanide Isomerization in the Second-Order Regime*[a]

| Activating gas | Relative efficiency ($\beta_c$) |
|---|---|
| $CH_3NC$ | (1.00) |
| $n\text{-}C_4H_{10}$ | 1.01 |
| $n\text{-}C_4F_{10}$ | 1.00 |
| $C_3H_8$ | 0.79 |
| $C_2H_6$ | 0.76 |
| $CH_4$ | 0.61 |
| $NH_3$ | 0.93 |
| CO | 0.46 |
| Xe | 0.232 |
| Ar | 0.279 |
| He | 0.243 |

[a]The data are from Reference 61.

Lindemann mechanism leads to the following master equation for each energy level:

$$\frac{dn_i}{dt} = \omega \sum_j P_{ij}n_j - \omega n_i - k_i n_i \qquad (47)$$

where $P_{ij}$ is the probability for conversion to the state indicated by $i$ from the state indicated by $j$. The quantities $k_i$ and $n_i$ are the unimolecular rate constant for state $i$ and number of molecules in state $i$, respectively.

Exact solutions to Eq. (47) are impossible because the individual $P_{ij}$s and their time dependence are not known for complex molecules containing many quanta of vibrational energy. However, a formal approximate solution to Eq. (47) indicates that after an initial short induction period the energy levels reach a steady state, $dn_i/dt = 0$.[62] If this steady state approximation is made, a stochastic solution to Eq. (47) is possible if a form for the $P_{ij}$s is assumed.[63,64] The stochastic analysis will yield a value of $\langle \Delta E \rangle$, the average amount of energy transferred per collision. The relative efficiency $\beta_p$ at low pressures may be determined by calculating the total reaction rate from the steady state distribution, $\sum_i k_i n_{ij}$, and dividing it by the strong collision reaction rate, Eq. (37).

The most often postulated models for $P_{ij}$ are stepladder, exponential, Poisson, and Gaussian distributions.[64] Each assumes that the number of *quanta* transferred per collision may be given as $\Delta E = m$ quanta and $\langle \Delta E \rangle = \bar{m}$ quanta. The stepladder model corresponds to a fixed step size for energy transfer where the step size is $\langle \Delta E \rangle$. The remaining three models allow for the transfer of large quanta of energy. For the exponential model the probability of removal of $m$ quanta is proportional to $\exp(-m/\bar{m})$. The probabilities for the Poisson and Gaussian distributions are $\bar{m}^m/m$ and $\exp[-(m-\bar{m})^2/2\sigma^2]$, respectively. Calculated values of $\beta_c$ using these models have been compared extensively with experimental determinations.[61,64,65] It appears that for the noble gases and small molecules with values of $\beta_c$ much less than unity the exponential model is the most appropriate, whereas the stepladder model gives the best fit to the experimental data for strong colliders. The stochastic calculations have also provided estimates of $\langle \Delta E \rangle$ necessary for the strong collision assumption to be valid.[61,64] In general, values of $\langle \Delta E \rangle$ larger than 5 kcal/mole are sufficient for the strong collision assumption. A weaker collider such as helium will normally have values of $\langle \Delta E \rangle \simeq 1$ kcal/mole. The specific value will depend on the molecule being activated and deactivated and the average energy of the reacting molecules, which is a function of pressure, temperature, and $E_0$.[64]

Chemical activation experiments have also provided information about the dynamics of intermolecular energy transfer in unimolecular reactions.[24] The deactivation and unimolecular isomerization or decomposition of the chemically activated molecule $A^*$ may be regarded as the following cascade process proceeding through $n$ energy levels, which start at $E^*$ and end at the first level below $E_0$:

$$A_1^* \xrightarrow{\;k_a(E_1^*)\;} \text{products}$$

$$\xrightarrow{\;\omega\;} A_2^* \xrightarrow{\;k_a(E_2^*)\;} \text{products}$$

$$\xrightarrow{\;\omega\;} A_3^* \cdots$$

$$\cdots$$

$$\cdots$$

$$\xrightarrow{\;\omega\;} A_{n+1}^* \xrightarrow{\;k_a(E_{n+1}^*)\;} \text{products}$$

$$\xrightarrow{\;\omega\;} A_n^*$$

Because $A_n^*$ has energy less than $E_0$, it represents a stable molecule. Collisional excitation may be included in the above scheme; however, it is seldom necessary. The strong collision assumption means the $A_1^* \rightarrow A_n^*$ transition occurs in one step.

What is predicted by this cascade mechanism[66] and found in many chemical activation experiments[24] is an increase in decomposition or isomerization relative to the strong collision prediction, because deactivation is a multistep process and there is a probability for unimolecular reaction after each of the deactivation steps until $A^*$ has energy less than $E_0$. The effect of the multistep deactivation may be seen in plots of $\omega D/S$ of Eq. (41) vs. $\omega$ or $D/S$, which show an increase at low pressures. A fit to the data using one of the stochastic models described above for the thermal systems yields a value for $\langle \Delta E \rangle$. An example of such a calculation is given in Fig. 10. The chemically activated molecule is $CH_3CF_3^*$ formed by $CH_3 + CF_3$ radical recombination.[67]

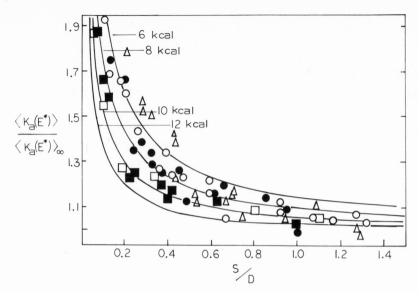

Fig. 10. Efficiency of various bath gases to deactivate chemically activated $CF_3CH_3$. The curves are calculated using the RRKM theory and assuming a stepladder model for deactivation; $(\Delta E)$ values are indicated.[67] (○)$C_2F_6$; (□)$C_6F_{14}$; (■)$C_8F_{18}$; (●)$CF_3N_2CH_3$; (△)$C_4F_8$.

The values of $\langle \Delta E \rangle$ were calculated using a stepladder model. There have been many studies similar to the one presented here. Some of the major contributions have come from the work of Rabinovitch,[68,69] Setser,[67,70] Troe,[71] and their co-workers. Some of the data are presented in Table 3. On the whole the chemical activation and thermal activation $\langle \Delta E \rangle$ values compare quite favorably. However, for the chemical activation experiments the quantity $\langle E_v^* \rangle - E_0$ is almost always larger than the deactivation step size $\langle \Delta E \rangle$. As a result, one collision is not sufficient for deactivation of A*. This leads to a failure of the strong collision assumption, which becomes more pronounced as $\langle \Delta E \rangle$ is decreased.

Some attempts have been made to interpret the $\langle \Delta E \rangle$ values by theoretical models of the collision dynamics. One possibility is deactivation occurring via a collision complex with statistical redistribution of vibrational energy. However, such a model does not adequately explain the changes in $\langle \Delta E \rangle$ with degree of excitation and isotopic substitution in A*.[72] For $CH_3NC$ deactivation a model that assumes the involvement of only the translational and rotational modes, with the necessary angular momentum considerations, appears to explain the collisional efficiencies of the various bath gases.[72] This model delineates three classes of deactivators, monoatomic, diatomic linear, and polyatomic nonlinear, in agreement with the experimental results.[61]

In reviewing our understanding of intermolecular dynamics, the strong collision assumption appears to be valid for large deactivating molecules. The

Table 3. *Vibrational Energy Decrements for Activated Molecules*[a]

| Activated molecule | Bath gas | $\langle \Delta E \rangle$ | Ref. |
|---|---|---|---|
| | Thermal activation | | |
| $CH_3NC$ | He | 1.3 | 72 |
| $\langle E_v^* \rangle = 39$ | Ar | 1.4 | |
| $E_0 = 38$ | Xe | 1.2 | |
| | $H_2$ | 1.3 | |
| | $N_2$ | 2.1 | |
| | $C_2H_4$ | 2.5 | |
| $C_2H_5NC$ | He | 1.4 | 65 |
| $\langle E_v^* \rangle = 39$ | Ar | 1.8 | |
| $E_0 = 38$ | Xe | 2.0 | |
| | Chemical activation | | |
| $C_2H_4Cl_2$ | He | 1.4 | 70 |
| $\langle E_v^* \rangle = 88$ | Ar | 4.0 | |
| $E_0 = 60$ | Xe | 5.0 | |
| | $N_2$ | 4.0 | |
| | $CF_4$ | 5 | |
| | $SF_6$ | 7.0 | |
| | $C_4F_8$ | 12 | |
| $CH_3CF_3$ | $C_2F_6$ | 7 | 67 |
| $\langle E_v^* \rangle = 102$ | $c\text{-}C_4F_8$ | 7 | |
| $E_0 = 68$ | $n\text{-}C_6F_{14}$ | 10 | |
| Dimethylcyclopropane | CO | 4.6 | 68 |
| $\langle E_v^* \rangle = 112$ | $cis\text{-}C_4H_{10}$ | 11.4 | |
| $E_0 = 60$ | | | |
| Cycloheptatriene | He | 0.3 | 71 |
| $\langle E_v^* \rangle = 113$ | Ar | 0.8 | |
| $E_0 = 50$ | Xe | 1.2 | |
| | $H_2$ | 0.6 | |
| | $N_2$ | 0.6 | |
| | CO | 1.1 | |
| | $CF_4$ | 2.6 | |
| | $SF_6$ | 1.9 | |
| | $n\text{-}C_4H_{10}$ | 4.8 | |

[a] Energies are in kilocalories per mole.

correctness of the assumption depends on the degree of excitation of $A^*$, and therefore the assumption of strong collisions is more restrictive for chemical than thermal activation. Stochastic models suggest that vibrational deactivation by atoms and diatoms may follow exponential distributions, and polyatomic deactivations are described by a stepladder model. However, exact solutions to the microscopic collision dynamics have not been performed.

## 4.2. Intramolecular Energy Transfer

The basic assumption of the RRKM theory is that vibrational energy redistribution is rapid. There have been several tests of the RRKM theory that have provided information about the rates of intramolecular energy relaxation. However, before discussing them, let us consider in rather simple terms what we expect if intramolecular energy distribution is not rapid.

Assume that the strong collision assumption is valid for thermal unimolecular reactions. As discussed above, this appears to be correct for most polyatomic molecules. The likelihood of any energy is then proportional to its equilibrium probability, and the states at that energy all have equal probabilities. If energy relaxation is faster than unimolecular reaction, the lifetime distribution for molecules with energy $E^*$ will be random, Eq. (3). However, as displayed in Fig. 5(e), if transitions between two (or more) groups of states are less probable than those leading to products, the molecule is intrinsically non-RRKM and will display a nonrandom lifetime distribution regardless of the random collisional preparation of vibrational states.

The situation where most of the states are badly coupled to the product channel is illustrated in Fig. 11. At very high pressures the impaired internal coupling will have no effect on $k_{uni}$ because only those states prepared near the critical configuration can react. In the pressure regime that is sufficient for the molecules activated in the correct part to react without any appreciable energy transfer between the two parts, the rate of decomposition will be that of a much simpler molecule, corresponding to the part that is reacting. At much lower pressures, when transfer of energy between the two parts becomes the most important source of activations for the decomposing part, the rate will become

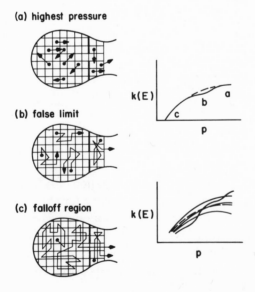

**(a) highest pressure**

**(b) false limit**

**(c) falloff region**

Fig. 11. Left: State occupation for an intrinsically non-RRKM molecule. Upper right: corresponding regions of the falloff curve. Lower right: some possibilities if apparent non-RRKM behavior is also present.[15]

first order again (non-RRKM false limit). At still lower pressures it will again decline until the reaction rate is the collisional preparation rate. If there is more than one intramoleculear bottleneck, the above process could be repeated several times. The effect of weak vibrational coupling was recognized several years ago by Rice.† He pointed out that at very high pressures a molecule with an intramolecular bottleneck would behave as if it had fewer atoms than are actually present.

The meaning of the non-RRKM false high-pressure limit may be seen from the following amended Lindemann mechanism:

$$A + M \xrightarrow{k_1^+} A^+ + M$$

$$A^+ + M \xrightarrow{k_2^+} A + M$$

$$A^+ \xrightarrow{\lambda} A^*$$

$$A + M \xrightarrow{k_1^*} A^* + M \qquad (48)$$

$$A^* + M \xrightarrow{k_2^*} A + M$$

$$A^* \xrightarrow{k} \text{products}$$

Here, $A^+$ represents the states badly coupled to the reaction coordinate. For the situation depicted in Fig. 11, $k_1^+ \gg k_1^*$. If we apply the steady state assumption to $A^+$ and $A^*$ in Eq. (48), we find

$$k_{uni} = \frac{kk_1^*[M]}{k_2^*[M] + k} + \frac{k\lambda k_1^+[M]}{(k_2^*[M] + k)(k_2^+[M] + \lambda)} \qquad (49)$$

The sum $k_1^*[M] + k_1^+[M]$ is the bimolecular rate for formation of A with energy $E^*$. The ratios $k_1^+/k_2^+$ and $k_1^*/k_2^*$ are each the equilibrium probabilities that A is produced with energy $E^*$, and $k_1^*/k_1^+$ is the ratio of molecular phase space volumes for the two states (* and +). For the cases $\omega \to \infty$ and $\omega \to 0$ and after Boltzmann averaging, Eq. (49) becomes Eq. (25), the RRKM expression for $k_{uni}$. The false high-pressure limit exists at $\lambda < k_2^+[M]$ and $k_2^*[M] < k$. Equation (49) then becomes

$$k_{uni} = k_1^*[M] + \lambda k_1^+/k_2^+ \qquad (50)$$

†The discussion by O. K. Rice is given in Einige Bemerkungen über Energie austausch innerhalb Molekülen und zwischen Molekülen bei Zusammenstoss, *Z. Phys. Chem. B* **7**, 226–233 (1930). This paper is discussed in L. S. Kassel, *Kinetics of Homogeneous Reactions*, Chemical Catalog Co., New York (1932), pp. 106–108, and in Reference 15.

Because the first term is small, the measured high-pressure rate in the false limit gives the intramolecular relaxation rate directly. In the region of the true high-pressure limit, $k_2^*[M] \rightarrow \infty$,

$$k_{\text{uni}} = \frac{kk_1^*[M]}{k_2^*[M] + k} \tag{51}$$

Equation (51) is a normal unimolecular rate expression for a molecule whose phase space volume is $k_1^*/(k_1^+ + k_1^*)$ times that of A. Because $k_1^+ > k_1^*$, Eq. (51) depicts the unimolecular behavior of a molecule containing fewer atoms than are actually present.

It must be emphasized that of the pressure ranges identified in Fig. 11, region (a) and most of region (b) are almost never explored in the laboratory. A conventional experimental curve offers no immediate clue to what type of high-pressure limit it approaches. The above description, Eqs. (49)–(51), is equivalent to one that results from assuming that intramolecular vibrational energy relaxation can be described by a master equation formalism.[15,22]

A much more difficult case to explain than intrinsic non-RRKM behavior of thermally excited molecules is apparent non-RRKM behavior. Here the molecular vibrational states are chosen nonrandomly, as is the situation in a large number of experiments, prominently including those based on chemical and photochemical activation. The unimolecular rate constant will then usually not agree with the RRKM theory at high pressures. The nature of the lifetime distribution $P(\tau)$ will depend on the nonrandomness of the excitation process, the rate of intramolecular energy relaxation, and whether the molecule also displays intrinsic non-RRKM behavior. Three possibilities are illustrated in Fig. 5(d).

It is instructive to formulate the unimolecular rate constant for nonrandomly monoenergetically excited molecules A in terms of the lifetime distribution $P(\tau)$. The quantity $\tau$ is the time required for A to undergo dissociation or isomerization in the absence of collisions. If we make the strong collision assumption, the probability that A avoids a collision for $\tau$ seconds is[1]

$$W(\tau) = \exp(-\omega\tau) \tag{52}$$

The unimolecular rate constant for A is equal to $\omega D/S$, Eq. (41), and in terms of the probabilities $P(\tau)$ and $W(\tau)$ it may be expressed as

$$k(E^*) = \int_0^\infty P(\tau) W(\tau) \, d\tau \Big/ \int_0^\infty P(\tau)[1 - W(\tau)] \, d\tau \tag{53}$$

The rate constant is represented as $k(E^*)$ to differentiate it from $k_a(E^*)$, the RRKM rate constant. Substitution of the random lifetime distribution, Eq. (3), for $P(\tau)$ gives $k(E^*)$ equal to $k_a(E^*)$.

For a nonrandom excitation process $k(E^*)$ will differ from the RRKM expression. The simplest case that identifies the three cases depicted in Fig. 5(d) results when the nonrandomly excited states $(+)$ relax to a statistical distribution $(*)$ with one rate, $\lambda$. If we also allow $A^+$ to undergo unimolecular reaction with rate constant $k_{NR}$, we have

$$A^+ \xrightarrow{k_{NR}} \text{product(s)}$$

$$A^+ \xrightarrow{\lambda} A^* \qquad\qquad (54)$$

$$A^* \xrightarrow{k} \text{product(s)}$$

and

$$P(\tau) = k_{NR}e^{-(k_{NR}+\lambda)\tau} + \frac{k\lambda}{k-k_{NR}-\lambda}(e^{-(k_{NR}+\lambda)\tau} - e^{-k\tau}) \qquad (55)$$

Inserting Eq. (55) into Eq. (53) and integrating, we have

$$k(E^*) = \frac{k_{NR}(k+\omega) + k\lambda}{k+\lambda+\omega} \qquad\qquad (56)$$

The measured rate constant is then a function of pressure and will equal $k$ only if $\lambda \gg (\omega + k)$ and $k_{NR}$. Apparent non-RRKM behavior may be detected by measuring $k(E^*)$ vs. pressure. If molecule A displays intrinsic non-RRKM behavior and there are multiple unimolecular paths for both $A^+$ and $A^*$, the treatment is basically the same as that given above but more difficult.

Equations (49)–(56) indicate ways of detecting non-RRKM behavior. In certain cases they also allow determination of the relaxation rate, $\lambda$. In the following section we shall discuss studies designed to evaluate rates of intramolecular energy relaxation. They include thermal and chemical activation and classical trajectory techniques.

### 4.2.1. Fall-off Experiments

One might imagine that every molecule has been carefully tested for intrinsic non-RRKM behavior. However, because of the experimental difficulties involved in attaining heterogeneous reaction at high pressures, such studies are very seldom performed. At present there are only two well-documented examples. For decomposition of nitryl chloride, $NO_2Cl \rightarrow NO_2 + Cl$, Dutton et al.[73] found no false limit in the unimolecular rate constant in the presence of 50–300 atm of $N_2$. The highest pressure corresponds to an average time between collisions of $\sim 2 \times 10^{-13}$ sec, which is nearly the same as one period of vibration of the reaction coordinate. Their result is rather conclusive evidence that $NO_2Cl$ does not display intrinsic non-RRKM behavior. In a similar experiment, Wilson et al.[74] studied the unimolecular decomposition of ethyl-cyclobutane in the presence of $N_2$ pressures from 7 to 170 atm. They found a slow decrease in the unimolecular rate constant as pressure was increased. This result would be expected if collisional preparation does not produce a truly random distribution but produces one with a shortage of states prepared near the critical configuration. We might therefore expect some shortage of lifetimes as $t \to 0$, which would explain the result of Wilson et al. However, they choose to explain it in terms of the volume of activation.

### 4.2.2. Chemical Activation Experiments

The first chemical activation experimental work that tested the RRKM postulate of rapid vibrational energy redistribution was done by Butler and Kistiakowsky.[59] By reacting singlet methylene radicals with propylene and cyclopropane they were able to produce chemically activated methylcyclopropane molecules with different initial energy distributions:

$$^1CH_2 + CH_3CH{=}CH_2 \to CH_3{-}\Delta^* \to \text{butenes}$$
$$^1CH_2 + \Delta \to CH_3{-}\Delta^* \to \text{butenes}$$

$$(57)$$

The experimental observation is that the ratios of butenes were the same for both methylcyclopropanes. The implication is that vibrational energy redistribution is rapid so that the initial nonrandom distributions relax before isomerization occurs; $\lambda \gg k + \omega$, Eq. (56). Because the isomerization rate constants were about $10^{10} \, \text{sec}^{-1}$, this result implied that vibrational energy redistribution occurred in $10^{-10}$ sec. A similar experiment, yielding the same conclusion, involves the addition of H atoms to butene-1 and *cis*-butene-2 to make chemically activated *sec*-butyl radicals.[75]

It should be possible to detect the initial chemical activation nonrandom distribution by measuring the unimolecular rate constant at high pressures. This technique has been applied to decomposition of *sec*-butyl radicals formed by H + *cis*-butene-2 reaction.[76] With use of $H_2$ as a deactivator it was found that $k(E^*)$ was invariant over a pressure range of 0.01–203 atm. At 203 atm the collision frequency is $\sim 5 \times 10^{12} \, \text{sec}^{-1}$, which yields an average interval between collisions of $2 \times 10^{-13}$ sec. The inference is that vibrational energy relaxation occurs on even a shorter time scale.

Direct evidence for apparent non-RRKM in chemical activation systems has been provided by Rabinovitch and co-workers. They used the following reaction to produce vibrationally excited hexafluorobicyclopropyl-$d_2$[16]:

$$^1CH_2 + \underset{\underset{CH_2}{\diagdown \diagup}}{CF_2{-}CF_2}{-}CF{=}CF_2 \to \underset{\underset{CH_2}{\diagdown \diagup}}{CF_2{-}CF}{-}\underset{\underset{CD_2}{\diagdown \diagup}}{CF{-}CF_2^*}$$

$$(58)$$

The molecule decomposes by $CF_2$ elimination, which should occur with equal probabilities from each ring if energy is randomized. However, at pressures in excess of 100 torr there is a measurable increase in the fraction of decomposition in the ring initially excited. Their model interpretation of the experimental measurements yielded an intramolecular relaxation rate of $\sim 10^{12}$ sec$^{-1}$. More recently they have investigated the series of chemically activated fluoroalkyl cyclopropanes[17]:

$$F_2 \overset{F}{\triangle}{-}R, \quad R = CF_3, C_3F_7, \text{ and } C_5F_{11}$$

with $D_2$

The chemically activated molecules are formed by $^1CD_2$ reaction with the appropriate fluorinated alkene. In all these cases apparent non-RRKM behavior was observed. The measured values of $k(E^*)$ are extremely dependent on pressure. The results are displayed in Fig. 12.

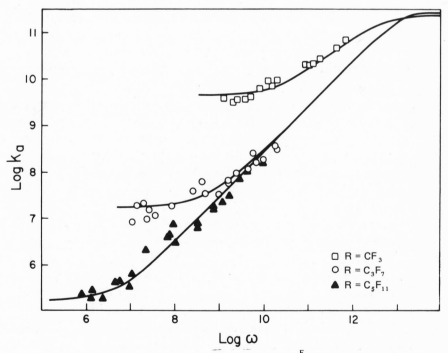

Fig. 12. Unimolecular rate constant $\omega$ for $F_2 \overset{F}{\triangle}{-}R$ decomposition.

with $D_2$

### 4.2.3. Classical Trajectory Calculations

Solutions to Hamiltonian classical equations of motion describing the nuclei in model systems have been quite useful in deducing efficiencies of intramolecular energy relaxation. Calculations have been performed with both random and nonrandom initial distributions of energy in the energized molecules. In a conservative system the potential energy $V(q_i)$ is a function of coordinates alone. The classical equations for this case are

$$p_i = \partial T / \partial \dot{q}_i \tag{59}$$

$$H = T(p_i, q_i) + V(q_i) \tag{60}$$

$$\dot{q}_i = \partial H / \partial p_i = \partial T / \partial p_i \tag{61}$$

$$\dot{p}_i = -\partial H / \partial q_i \tag{62}$$

The molecular trajectory is computed by numerical integration of Eqs. (61) and (62).

The earliest studies of this kind used fictitious models in that they did not represent any particular molecules. Thiele and Wilson[77] studied the dissociation of linear, symmetric, triatomic molecules having Morse bond-stretching potentials. A similar situation for four-atom linear molecules has been investigated by Harter et al.[78] Hung and Wilson[79] and Hung[80] have studied rotating versions of the above models. The effectiveness of nonrandom and random initial conditions was explored by Baetzold and Wilson.[81] The most comprehensive study is that of Bunker on model triatomics.[82]

The results of the above trajectory calculations illustrated several important points. It was found that the normal mode description, which is implicit in the Slater theory, is rather inadequate for dissociating molecules. Harmonic molecules do not have a metrically indecomposable phase space; some molecules can never dissociate even though sufficiently energized. This effect still exists if external rotation is included. However, it is removed by inclusion of anharmonicity.

The Monte Carlo trajectory studies of Bunker[82] gave additional information regarding the lifetime distributions for representative anharmonic triatomic molecules at constant energy. Lifetime distributions can be calculated by determining the times required for a random selection of starting conditions to decompose. Because the Hamiltonian is not separable, linearly dependent random sampling procedures are not applicable for choosing the initial conditions. The method employed here was to use an extended rejection technique to select points at random in multidimensional $(p, q)$ space, in a region bounded by the hypersurfaces on which $H$ and $H + dH$ are constant.

Collection of the Monte Carlo lifetime distributions into a histogram with time divisions of $\Delta\tau$ gives a convenient comparison with the RRKM theory. If the distribution is random,

$$\ln(N/N_0 \, \Delta\tau) = \ln k_a(E^*) - k_a(E^*)\tau \tag{63}$$

where $N$ is the number of lifetimes in each division and $N_0$ is the total number of trajectories studied for the particular model and energy. The intercept $\ln k_a(E^*)$ can be obtained by inspection. After anharmonicity and rotational effects were calculated correctly,[19] the agreement between the Monte Carlo and the RRKM $k_a(E^*)$s is about $\pm 50\%$ for the bent triatomics and $\pm 20\%$ for the linear ones.

Although most of the model triatoms studied by Bunker[82] display random lifetimes, some of the models with widely separated frequencies and disparate masses do not. Examples of random and nonrandom distributions are given in Fig. 13. The obvious meaning of this finding is that there is slow exchange of vibrational energy between some of the various molecular vibrations. If intramolecular relaxation can be described by a master equation formalism,[82] the lifetime distribution can be written as

$$P(\tau) = \sum_i c_i e^{-k_i \tau} \tag{64}$$

$$\sum_i c_i = k_a(E^*) \tag{65}$$

$$\sum_i (c_i/k_i) = 1 \tag{66}$$

The above model gives excellent agreement with the nonrandom distributions. The sizes of the intramolecular relaxation rate constants $c_i$ vary from $5 \times 10^{10}$ to $10^{13} \, \mathrm{sec}^{-1}$.

The latest generation of trajectory studies has pertained to specific unimolecular reactions. Using the available information for force constants, bond energies, activation energies, and critical configuration geometries, models for the potential energies have been chosen in an *a priori* fashion. The first study of

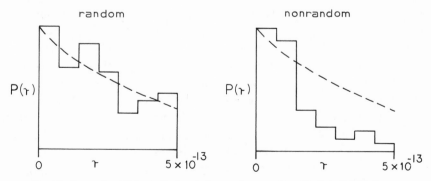

Fig. 13. Examples of Monte Carlo, random and nonrandom, lifetime distributions.

this kind was for $CH_3NC$ isomerization.[15,83] The simplest model used was

$$V = \sum_{i=1}^{5} D_i\{1 - \exp[-\beta_i(r_i - r_i^0)]\}^2 + \tfrac{1}{2}k_\phi \sum_{i=1}^{3} (\phi_i - \phi_i^0)^2$$

$$+ \tfrac{1}{2}k_x \sum_{i=1}^{3} (x_i - x_i^0)^2 + \sum_{i=0}^{5} \alpha_i \cos i\theta \qquad (67)$$

It has five Morse bonds corresponding to the $N\equiv C$ group, the three C—H, and a linkage from the methyl C to the center of the NC group. Harmonic forces resist the distortion of the H—C—H angles $x$ and H—C— (NC center) angles $\phi$. The truncated Fourier cosine series expresses the variation of potential energy along the reaction coordinate, which is the angle $\theta$ between NC and the bond from the NC center to the methyl C. The Fourier coefficients were chosen to give a 37.8-kcal/mole potential barrier to isomerization to $CH_3CN$, with an exoergicity of 15.3 kcal/mole, and to approximately match the spectroscopic bending frequencies when $\theta = 0°$ and $180°$. The potential barrier was chosen at $\theta = 90°$. A recent *ab initio* SCF calculation places it at $100.8°$.[84] The more complicated model included quartic bending forces for the $\phi$s and $x$s and repulsions between NC an the H's. The repulsions introduce a $CH_3$—(NC) torsional potential.

Three different techniques were used to approximate random sampling. For a molecule of this size, with a 30-dimensional internal coordinate phase space, the multidimensional rejection technique is too time consuming to be practical. The three approximate methods were (1) mode sampling, which apportioned energy on an atom-by-atom and vibrational mode-by-mode basis; (2) orthant sampling, in which one of the 30 orthants is chosen at random; a random phase space vector is erected with it, and the vector's length is scaled until the desired energy is reached; and (3) progressive sampling, in which an initial phase point is chosen at an angle $\Psi < 90°$ from a phase point on a trajectory initiated by one of the above two methods. Progressive sampling produces a much finer sampling of initial states than orthant sampling, because in spaces of high dimensionality random vectors like those chosen in orthant sampling will nearly always have values of $\Psi$ equal to $90°$.

Monte Carlo lifetimes were calculated at 200, 100, and 70 kcal/mole using these three sampling methods. In light of the thermal behavior of $CH_3NC$ the Monte Carlo results were quite surprising. After a short time transient was separated from the distributions, the Monte Carlo rate constants for the three sampling methods were found to differ with each other and with the RRKM prediction. The obvious conclusion was that the reaction coordinate for $CH_3NC$ is badly coupled to the remaining vibrational modes as depicted in Fig. 11, and therefore the Monte Carlo rate constants are the intramolecular relaxation rates, Eqs. (48)–(50).

*Table 4. Monte Carlo and RRKM Rate Constants*

| Molecule and path | Rate constants[a] | |
|---|---|---|
| | Theoretical (RRKM) | Calculated (Monte Carlo) |
| H—C≡C—Cl (H–C) | $7.0 \times 10^{11}$ | $4.9 \pm 0.7 \times 10^{11b}$ <br> $5.8 \pm 0.8 \times 10^{11c}$ |
| H—C≡C—Cl (Cl–C) | $4.9 \times 10^{11}$ | $6.8 \pm 0.9 \times 10^{11b}$ <br> $7.1 \pm 0.9 \times 10^{11c}$ |
| H—C≡C—H | $14.7 \times 10^{11}$ | $9.6 \pm 1.8 \times 10^{11}$ |
| Cl—C≡C—Cl | $12.8 \times 10^{11}$ | $13.3 \pm 1.7 \times 10^{11}$ |

[a] The rate constants are in units of $sec^{-1}$ (Reference 85).
[b] The rate constant is calculated using orthant sampling.
[c] The rate constant is calculated using progressive sampling.

To see if the result for $CH_3NC$ is a general phenomena, a trajectory study was performed for H—C≡C—H, H—C≡C—Cl, and Cl—C≡C—Cl decomposition.[85] With use of a model potential with three Morse functions and two harmonic bends, lifetime distributions were determined at 200 kcal/mole for both random and nonrandom initial conditions. An earlier linear tetraatomic trajectory study[78] and quantum yield measurements of fluorescence in H—C≡C—Cl[86] suggested that intramolecular relaxation might be slow in these molecules. For random excitation both progressive and orthant sampling

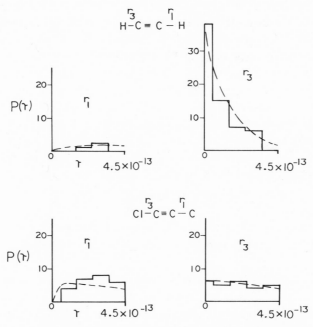

Fig. 14. $P(t) \times 10^{11}$ $sec^{-1}$ vs. $\tau$ for H—C≡C—H and Cl—C≡C—Cl computed by classical trajectories. Molecules were excited nonrandomly.[85]

methods were used. After separation of an initial transient, the lifetime distributions were exponential, and the Monte Carlo rate constants agreed to within $\pm 30\%$ of the RRKM prediction. The comparison is given in Table 4.

The effect of a nonrandom initial distribution was also investigated. This was done by sampling only 12 dimensions of the 18-dimensional phase space using the orthant sampling method. As a result little relative energy resided initially in either a H—C or Cl—C stretching coordinate. An example of these nonrandom distributions is shown in Fig. 14. The dashed line is a fit to the data using a mechanism similar to Eq. (54). Relaxation rates were derived by assuming that $k$ equals the RRKM prediction and by equating $k_{NR}$ with the $\tau = 0$ intercept of the lifetime distribution for the reactive bond receiving initial excitation. The values for $\lambda$ vary from $5 \times 10^{11}$ to $>3 \times 10^{13}$ sec$^{-1}$, with intramolecular relaxation fastest in Cl—C≡C—Cl and slowest in H—C≡C—H.

### 4.3. Energy Partitioning in Unimolecular Fragments

The RRKM theory may be used to predict the translational energy distribution of the unimolecular fragments at the critical configuration. For unimolecular reactions with no potential energy release other than that resulting from orbital angular momentum, the translational energy distribution in the critical configuration is directly related to the kinetic energies of the decomposition products. If a molecule is monoenergetically excited, the rate constant for decomposition when the separating fragments have relative translational energy in the interval $E_t^+ \to E_t^+ + dE_t^+$ is

$$d[k(E_t^+)] = \frac{\nu}{2} \frac{N(E_t^+)N(E^+ - E_t^+)\,dE_t^+}{N(E_v^*)} \tag{68}$$

where $N(E_t^+)$, $N(E^*)$, and $\nu$ are as defined previously, Eqs. (5)–(10). The quantity $N(E^+ - E_t^+)$ is the density associated with the active states in the critical configuration. The probability of a particular $E_t^+$ is then

$$P(E_t^+)\,dE_t^+ = d[k(E_t^+)]\Big/\int_0^{E^+} d[k(E_t^+)] \tag{69}$$

After cancelling terms and integrating, Eq. (69) becomes

$$P(E_t^+) = N(E^+ - E_t^+)\Big/\sum_{E_v^+=0}^{E^+} P(E_v^+) \tag{70}$$

Because it is reasonable to assume that the active rotational energy $E_r^+$ will ultimately result in relative kinetic energy, the total kinetic energy of the decomposition products is $E_t^+ + E_r^+$, and its probability is $P(E_t^+)$. For the usual unimolecular reaction there is a distribution of rotational energies, and we have

$$P(E_t^+ + E_r^+) = \int P(E_t^+)P(E_r^+)\, dE_r^+ \tag{71}$$

The appropriate form for $P(E_r^+)$ depends on the particular system under question. The reader is referred to the literature for some specific examples.[21,87–89]

Equations (70) and (71) can be used to predict the kinetic energy distribution for the general situation of potential energy release in addition to that resulting from $E_r^+$ if the assumption of no "exit channel" coupling is made. This hypothesis will be valid if there is no translation–vibration or rotation–vibration energy transfer as the decomposition fragments separate from the critical configuration. If there is strong exit channel coupling, the products' energy spectrum will be determined by the dynamical properties of the potential energy surface.

Measurements of kinetic energy distributions following unimolecular reaction have been made using molecular beams[3–9] and mass spectrometry.[90,91] Most of the studies have pertained to ionic decompositions. The most extensive study of neutral reactions is that of Lee et al.[3–9] using crossed molecular beam techniques. By reacting F atoms with various olefins, they prepared a series of chemically activated molecules that decomposed by either H·, Cl·, or ·$CH_3$ emission. Both angular distributions and recoil energy distributions were measured for the reactions. Each of the systems was characterized by an isotropic c.m. angular distribution, consistent with the formation of long-lived complexes. However, only for Cl atom emission did the relative translational energy distributions approximate those predicted by Eq. (71). A comparison with the RRKM prediction for $F + C_2H_4 \rightarrow C_2H_3F + H$ is shown in Fig. 15. The immediate question is whether this lack of agreement is

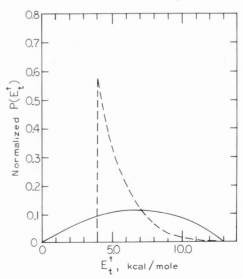

Fig. 15. Translational energy distribution for the $F + C_2H_4 \rightarrow H + C_2H_3F$ reaction; solid line, experimental result; dashed line, RRKM prediction.[3]

due to a failure of the RRKM theory or results from exit channel coupling. For H and $CH_3$ elimination the potential barrier in the exit channel may be as high as 3 and 8 kcal/mole, respectively. In the RRKM calculation shown in Fig. 15 the implicit assumption is that all this energy results in product translational energy. Is such an assumption valid? Release of some of the potential energy to $C_2H_3F$ as the products depart would explain the measured distribution. Our current knowledge of potential energy surfaces is not detailed enough to choose between these alternatives and that chemically activated $C_2H_4F$ exhibits non-RRKM behavior remains a real possibility.

Some light has been shed on this problem by Moehlmann et al.,[13] who have made a study of the reactions

$$F + C_2H_3X \rightarrow X + C_2H_3F, \qquad X = H, Cl, Br, and CH_3 \qquad (72)$$

by arrested relaxation infrared chemiluminescence. Measurements of infrared emission from $C_2H_3F$ indicated that energy was randomized for $X = Cl$ and Br but not for $X = H$ and $CH_3$. The pointed out that the existence of a barrier in the exit channel can lead to a nonstatistical product energy distribution, even though a statistical distribution exists in the critical configuration. Therefore, the lack of a barrier for Cl and Br elimination allows the product vibrational energy to be distributed statistically. The question does arise as to why $C_2H_3XF$ can randomly redistribute its excess energy and $C_2H_3F$ cannot. One possible answer would be the much lower degree of vibrational excitation in $C_2H_3F$, which makes the normal mode approximation more appropriate and as a result intramolecular energy transfer much slower.

Another series of reactions for which energy partitioning has been measured is three- and four-centered HX ($X = Cl, F$) elimination from halogenated alkanes. The difference between these two reactions is that elimination occurs from the same carbon in the three-centered process, e.g., $CF_3H \rightarrow HF + CF_2$ and $CH_2Cl - CDCl_2 \rightarrow CH_2Cl - CCl + DCl$, and from adjacent carbons in the four-centered step, e.g., $CCl_3CH_3 \rightarrow HCl + CCl_2CH_2$. Translational energy distributions following HCl elimination from alkyl halide positive ions has been measured by mass spectrometry.[91] Infrared chemiluminescence measurements have revealed the degree of vibrational excitation in both HF and HCl.[10,11] The energy partitioning to the product olefins has been estimated by observing their subsequent unimolecular reactions.[54,55]

So far the following picture has emerged with respect to these elimination reactions. For the three-centered process the energy partitioning is purely statistical. This result agrees with the RRKM prediction and might be expected because very little potential energy is released in passing from the elimination barrier to the carbene. Dynamical considerations appear to be much more important for four-centered elimination. The degree of vibrational excitation of HX is still statistical; however, $\gg 25\%$ of the approximately 50 kcal/mole of potential energy is released as relative translational or rotational energy. This fraction is much greater than the RRKM prediction, Eq. (71). Therefore, these four-centred processes appear to be a good example of strong exit channel coupling. Although it has yet to be tested, the following two-step model suggested by Kim and Setser[55] seems to present a meaningful representation of the unimolecular dynamics for four-centered elimination. The first step is the transfer of H to X because its motion is rapid compared to that of the X atom. The HX elimination then occurs as X recoils from the olefin, releasing a relatively large fraction of the energy to relative translation.

## 4.4. Status of the RRKM Theory

We shall conclude this section on unimolecular dynamics by making a critical review of the applicability of the RRKM theory. The above studies show that the RRKM theory cannot be used to interpret either product energy distributions when there is strong exit channel coupling or the initial lifetimes of molecules excited nonrandomly. However, what do they say about the validity of applying the RRKM theory to the more normal situations of chemical activation with long unimolecular lifetimes $>10^{-10}$ sec and thermal activation?

In general the classical trajectory calculations support the RRKM theory for unimolecular dissociation reactions. The intramolecular vibrational energy relaxation rates are greater than $10^{11}$ sec$^{-1}$ for both the triatomic and tetra-atomic models. Only under very unusual circumstances do we expect intrinsic non-RRKM behavior. The most likely candidates would be small molecules with large frequency or mass differences. Apparent non-RRKM behavior should be measurable at very high pressures. It should be noted that $NO_2Cl$ dissociation appeared to be RRKM in the suprahigh pressure regime.

The situation is much different for $CH_3NC$ isomerization. Here the classical trajectory calculations predict intrinsic non-RRKM behavior, whereas the thermal experiments are well described by the RRKM theory. If one assumes that the classical calculation adequately models reality, $CH_3NC$ is then a molecule characterized by an internal bottleneck for intramolecular energy relaxation (Fig. 11). The measured high-pressure thermal rate constant is then the rate of vibrational energy relaxation, Eq. (50) and region (b) of Fig. 11. If this is the case, the critical configuration deduced from the high-pressure Arrhenius parameters would not resemble that of the RRKM theory.

In deciding whether $CH_3NC$ is intrinsically non-RRKM we should ask the following two questions of the thermal measurements: (1) Is the derived critical configuration realistic, and (2) can the thermal falloff measurements for methyl and ethyl isocyanide and their isotopic variants be explained by an adequate non-RRKM theory? As described previously, the critical configuration chosen by Rabinovitch et al. for isocyanide isomerization is a cyclic species. The major frequency alteration in the critical configuration is the lowering of the C—N stretching normal mode by a factor of $\sim 1.6$. This results in a "tight" critical configuration with an $A$ factor of $10^{13.6} \sec^{-1}$. Is this the order of magnitude we expect for the $A$ factor? This question has not been answered and probably will not be until a complete potential energy surface has been calculated for $CH_3NC$ isomerization. The RRKM $A$ factor may then be calculated from the molecular force constants at the potential energy barrier. The *ab initio* calculations of Schaeffer et al. for the $CH_3NC \rightarrow CH_3CN$ potential surface shows that $CH_3$ and $N \equiv C$ have nearly free internal rotation in the critical configuration. This result does place some doubt on the tight cyclic structure derived from the thermal $A$ factor. However, the $A$ factor of $10^{13.6} \sec^{-1}$ does appear to be consistent with other types of isomerizations.[25]

One can only conjecture about the possibility of a non-RRKM theory interpreting all the thermal isocyanide isomerization rate data. It must first be able to predict the same false limit for all the isocyanides. This means that the rate of intramolecular energy relaxation into the reaction coordinate, as depicted in Fig. 11, is the same for each of the isocyanides. This would occur if the coupling with the product channel were the same for all the isocyanides. The theory must also be able to predict the shape of the unimolecular falloff and its variation with isotopic substitution and molecular complexity.

The current dilemma regarding the actual behavior of $CH_3NC$ will undoubtedly be resolved in the future. A solution could be provided by the study of $CH_3NC$ isomerization at suprahigh pressures or a determination of the critical configuration structure by *ab initio* techniques. If it turns out that $CH_3NC$ is indeed a non-RRKM molecule, the thermal measurements should provide some of the best tests of non-RRKM theories.

From the above discussion we see that identical high-pressure thermal unimolecular rate constants for a homologous series of reactants either implies identical critical configuration structures if the RRKM theory is valid or similar intramolecular relaxation rates if there is a bottleneck, as in Fig. 11. Therefore, the finding of identical $A$ factors for a series of compounds that undergo a specific reaction type, such that $E_0$ remains unchanged, is not a sufficient test of the RRKM theory. Although it has seldom been suggested, the small $A$ factors for many unimolecular reactions could refer to weak coupling between the reaction coordinate and the remaining vibrational modes, instead of tight critical configurations.

The primary experimental tests indicating rapid intramolecular relaxation have been those utilizing chemical activation techniques to detect apparent non-RRKM behavior. The argument is that if vibrational energy relaxation is rapid between the molecular vibrations so that the initial nonrandom distribution quickly becomes random, the coupling with the reaction coordinate must be strong. Therefore, if we do not see apparent non-RRKM behavior, it is reasonable to assume that the molecule does not display intrinsic non-RRKM behavior. This argument assumes that the coupling between the molecular vibrations is the same as with the reaction coordinate. One has then eliminated the possibility depicted in Fig. 11, weak coupling with the reaction coordinate. The existence or nonexistence of apparent non-RRKM behavior does not make a direct statement about the possibility of intrinsic non-RRKM behavior.

## 5. New Directions

Some of the recent theoretical studies of unimolecular reactions are related to the concept of ergodicity. To quote Maxwell,† the ergodic hypothesis " ... is that the system, if left to itself in its actual state of motion, will, sooner or later, pass through every phase which is consistent with the equation of energy." If the system is considered to be an isolated molecule, ergodicity has immediate relevance to unimolecular dynamics. For a nonergodic molecule the phase space would be separable (metrically indecomposable), and some molecules would never be able to undergo isomerization or dissociation even though sufficiently energized. Such molecules would definitely be non-RRKM. If possible, we would then like to determine the energy necessary for a molecule, represented by a set of coupled anharmonic oscillators, to be ergodic.

One of the first studies of this type is that of Fermi et al.,[92] who simulated the planar motion of a string of 64 particles, bound by anharmonic forces, rigidly supported at the ends. Instead of finding rapid redistribution of energy among the normal modes, they found that most of it remained concentrated in a few modes. This result indicated that more than anharmonicity is necessary for a system to be ergodic.

An answer to the phenomena found by Fermi et al. has been explained by the work of Kolmogrov et al.[93] They have proved rigorously a theorem that states that if there are no resonances among a set of harmonic oscillators, the addition of an anharmonic perturbation that is sufficiently small compared to the total energy does not make the system ergodic.

Numerical studies of the Hamiltonians

$$H = \tfrac{1}{2}(p_1^2 + p_2^2 + q_1^2 + q_2^2) + (q_1^2 q_2 - \tfrac{1}{3} q_2^3) \tag{73}$$

and

$$H = \tfrac{1}{2}(p_1^2 + p_2^2) + a(q_1^2 + q_2^2) - \lambda q_1^2 q_2$$

have also shown the tendency for nonergodic behavior at low energies and an approach to ergodicity as the energy is increased.[94,95] Thus there appears to be a critical energy above which trajectories will be ergodic. For the Hamiltonians described by Eqs. (73) and (74) Mo has shown that this critical energy is 0.75 of that necessary for dissociation.[96] In the future we should expect the extension

---

†J. L. Lebowitz and O. Penrose, Modern ergodic theory, *Phys. Today* **26**(2), 23 (1973).

of this work to more complicated Hamiltonians so the relationship between the critical energy for ergodic behavior and the threshold for unimolecular reaction becomes known for real molecular systems. Nordholm and Rice have studied intramolecular vibrational energy relaxation using the idea of quantum ergodicity.[97,98]

One of the major limitations of using the concept of ergodicity to describe intramolecular energy relaxation is that it is time independent. That every state will ultimately dissociate is not sufficient for the RRKM theory to be applicable. A molecule that is ergodic is not necessarily an RRKM molecule. For a randomly excited molecule to have a random lifetime, intramolecular energy relaxation into the reaction coordinate has to be faster than unimolecular reaction. The example shown in Fig. 11 may exhibit ergodic behavior even though it is intrinsically non-RRKM.

At the present most theories of intramolecular energy relaxation are based on the master equation formalism.[22,82,99-102] One of the most recent treatments of this type is that of Gelbart et al.[100] Their model predicts intrinsic non-RRKM behavior for weak vibrational exchange mechanisms and also explains apparent non-RRKM behavior because energy transfer between each of the molecular vibrations is treated individually. Because the extent of vibrational coupling is not known for actual molecules, comparisons have not been made with real molecular systems. However, stochastic treatments do provide an alternative choice to the RRKM theory, and they may be used more extensively in the future.

Another direction for future research will be the continuing experimental study of molecular dynamics and the interpretation of the measurements using classical or semiclassical[103] trajectory calculations on *ab initio* potential energy surfaces. Some important experimental data are already available, e.g., $F + C_2H_4$ and $CF_3H$ energy partitioning.[3,11] Such experimental and theoretical studies will provide rates of intramolecular energy relaxation and predictions about the statistical or nonstatistical behavior of molecules.

ACKNOWLEDGMENTS

I wish to thank Professors Chris Sloane and Don Bunker for very meaningful discussions during the preparation of this chapter. An acknowledgment is due to my wife, Betty Hase, who prepared the illustrations.

## References

1. N. B. Slater, *Theory of Unimolecular Reactions*, Cornell University Press, Ithaca, N.Y. (1959).

2. R. A. Marcus, Unimolecular dissociations and free radical recombination reactions, *J. Chem. Phys.* **20**, 359–364 (1952).
3. J. M. Parson and Y. T. Lee, Crossed molecular beam study of $F + C_2H_4$, $C_2D_4$, *J. Chem. Phys.* **56**, 4658–4666 (1972).
4. J. M. Parson, K. Shobatake, Y. T. Lee, and S. A. Rice, Unimolecular decomposition of the long-lived complex formed in the reaction $F + C_4H_8$, *J. Chem. Phys.* **59**, 1402–1415 (1973).
5. K. Shobatake, J. M. Parson, Y. T. Lee, and S. A. Rice, Unimolecular decomposition of long-lived complexes of fluorine and substituted mono-elfins, cyclic olefins, and dienes. *J. Chem. Phys.* **59**, 1416–1426 (1973).
6. K. Shobatake, J. M. Parson, Y. T. Lee, and S. A. Rice, Laboratory angular dependence and the recoil–energy spectrum of the products of the reaction $F + C_6D_6 \rightarrow D + C_6D_5F$, *J. Chem. Phys.* **59**, 1427–1434 (1973).
7. K. Shobatake, Y. T. Lee, and S. A. Rice, Reactions of F atoms and aromatic and heterocyclic molecules: Energy distribution in the reaction complex, *J. Chem. Phys.* **59**, 1435–1448 (1973).
8. K. Shobatake, Y. T. Lee, and S. A. Rice, Crossed molecular beams study of the reaction $F + (C_2H_2Cl_2) \rightarrow Cl + (C_2H_2ClF)$, *J. Chem. Phys.* **59**, 6104–6111 (1973).
9. J. M. Parson, K. Shobatake, Y. T. Lee, and S. A. Rice, Substitution reactions of fluorine atoms with unsaturated hydrocarbons: Crossed molecular beam studies of unimolecular decomposition, *Discuss. Faraday Soc.* **55**, 344–356 (1973).
10. P. N. Clough, J. C. Polanyi, and R. T. Taguchi, Vibrational energy distribution in HF formed by elimination from activated $CH_3CF_3$ and $CH_2CF_2$, *Can. J. Chem.* **48**, 2919–2930 (1970).
11. H. W. Chang, D. W. Setser, and M. J. Perona, Comparison of energy partitioning from three-centered processes. Bimolecular transfer and unimolecular elimination reactions, *J. Phys. Chem.* **75**, 2070–2072 (1971).
12. J. G. Moehlmann and J. D. McDonald, Vibrational energy distribution of chemically activated cyclooctanone, *J. Chem. Phys.* **59**, 6683–6684 (1973).
13. J. G. Moehlmann, J. T. Gleaves, J. W. Hudgens, and J. D. McDonald, Infrared chemiluminescence studies of the reaction of fluorine atoms with monosubstituted ethylene compounds, *J. Chem. Phys.* **60**, 4790–4799 (1974).
14. J. T. Gleaves and J. D. McDonald, Infrared chemiluninescence studies of hydrogen halide elimination reactions, *J. Chem. Phys.* **62**, 1582–1583 (1975).
15. D. L. Bunker and W. L. Hase, On non-RRKM unimolecular kinetics: Molecules in general, and $CH_3NC$ in particular, *J. Chem. Phys.* **59**, 4621–4632 (1973).
16. J. D. Rynbrandt and B. S. Rabinovitch, Intramolecular energy relaxation. Nonrandom decomposition of hexafluorobicyclopropyl, *J. Phys. Chem.* **75**, 2164–2171 (1971).
17. J. F. Meagher, K. J. Chao, J. R. Barker, and B. S. Rabinovitch, Intramolecular vibrational energy relaxation. Decomposition of a series of chemically activated fluoralkyl cyclo-propanes, *J. Phys. Chem.* **78**, 2535–2543 (1974).
18. R. A. Marcus, Dissociation and isomerization of vibrationally excited species. III, *J. Chem. Phys.* **43**, 2658–2661 (1965).
19. D. L. Bunker and M. Pattengill, Monte Carlo calculations. VI. A re-evaluation of the RRKM theory of unimolecular reaction rates, *J. Chem. Phys.* **48**, 772–776 (1968).
20. E. V. Waage and B. S. Rabinovitch, Centrifugal effects in reaction rate theory, *Chem. Rev.* **70**, 377–387 (1970).
21. D. L. Bunker, RRKM theory, the $CH_3NC$ paradox, and the decomposition of hot-atom substitution products, *J. Chem. Phys.* **57**, 332–335 (1972).
22. D. L. Bunker, *Theory of Elementary Gas Reaction Rates*, Pergamon Press, Inc., Elmsford, N.Y. (1966).
23. K. J. Laidler, *Theories of Chemical Reaction Rates*, McGraw-Hill Book Company, New York (1969).
24. D. W. Setser, *in: MTP International Review of Sciences, Physical Chemistry* (J. Polanyi, ed.), Vol. 9, pp. 1–43, University Park Press, Baltimore (1972).
25. P. J. Robinson and K. A. Holbrook, *Unimolecular Reactions*, John Wiley & Sons, Inc. (Interscience Division), New York (1972).

26. W. Forst, *Theory of Unimolecular Reactions,* Academic Press, Inc., New York (1973).
27. S. Glasstone, K. J. Laidler, and H. Eyring, *The Theory of Rate Processes,* McGraw-Hill Book Company, New York (1941).
28. W. Forst, Methods for calculating energy level densities, *Chem. Rev.* **71**, 339–356 (1971).
29. D. C. Tardy, B. S. Rabinovitch, and G. Z. Whitten, Vibration–rotation energy-level density calculations, *J. Chem. Phys.* **48**, 1427–1429 (1968).
30. W. L. Hase, Theoretical configuration for ethane decomposition and methyl radical recombination, *J. Chem. Phys.* **57**, 730–733 (1972).
31. W. H. Wong and R. A. Marcus, Concept of minimum state density in the activated complex theory of bimolecular reactions, *J. Chem. Phys.* **55**, 5625–5629 (1971).
32. H. S. Johnston, *Gas Phase Reaction Rate Theory,* The Ronald Press Company, New York (1966).
33. M. Quack and J. Troe, Specific rate constants of unimolecular processes. II. Adiabatic channel model, *Ber. Bunsenges. Phys. Chem.* **78**, 240–252 (1974).
34. R. C. Baetzold and D. J. Wilson, Classical unimolecular rate theory. II. Effect of the distribution of initial conditions, *J. Phys. Chem.* **68**, 3141–3145 (1964).
35. E. V. Waage and B. S. Rabinovitch, Simple and accurate approximation for the centrifugal factor in RRKM theory, *J. Chem. Phys.* **52**, 5581–5584 (1970).
36. H. S. Johnston and P. Goldfinger, Theoretical interpretation of reactions occuring in photoclorination, *J. Chem. Phys.* **37**, 700–709 (1962).
37. K. J. Mintz and R. J. Cvetanovic, Arrhenius A factors of unimolecular decomposition of alcohols formed by insertion of $O(^1D_2)$ atoms into the C—H bonds of paraffins, *Can. J. Chem.* **51**, 3386–3393 (1973).
38. W. L. Hase, R. L. Johnson, and J. W. Simons, The decomposition of chemically activated *n*-butane, isopentane, neohexane, and *n*-pentane and the correlation of their decomposition rates with radical recombination rates, *Int. J. Chem. Kinet.* **4**, 1–35 (1972).
39. F. B. Growcock, W. L. Hase, and J. W. Simons, Kinetics of chemically activated ethane, *Int. J. Chem. Kinet.* **5**, 77–92 (1973).
40. W. L. Hase, C. J. Mazac, and J. W. Simons, Decomposition kinetics of chemically activated dimethylsilane and ethylsilane, *J. Am. Chem. Soc.* **95**, 3454–3459 (1973).
41. H. O. Pritchard, R. G. Sowden, and A. F. Trotman-Dickenson, Studies in energy transfer. II. The isomerization of cyclopropane—A quasi-unimolecular reaction, *Proc. R. Soc. London Ser. A* **217**, 563–571 (1953).
42. T. S. Chambers and G. B. Kistiakowsky, Kinetics of the thermal isomerization of cyclopropane, *J. Am. Chem. Soc.* **56**, 399–405 (1934),
43. M. C. Lin and K. J. Laidler, Fall-off behavior and kinetic isotope-effects in reactions of cyclic hydrocarbons, *Trans. Faraday Soc.* **64**, 927–944 (1968).
44. P. Jeffers, D. Lewis, and M. Sarr, Cyclopropane structural isomerization in shock waves, *J. Phys. Chem.* **77**, 3037–3041 (1973).
45. F. W. Schneider and B. S. Rabinovitch, The thermal unimolecular isomerization of methyl isocyanide. Fall-off behavior, *J. Am. Chem. Soc.* **84**, 4215–4230 (1962).
46. F. W. Schneider and B. S. Rabinovitch, The unimolecular isomerization of methyl-$d_3$ isocyanide. Statistical-weight inverse secondary intermolecular kinetic isotope effects in nonequilibrium thermal systems, *J. Am. Chem. Soc.* **85**, 2365–2370 (1963).
47. B. S. Rabinovitch, P. W. Gilderson, and F. W. Schneider, The thermal unimolecular isomerization of methyl-$d_1$ isocyanide. Fall-off and inverse isotope effect, *J. Am. Chem. Soc.* **87**, 158–160 (1965).
48. K. M. Maloney and B. S. Rabinovitch, The thermal isomerization of isocyanide. Variation of molecular parameters, Ethyl isocyanide, *J. Phys. Chem.* **73**, 1652–1666 (1969).
49. K. M. Maloney, S. P. Pavlou, and B. S. Rabinovitch, Kinetic isotope effects in nonequilibrium thermal unimolecular systems. Ethyl isocyanide-$d_5$, *J. Phys. Chem.* **73**, 2756–2760 (1969).
50. J. V. Michael and G. N. Suess, Application of RRKM theory to the chemical and thermal activation of ethyl radicals, *J. Chem. Phys.* **58**, 2807–2812 (1973).
51. K. J. Laidler and J. C. Polanyi, Theories of the kinetics of bimolecular reactions, *in: Progress in Reaction Kinetics* (G. Porter, ed.), Vol. 3, pp. 1–61, Pergamon Press, Inc., Elmsford, N.Y. (1965).

52. E. Tschuikow-Roux, Critical bond length in radical combination and unimolecular dissociation reactions, *J. Phys. Chem.* **72**, 1009–1011 (1968).
53. C. W. Larson and B. S. Rabinovitch, Competitive unimolecular decomposition of alkyl radicals. Tertiary butyl rupture, *J. Chem. Phys.* **52**, 5181–5183 (1970).
54. K. Dees, D. W. Setser, and W. G. Clark, The reactions of methylene with 1,2-dichloroethane and nonequilibrium unimolecular HCl elimination from 1,3-dichloropropane, 1,4-dichlorobutane, and 1-chloropropane, *J. Phys. Chem.* **75**, 2231–2240 (1971).
55. K. C. Kim and D. W. Setser, Unimolecular reactions and energy partitioning. Three- and four-centered elimination reactions of chemically activated 1,1,2-trichloroethane-$d_0$-$d_1$ and $d_2$, *J. Phys. Chem.* **78**, 2166–2179 (1974).
56. F. S. Rowland, *in: MTP International Review of Sciences, Physical Chemistry* (J. Polanyi, ed.), Vol. 9, pp. 109–133, University Park Press, Baltimore (1972).
57. K. A. Krohn, N. J. Parks, and J. W. Root, Chemistry of nuclear recoil [18]F atoms. VI. Approximate energetics and molecular dynamics in $CH_3CF_3$, *J. Chem. Phys.* **55**, 5785–5794 (1971).
58. C. C. Chou and W. L. Hase, Rice–Ramsperger–Kassel–Marcus theory applied to decomposition of hot atom substitution products. $c$-$C_4H_7T$ and $c$-$C_4D_7T$, *J. Phys. Chem.* **78**, 2309–2315 (1974).
59. J. N. Butler and G. B. Kistiakowsky, Reactions of methylene. IV. Propylene and cyclopropane, *J. Am. Chem. Soc.* **82**, 759–765 (1960).
60. A. F. Trotman-Dickenson, *Gas Kinetics*, Butterworths, London (1955).
61. S. C. Chan, B. S. Rabinovitch, J. T. Bryant, L. D. Spicer, T. Fujimoto, Y. N. Lin, and S. P. Pavlou, Energy transfer in thermal methyl isocyanide isomerization. A comprehensive investigation, *J. Phys. Chem.* **74**, 3160–3176 (1970).
62. W. G. Valance and E. W. Schlag, Nonequilibrium effects in unimolecular reaction theory, *J. Chem. Phys.* **45**, 4280–4288 (1966).
63. V. J. Troe and H. G. Wagner, Unimolecular reactions in thermal systems, *Ber. Bunsenges. Phys. Chem.* **71**, 937–979 (1967).
64. D. C. Tardy and B. S. Rabinovitch, Collisional energy transfer in thermal unimolecular systems. Dilution effects and fall-off region, *J. Chem. Phys.* **48**, 1282–1301 (1968).
65. S. P. Pavlou and B. S. Rabinovitch, Energy transfer in thermal isocyanide isomerization. Noble gases in the ethyl isocyanide system, *J. Phys. Chem.* **75**, 1366–1374 (1971).
66. M. Hoare, Steady state unimolecular processes in multilevel systems, *J. Chem. Phys.* **38**, 1630–1635 (1963).
67. H. W. Chang, N. L. Craig, and D. W. Setser, Nonequilibrium unimolecular reactions and collisional deactivation of chemically activated fluoroethane and 1,1,1-trifluoroethane, *J. Phys. Chem.* **76**, 954–963 (1972).
68. J. D. Rynbrandt and B. S. Rabinovitch, Collisional transition probability distributions for deactivation of vibrationally excited dimethylcyclopropane, *J. Phys. Chem.* **74**, 1679–1685 (1970).
69. J. H. Georgakakos and B. S. Rabinovitch, Collision transfer of vibrational energy from highly excited polyatomics. Transfer probabilities and cross sections for inefficient bath gases, *J. Chem. Phys.* **56**, 5921–5930 (1972).
70. D. W. Setser and E. E. Siefert, Vibrational energy transfer probabilities of highly vibrationally excited dichloroethane with argon, krypton, xenon, and sulfur hexafluoride, *J. Chem. Phys.* **57**, 3623–3628 (1972).
71. Von S. H. Luu and J. Troe, Photoisomerization of cycloheptatriene, II. Temperature dependence of collisional energy transfer, *Ber. Bunsenges. Phys. Chem.* **78**, 766–773 (1974).
72. Y. N. Lin and B. S. Rabinovitch, A simple quasi-accommodation model of vibrational energy transfer. Low pressure thermal methyl isocyanide isomerization, *J. Phys. Chem.* **74**, 3151–3159 (1970).
73. M. L. Dutton, D. L. Bunker, and H. H. Harris, Two familiar gas reactions at suprahigh pressure, *J. Phys. Chem.* **76**, 2614–2617 (1972).
74. J. Aspen, N. A. Khawaja, J. Reardon, and D. J. Wilson, Pyrolysis of ethylcyclobutane in the gas phase at high pressures, *J. Am. Chem. Soc.* **91**, 7580–7582 (1969).

75. R. E. Harrington, B. S. Rabinovitch, and H. M. Frey, Decomposition of activated sec-butyl radicals from different sources and unimolecular reaction theory, *J. Chem. Phys.* **33**, 1271–1272 (1960).

76. I. Oref, D. Schuetzle, and B. S. Rabinovitch, Unimolecular decomposition and intramolecular energy relaxation in the suprahigh-pressure region, *J. Chem. Phys.* **54**, 575–578 (1971).

77. E. Thiele and D. J. Wilson, Anharmonicity in unimolecular reactions, *J. Chem. Phys.* **35**, 1256–1263 (1961).

78. R. J. Harter, E. B. Alterman, and D. J. Wilson, Anharmonic effects in unimolecular rate theory. Vibrations and collisions of simple polyatomic systems, *J. Chem. Phys.* **40**, 2137–2145 (1964).

79. N. C. Hung and D. J. Wilson, Anharmonic effects in unimolecular rate theory. Dynamics of a rotating anharmonic triatomic molecule, *J. Chem. Phys.* **38**, 828–831 (1963).

80. N. C. Hung, Rotational–vibrational energy transfer. Dynamics of a rotating anharmonic four-atom molecule, *J. Chem. Phys.* **57**, 5202–5215 (1972).

81. R. C. Baetzold and D. J. Wilson, Classical unimolecular rate theory. III. Effect of initial conditions on lifetime-distributions, *J. Chem. Phys.* **43**, 4299–4303 (1965).

82. D. L. Bunker, Monte Carlo calculations. IV. Further studies of unimolecular decomposition, *J. Chem. Phys.* **40**, 1946–1957 (1964).

83. H. H. Harris and D. L. Bunker, Methyl isocyanide is probably a non-RRKM molecule, *Chem. Phys. Lett.* **11**, 433–436 (1971).

84. D. H. Loskiw, C. F. Bender, and H. F. Schaefer III, Some features of the $CH_3NC \rightarrow CH_3CN$ potential surface, *J. Chem. Phys.* **57**, 4509–4511 (1972).

85. W. L. Hase and Da-Fei Feng, Classical trajectory study of the unimolecular decomposition of $H-C\equiv C-Cl$, $H-C\equiv C-H$ and $Cl-C\equiv C-Cl$, *J. Chem. Phys.* **61**, 4690–4699 (1974).

86. K. Evans, R. Scheps, S. A. Rice, and D. Heller, Primary photochemical and photophysical processes in chloro- and bromoacetylene, *Chem. Soc. Faraday Trans. 2* **69**, 856–880 (1973).

87. C. E. Klots, Quasi-equilibrium theory of ionic fragmentation: Further considerations, *Z. Naturforsch. Teil A* **27**, 553–561 (1972).

88. S. A. Safron, N. D. Weinstein, D. R. Herschbach, and J. C. Tully, Transition state theory for collision complexes: Product translational energy distributions, *Chem. Phys. Lett.* **12**, 564–568 (1972).

89. R. A. Marcus, On the theory of energy distributions of products of molecular beam reactions involving transient complexes, *J. Chem. Phys.* **62**, 1372–1384 (1975).

90. E. L. Spotz, W. A. Seitz, and J. L. Franklin, Translational energy of fragments of ion decomposition and totality of states functions, *J. Chem. Phys.* **51**, 5142–5148 (1969).

91. K. C. Kim, J. H. Beynon, and R. G. Cooks, Energy partitioning by mass spectrometry: chloroalkanes and chloroalkenes, *J. Chem. Phys.* **61**, 1305–1314 (1974).

92. E. Fermi, J. Pasta, and S. Ulam, Studies of non linear problems, in: *Enrico Fermi: Collected Papers*, Vol. II, pp. 978–988, University of Chicago Press, Chicago (1965).

93. G. H. Walker and J. Ford, Amplitude instability and ergodic behavior for conservative nonlinear oscillator systems, *Phys. Rev. A* **188**, 416–432 (1969).

94. B. Barbanis, On the isolating character of the third integral in a resonance case, *Astron. J.* **71**, 415–424 (1966).

95. M. Henon and C. Heiles, The applicability of the third integral of motion: Some numerical experiments, *Astron. J.* **69**, 73–79 (1964).

96. K. C. Mo, Theoretical prediction for the onset of widespread instability in conservative nonlinear oscillator systems, *Physica* **57**, 445–454 (1972).

97. K. S. J. Nordholm and S. A. Rice, Quantum ergodicity and vibrational relaxation in isolated molecules, *J. Chem. Phys.* **61**, 203–223 (1974).

98. S. Nordholm and S. A. Rice, Quantum ergodicity and vibrational relaxation in isolated molecules. II. $\lambda$-Independent effects and relaxation to the asymptotic limit, *J. Chem. Phys.* **61**, 768–779 (1974).

99. J. W. Brauner and D. J. Wilson, Intramolecular energy transfer in unimolecular reactions. II. A weakly-coupled-oscillators model, *J. Phys. Chem.* **67**, 1134–1138 (1963).

100. W. M. Gelbart, S. A. Rice, and K. F. Freed, Stochastic theory of vibrational relaxation and dissociation, *J. Chem. Phys.* **52**, 5718–5732 (1970).

101. W. M. Gelbart, S. A. Rice, and K. F. Freed, Random matrix theory and the master equation for finite systems, *J. Chem. Phys.* **57**, 4699–4712 (1972).
102. K. G. Kay, Theory of vibrational relaxation in isolated molecules, *J. Chem. Phys.* **61**, 5205–5220 (1974).
103. W. H. Miller, The semiclassical nature of atomic and molecular collisions, *Acc. Chem. Res.* **4**, 161–167 (1971).

# 4

# Semiclassical Methods in Molecular Collision Theory

## M. S. Child

### 1. Introduction

The connection between classical and quantum mechanics has been of continuing interest since the time of Bohr and de Broglie, and the reader is referred to a number of established texts.[1-3] However, it is only in recent years that this connection has been applied to the solution of molecular scattering problems. The aim is to exploit the relative tractability of the classical equations of motion to obtain quantum mechanically accurate transition probabilities and collision cross sections.

A simple, almost trivial, example is the diffraction of a free particle described by the wave function

$$\psi(\mathbf{q}) = \exp(i\mathbf{p} \cdot \mathbf{q}/\hbar) \tag{1}$$

which shows that knowledge of the classical momentum is sufficient to determine the quantum mechanical diffraction pattern. The more general case, which has close analogies with the optics of a medium with variable refractive index,[4-5] requires that the momentum be allowed to vary with position, with knowledge of the function $\mathbf{p}(\mathbf{q})$ derived from the classical trajectory. This is accommodated at the simplest level by introduction of the *primitive semiclassical form*

$$\psi(\mathbf{q}) = A(\mathbf{q}) \exp\left(\frac{i}{\hbar}\int \mathbf{p} \cdot d\mathbf{q}\right) \tag{2}$$

_M. S. Child_ • Theoretical Chemistry Department, University of Oxford, Oxford, England

in place of Eq. (1). Here the exponent may be recognized as the classical action for the motion, and the amplitude $A(\mathbf{q})$ is constrained by the conservation of flux. A description of this type, or one of its analogues in another representation, proves adequate in accounting for various fluctuations in the quantal transition probability and other diffraction effects and may even be adapted to allow for changes in electronic state, or surface-hopping processes, as discussed in Chapter 5.

Problems arise, however, near what are termed the *caustics* of the classical motion, which are the envelopes of the relevant families of classical paths. The difficulty, because the motion is bounded by its caustics, is that one of the classical momenta must change sign. Hence it becomes impossible to reconcile the existence of a precise trajectory defined by the function $\mathbf{p}(\mathbf{q})$ required for evaluation of the integral in Eq. (2) with the requirements of the uncertainty principle. At points far from the caustics ambiguity in the direction of motion is sufficient to allow an adequate momentum uncertainty, but the problem becomes acute as the relevant momentum component tends toward zero. Consequently the amplitude $A(\mathbf{q})$ in Eq. (2) diverges, and the primitive semiclassical description becomes invalid.

Nevertheless, it is still possible to use knowledge of the classical motion to construct so-called "uniform approximations" that take different functional forms according to the topologies of the caustics concerned.[3,6–12] These approximations have the particular value of extending the theory to cover *classically forbidden* events by utilizing the solutions of the classical equations in complex space and time. Indeed, sufficiently far outside the caustics of the motion, the uniform approximation is designed to revert to a classically forbidden primitive form analogous to Eq. (2) but with the real momentum $\mathbf{p}$ replaced by a purely imaginary value $i|\mathbf{p}|$:

$$\psi(\mathbf{q}) = A(\mathbf{q}) \exp\left(-\int |\mathbf{p}| \cdot d\mathbf{q}\right) \tag{3}$$

The simplest example of this type is the familiar JWKB description of quantum mechanical tunneling,[13] but the analogues of Eq. (3) in other representations provide similar insight into the quantum mechanical description of many other classically forbidden events. Elastic scattering beyond the rainbow angle[8,9,14,15] and thermal energy vibrational relaxation may be cited as important examples.

This chapter is designed to illustrate the fundamental principles in the development of the theory. In Section 2 we shall cover the JWKB or primitive semiclassical description of motion in one degree Cartesian of freedom and the corresponding uniform approximations. This is followed by the semiclassical theory of transformations from one system of variables to another. Transformations to the energy–time and action–angle representations are treated in detail, because they allow particularly simple descriptions of the unperturbed asymptotic motions of the system. The above elements are then combined in Section 3 to give a unified semiclassical theory, which is applied to elastic and simple inelastic scattering events. Constraints of space allow for the inclusion of only one of two possible formulations, but excellent accounts of the other available in the literature.[16] This unified theory employs exact solutions of the classical equations of motion. In Section 4 we shall be concerned with other semiclassical methods that employ various approximations to the exact classical trajectory.

## 2. Elements of the Theory

### 2.1. The JWKB Approximation

The JWKB approximation[17–18] illustrates the simplest and one of the most pervasive arguments in the semiclassical theory. It is derived as what is in effect an asymptotic[19] solution of the Schrödinger equation

$$\left[\frac{d^2}{dq^2}+\frac{p^2(q)}{\hbar^2}\right]\psi=0 \tag{4}$$

where $p(q)$ is the classical momentum function given by

$$p^2(q)=2m[E-V(q)] \tag{5}$$

The approximation is obtained from the expression

$$\psi(q)=\exp[iS(q)/\hbar] \tag{6}$$

by expanding $S(q)$ in powers of $\hbar$

$$S(q)=S_0(q)+\hbar S_1(q)+\hbar^2 S_2(q)+\cdots \tag{7}$$

The first two terms, obtained by equating the coefficients of $\hbar^{-2}$ and $\hbar^{-1}$ in turn in Eq. (4), yield[17]

$$S_0(q)=\pm\int_{q_0}^{q} p(q')\,dq' \tag{8}$$

with $q_0$ arbitrary, and

$$S_1(q)=\frac{i}{2}\ln(dS_0/dq)=\frac{i}{2}\ln p(q) \tag{9}$$

Hence

$$\psi(q)\simeq[p(q)]^{-1/2}\left\{A_+\exp\left[\frac{i}{\hbar}\int_{q_0}^{q}p(q')\,dq'\right]+A_-\exp\left[-\frac{i}{\hbar}\int_{q_0}^{q}p(q')\,dq'\right]\right\} \tag{10}$$

or if $p(q)$ is purely imaginary, as when $q$ lies in the classically inaccessible region,

$$\psi(q) \simeq |p(q)|^{-1/2} \left\{ B_+ \exp\left[ \frac{1}{\hbar} \int_{q_0}^q |p(q')| \, dq' \right] + B_- \exp\left[ -\frac{1}{\hbar} \int_{q_0}^q p(q') \, dq' \right] \right\} \tag{11}$$

Solutions of this type will be termed primitive semiclassical forms.

Equations (10) and (11) have the following physical interpretation. In each case the phase of the wave function depends on the classical action $\int p \, dq$, which becomes imaginary when $q$ is classically forbidden. The first and second terms in Eq. (10) correspond to motion in the positive and negative directions in $q$, whereas those in Eq. (11) are the increasing and decreasing exponential factors required to account for quantum mechanical tunneling. Finally, the preexponential factor ensures a constant current density

$$j = -\frac{i\hbar}{2m} \left( \psi^* \frac{d\psi}{dq} - \psi \frac{d\psi^*}{dq} \right) \tag{12}$$

The validity of this approximation depends on neglect of terms in $\hbar^{-1}$ when solving for $S_0(q)$, with the implication that

$$|\hbar(d^2 S_0/dq^2)| \ll |dS_0/dq|^2$$

or

$$\frac{d}{dq} \left[ \frac{\hbar}{p(q)} \right] \ll 1 \tag{13}$$

Breakdown is therefore expected at the classical turning points because $p(q) \to 0$. This failure of the approximation may be traced to the uncertainty principle. Consider for simplicity the case of a linear potential

$$V(q) = E - F(q - a) \tag{14}$$

giving rise to an uncertainty

$$\Delta p \simeq 2[2mF(q-a)]^{1/2} \tag{15}$$

associated with the ambiguity of sign in $p(q)$ derived from Eq. (5). Clearly $\Delta p$ vanishes at the precise point $q = a$, but Eq. (13) is readily seen to be consistent with Eq. (15) at a distance

$$\Delta q = q - a \tag{16}$$

from this point provided that

$$\Delta p \, \Delta q \gg \hbar \tag{17}$$

The conclusion is that a wave function constructed by exponentiation of the classical action may be expected to remain valid provided the associated classical uncertainties such as those given by Eqs. (15) and (16) are consistent with the uncertainty principle. Similar JWKB-like approximations are described below for the general multidimensional problem, but they again become invalid at the caustics of the motion, which are the generalizations of the classical turning points.

## 2.2. Uniform Approximations

Two types of uniform approximation have been developed to improve these JWKB or primitive semiclassical forms. Both involve a variable transformation to map the problem at hand onto one with a known solution and the same classical turning point structure. The first, which is mainly applicable in one dimension, is a "uniform" solution to the Schrödinger equation itself; the second is concerned with the uniform evaluation of integrals that arise from the solutions of multidimensional equations of motion.

### 2.2.1. Uniform Solutions of the Schrödinger Equation[6,10,12]

Given the equation

$$\left[\frac{d^2}{dq^2}+\frac{p^2(q)}{\hbar^2}\right]\psi=0 \tag{18}$$

one seeks a transformation $Q(q)$ in terms of which $\psi(q)$ may be mapped onto the known solution of a suitable comparison equation,

$$\left[\frac{d^2}{dQ^2}+\frac{p^2(Q)}{\hbar^2}\right]\Psi=0 \tag{19}$$

The forms of the primitive JWKB expressions

$$\psi(q)\sim[p(q)]^{-1/2}\exp\left[\pm\frac{i}{\hbar}\int_{q_0}^{q}p(q')\,dq'\right]$$

$$\Psi(Q)\sim[P(Q)]^{-1/2}\exp\left[\pm\frac{i}{\hbar}\int_{Q_0}^{Q}P(Q')\,dQ'\right] \tag{20}$$

suggest an association between the solutions of the form

$$\psi_{\mathrm{app}}(q)=[P(Q)/p(q)]^{1/2}\Psi[Q(q)] \tag{21}$$

with the transformation $Q(q)$ implicitly defined by

$$\int_{q_0}^{q}p(q')\,dq'=\int_{Q_0}^{Q}P(Q')\,dQ' \tag{22}$$

or the equivalent form,

$$(dQ/dq)=p(q)/P(Q) \tag{23}$$

$\psi_{app}(q)$ defined in this way again depends only on knowledge of the classical momentum function $p(q)$, but the functional form now depends on the properties of the standard function $\Psi(Q)$.

The validity of this approximation is readily assessed by combining Eqs. (19) and (21)–(23) to obtain

$$\left[\frac{d^2}{dq^2} + \frac{p^2(q)}{\hbar^2} + \varepsilon(q)\right]\psi_{app} = 0 \tag{24}$$

where

$$\varepsilon(q) = -\left(\frac{dQ}{dq}\right)^{1/2} \frac{d^2}{dq^2}\left(\frac{dQ}{dq}\right)^{-1/2} \tag{25}$$

Hence $\psi_{app}$ will be a good approximation to Eq. (18) if

$$|\hbar^2 \varepsilon(q)/p^2(q)| \ll 1 \tag{26}$$

With a judicious choice of comparison equation, this is a much weaker inequality than the JWKB criterion (13).

The most important constraint on the comparison function $P(Q)$ in Eq. (19) is that it should have the same structure of zeros and singularities as $p(q)$ in the equation to be solved. Otherwise $\varepsilon(q)$ must diverge due to disappearance or divergence of $(dQ/dq)$ given by Eq. (23). A valid mapping, which ensures that the function $Q(q)$ passes through the pairs of special points $(q_r, Q_r)$, $r = 1 \cdots n$, say, is obtained by setting the lower integration limits in Eq. (22) to coincide with each of the pairs $(q_r, Q_r)$ in turn. This means that the comparison function $P(Q)$ must contain $n-1$ variable parameters in order that the equations

$$\int_{q_r}^{q_s} p(q')\,dq' = \int_{Q_r}^{Q_s} P(Q')\,dQ' \tag{27}$$

be simultaneously satisfied.

The simplest case, with a single classical turning point $q_0$, may be used to illustrate the technique. The obvious comparison function is simply

$$P^2(Q) = \hbar^2 Q \tag{28}$$

so that Eq. (19) reduces to the Airy equation[20]

$$\left(\frac{d^2}{dQ^2} + Q\right)\Psi = 0 \tag{29}$$

with the general solution

$$\Psi(Q) = a\,\mathrm{Ai}(-Q) + b\,\mathrm{Bi}(-Q) \tag{30}$$

and the function $Q(q)$ is determined, according to Eq. (22), by

$$\frac{1}{\hbar} \int_{q_0}^{q} p(q') \, dq' = \int_{0}^{Q} Q'^{1/2} \, dQ' = \frac{2}{3} Q^{3/2} \tag{31}$$

or

$$Q(q) = \left[\frac{3}{2\hbar} \int_{q_0}^{q} p(q') \, dq'\right]^{2/3} \tag{32}$$

Finally, the branch of $p(q')$ is chosen so that

$$p(q') = e^{\pm(3\pi/2)i} |p(q')| \tag{33}$$

in the nonclassical $[p^2(q') < 0]$ region in order to ensure that $Q$ is everywhere real, with positive and negative values for classical and nonclassical values of $q'$, respectively.

The usual requirement is that the wave function decrease exponentially into the classically forbidden region, and this is achieved by setting $b = 0$ in Eq. (30). Hence, according to Eqs. (21) and (30),

$$\psi(q) \simeq a[\hbar^2 Q(q)/p^2(q)]^{1/4} \text{Ai}[-Q(q)] \tag{34}$$

The relation between this uniform approximation and the previous JWKB form merits close investigation in view of its importance for later discussion. The first point that may be emphasized by adopting the slightly cumbersome representation[20]

$$\text{Ai}(-Q) = \frac{1}{2\pi\hbar} \int_{-\infty}^{\infty} \exp\left(\frac{iP^3}{3\hbar^3} - i\frac{PQ}{\hbar}\right) dP \tag{35}$$

is that $\text{Ai}(-Q)$ may be regarded as a superposition of all possible momentum eigenstates, $\exp(-iPQ/\hbar)$, whereas in the classical situation there are just two possible values of momentum, $\pm P(Q)$, at any given $Q$. The semiclassical connection between these extremes is obtained by approximating the integral by the method of stationary phase or steepest descent.[19,21] This depends on the argument that rapid oscillations in the integrand will lead to almost complete cancellation except in regions where the exponent is stationary with respect to $P$. These points

$$P = \pm\hbar Q^{1/2} \tag{36}$$

are seen to coincide exactly with the classical momenta given by Eq. (28). For positive (classically accessible) values of $Q$ the approximation is completed by making a quadratic approximation to the exponent about the roots of Eq. (36) and displacing the integration contour to follow the directions of steepest descent shown in Fig. 1(a). When $Q$ is negative (classically inaccessible) the argument differs only to the extent that the roots of Eq. (36) are imaginary and that the path of steepest descent passes through only one of them, as shown in

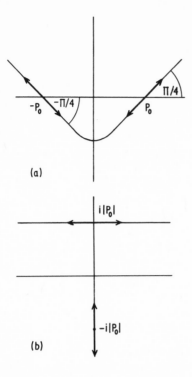

Fig. 1. Integration contours appropriate to stationary phase points in (a) the classical and (b) the nonclassical region. The arrows indicate the directions of steepest descent.

Fig. 1(b).[21] The validity of the resulting asymptotic forms,[20]

$$\text{Ai}(-Q) \sim \begin{cases} \pi^{-1/2}Q^{-1/4}\sin(\tfrac{2}{3}Q^{3/2}+\pi/4), & \text{as } Q \to \infty \\ \tfrac{1}{2}\pi^{-1/2}|Q|^{-1/4}\exp(-\tfrac{2}{3}|Q|^{3/2}), & \text{as } Q \to -\infty \end{cases} \tag{37}$$

increases with increasing $|Q|$, because the widths of the Gaussian approximations to the integrand employed depend inversely on $|Q|$. Hence these forms break down as $Q \to 0$, and the full Airy function is required to bridge this classical threshold region.

This general behavior is characteristic of many more complicated systems for which the full quantal solution may be expressed as a superposition over the continuum of classical solutions, whereas the number of classical trajectories is finite. This situation is most readily apparent in the Feynman path integral formulation of quantum mechanics.[22] For certain values of the relevant parameters (for example, the coordinate $Q$ in the above discussion), the quantal superposition may be reduced by the stationary phase approximation to a combination of primitive semiclassical forms, but this requires that phase differences between the classical paths be large. Situations involving a confluence of classical paths, which inevitably occur at the caustics of the motion, can be handled only by the appropriate uniform approximation. The Airy function is applicable to cases of two interfering paths, specified by variables in the range $(-\infty, \infty)$.

One immediate practical conclusion from Eqs. (34) and (37) is that the physically acceptable solution of Eq. (18) has one or the other of the following asymptotic approximations:

$$\psi(q) \sim \begin{cases} C[p(q)]^{-1/2} \sin\left[\dfrac{1}{\hbar}\displaystyle\int_{q_0}^{q} p(q')\,dq' + \pi/4\right], & \text{for } q \gg q_0 \\[2ex] \dfrac{1}{2}C|p(q)|^{-1/2} \exp\left[-\dfrac{1}{\hbar}\displaystyle\int_{q}^{q_0} |p(q')|\,dq'\right], & \text{for } q \ll q_0 \end{cases} \tag{38}$$

or

$$\psi(q) \sim \begin{cases} C'[p(q)]^{-1/2} \sin\left[\dfrac{1}{\hbar}\displaystyle\int_{q}^{q_0} p(q')\,dq' + \pi/4\right], & \text{for } q \ll q_0 \\[2ex] \dfrac{1}{2}C'|p(q)|^{-1/2} \exp\left[-\dfrac{1}{\hbar}\displaystyle\int_{q_0}^{q} |p(q')|\,dq'\right], & \text{for } q \gg q \end{cases} \tag{39}$$

according to whether $q > q_0$ or $q < q_0$ is classically accessible. These are of course both particular combinations of the primitive semiclassical expressions given by Eqs. (10) and (11).

The implication, in a bound state problem, with turning points $q_1$ and $q_2$, is that the same wave function must be described both by Eq. (38) and by Eq. (39), with $q_0$ replaced by $q_1$ and $q_2$, respectively, and with $C' = \pm C$. This requires that the sum of the two phases be equal to an integral multiple of $\pi$; in other words,

$$\frac{1}{\hbar}\int_{q_1}^{q_2} p(q')\,dq' = \left(n + \frac{1}{2}\right)\pi \tag{40}$$

This Bohr quantization condition[13] may also be derived from Eq. (27) by mapping the general problem onto that of a harmonic oscillator.[6]

Another important application of Eq. (38) concerns the elastic scattering phase shift derived from the radial equation

$$\left\{\frac{d^2}{dr^2} + \frac{2m}{\hbar^2}\left[E - V(R) - \frac{l(l+1)\hbar^2}{2mR^2}\right]\right\}\psi_l = 0 \tag{41}$$

A confluence between the turning point $a_l$ and the origin must first be removed by the Langer transformation[10,23]

$$R = e^q, \qquad \psi_l(R) = e^{q/2}\phi_l(q) \tag{42}$$

after which it is readily verified by use of Eq. (38) that

$$\psi_l(R) \sim C[p_l(R)]^{-1/2} \sin\left[\frac{1}{\hbar}\int_{a_l}^{R} p_l(R')\,dR' + \pi/4\right], \qquad \text{as } R \to \infty \tag{43}$$

where

$$p_l^2(R) = 2m[E - V(R) - (l+\tfrac{1}{2})^2\hbar^2/2mR^2] \tag{44}$$

Hence after comparison with the standard form

$$\psi_l(R) \sim A\ \sin(pR/\hbar - l\pi/2 + \eta_l), \qquad \text{as } R \to \infty \tag{45}$$

it follows in the uniform approximation that the phase shift is given by

$$\eta_l = \lim_{R \to \infty} \left\{ \frac{1}{\hbar} \left[ \int_{a_l}^{R} p_l(R')\ dR' - pR \right] + \left( l + \frac{1}{2} \right) \pi/2 \right\} \tag{46}$$

Finally, it may be noted that Eqs. (34), (38), and (39) have been extensively used[24] to give analytical approximations to the integrals generated by distorted wave and exponential[25] solutions to the close coupled equations and in other direct approaches to the scattering problem, of which the methods of Johnson[26] and Eu[27] are particularly relevant. Mention must also be made of applications of the associated equation technique to barrier penetration and curve crossing problems,[28–30] with particular reference to resonant scattering applications.[31]

### 2.2.2. Uniform Integral Evaluation

The physical significance of the following integral evaluation technique due to Chester et al.[7] will become apparent later. It involves the mapping of a general exponential integrand onto one for which the integral can be evaluated explicitly, subject to the requirement that the integration limits and the number of stationary phase points coincide in the two cases. Three examples will be given to illustrate the method.

The first two examples take the form

$$I(\nu) = \int_{-\infty}^{\infty} g(x)\ \exp[if(x, \nu)]\ dx \tag{47}$$

where $\nu$ is a parameter and $f(x, \nu)$ is assumed to have one and two stationary phase points, respectively. In the former case the obvious mapping is onto a complete Fresnel integral,[20] and this is acheived by the identity

$$f(x, \nu) = t^2 + \phi(\nu) \tag{48}$$

with $\phi(\nu)$ fixed by the requirement that the stationary points $x_0$ and $t_0$ on the two sides of Eq. (48) coincide:

$$x_0 \leftrightarrow t_0 = 0 \tag{49}$$

Hence

$$\phi(\nu) = f(x_0, \nu) \tag{50}$$

and

$$I(\nu) = \int_{-\infty}^{\infty} g(x)(dx/dt)\ \exp\{i[t^2 + \phi(\nu)]\}\ dt \tag{51}$$

Because Eq. (49) fixes only one point on the curve $x(t)$, the preexponent in Eq. (51) must be approximated by its value at that point, with $dx/dt$ determined by taking the second derivative of Eq. (48). Thus

$$I(\nu) \simeq g(x_0)(dx/dt)_0 e^{i\phi(\nu)} \int_{-\infty}^{\infty} e^{it^2} dt$$

$$= g(x_0)(2\pi/f_0'')^{1/2} \exp[if(x_0, \nu) + i\pi/4] \tag{52}$$

where $f''$ denotes the second derivative with respect to $x$. This is of course simply the stationary phase approximation.

In the case of two points of stationary phase at $x_a$, $x_b$ the mapping in Eq. (48) is replaced by[8,32]

$$f(x, \nu) = \tfrac{1}{3}t^3 - \zeta(\nu)t + \phi(\nu) \tag{53}$$

with $\zeta(\nu)$ and $\phi(\nu)$ determined by the coincidence constraints

$$\begin{aligned} x_a \leftrightarrow t_a &= \zeta^{1/2} \\ x_b \leftrightarrow t_b &= -\zeta^{1/2} \end{aligned} \tag{54}$$

so that

$$\begin{aligned} f_a &= \phi - \tfrac{2}{3}\zeta^{3/2} \\ f_b &= \phi + \tfrac{2}{3}\zeta^{3/2} \end{aligned} \tag{55}$$

Note that reference to the explicit dependence of $\zeta$ and $\phi$ on $\nu$ has been dropped for notational convenience. The analogue of Eq. (51) is therefore

$$I(\nu) = \int_{-\infty}^{\infty} g(x)(dx/dt) \exp[i(t^3/3 - \zeta t + \phi)] dt \tag{56}$$

and, with two known points on the curve $x(t)$, a linear expansion for the preexponent is permissible,

$$g(x)(dx/dt) = \alpha + \beta t \tag{57}$$

It follows that

$$I(\nu) \simeq \int_{-\infty}^{\infty} (\alpha + \beta t) \exp\left[i\phi(\nu) + \frac{it^3}{3} - i\zeta(\nu)t\right] dt$$

$$\sim 2\pi e^{i\phi(\nu)} \{\alpha \operatorname{Ai}[-\zeta(\nu)] + i\beta \operatorname{Ai}'[-\zeta(\nu)]\}$$

$$= \pi^{1/2} e^{i\phi(\nu)} \{(P_a^{1/2} + P_b^{1/2})\zeta^{1/4}(\nu)\operatorname{Ai}[-\zeta(\nu)]$$

$$- i(P_a^{1/2} - P_b^{1/2})\zeta^{-1/4}(\nu)\operatorname{Ai}'[-\zeta(\nu)]\} \tag{58}$$

where

$$P_i = 2\pi g_i^2/|f_i''| \tag{59}$$

The modulus is required in Eq. (59) because, according to Eqs. (53) and (54), $f(x, \nu)$ is stationary at $x_a$ and $x_b$, with $f_a < f_b$; hence $f_a'' > 0$ and $f_b'' < 0$. Equation (58) is the desired uniform approximation.

It is evident from Eq. (37) and their derivatives with respect to $-Q$ that

$$I(\nu) \sim P_a^{1/2} e^{if_a + i\pi/4} + P_b^{1/2} e^{if_b - i\pi/4}, \qquad \text{for } \zeta(\nu) \gg 1 \tag{60}$$

In other words, for *classically accessible* values of the parameter $\nu$ for which $\zeta(\nu) \gg 1$, Eq. (58) reduces to a combination of stationary phase components. Equally for *classically forbidden* values of $\nu$ such that $\zeta(\nu) \ll -1$, it follows on choosing the branch $\zeta = |\zeta| e^{i\pi}$ that

$$I(\nu) \sim P_a^{1/2} \exp\left[i\left(\phi + \frac{2i}{3}|\zeta|^{3/2}\right)\right], \qquad \text{for } \zeta(\nu) \ll -1$$

$$\sim P_a^{1/2} \exp(if_a) \tag{61}$$

whereas the choice $\zeta = |\zeta| e^{-i\pi}$ simply substitutes $b$ for $a$ in Eq. (61). In either case the stationary phase value of $t$, given Eq. (54), is purely imaginary and hence corresponds to a complex value of the physical variable $x$. If the function $f(x, \nu)$ is real, each such point must be accompanied by its complex conjugate; hence the imaginary part of $f(x, \nu)$ will be positive at one point and negative at the other. The stationary phase evaluation of the Airy function described in the previous section shows that the surviving term in Eq. (61) always corresponds to the former. Thus the resulting contribution to the integral is always exponentially small.

The final case is of the form[9,33]

$$I(\nu) = \int_0^{2\pi} g(x) \exp[if(x, \nu)] \, dx \tag{62}$$

with $f(x, \nu)$ as the combination of a term periodic in $x$ and an integral multiple of $x$. The transformation $x(t)$ is then defined by

$$f(\nu, x) = \phi(\nu) - \zeta(\nu) \cos t - mt \tag{63}$$

with stationary points given by

$$\begin{aligned} x_a \leftrightarrow t_a &= \pi/2 - \arccos(m/\zeta) \\ x_b \leftrightarrow t_b &= \pi/2 + \arccos(m/\zeta) \end{aligned} \tag{64}$$

so that $\phi(\nu)$ and $\zeta(\nu)$ in Eq. (63) are given by

$$\begin{aligned} f_a &= \phi - m\pi/2 - (\zeta^2 - m^2)^{1/2} + m \arccos(m/\zeta) \\ f_b &= \phi - m\pi/2 + (\zeta^2 - m^2)^{1/2} - m \arccos(m/\zeta) \end{aligned} \tag{65}$$

The approximation

$$g(x)(dx/dt) = \alpha \cos x + \beta \sin x \tag{66}$$

is readily shown in this case to yield[32]

$$I = (\pi/2)^{1/2} e^{i\phi} [(P_a^{1/2} + P_b^{1/2})(\zeta^2 - m^2)^{1/4} J_m(\zeta)$$
$$+ i(P_a^{1/2} - P_b^{1/2})\zeta(\zeta^2 - m^2)^{-1/4} J_m'(\zeta)] \quad (67)$$

where $J_m(\zeta)$ and $J_m'(\zeta)$ denote the $m$th-order Bessel function and its first derivative, and $P_a$ and $P_b$ are again defined by Eq. (59).

Other uniform approximations for more complicated systems have been collected by Connor.[11]

## 2.3. Canonical Transformations

The discussion above has been exclusively in the Cartesian $(q, p)$ system, but the theory that follows cannot be fully appreciated without some knowledge of the relation between classical canonical transformations to other conjugate systems $(q, p)$ and the unitary transformations of quantum mechanics.[1,34,35]

In classical mechanics a transformation from one set of coordinates and momenta $(q, p)$ to another set $(\tilde{q}, \tilde{p})$ is termed canonical if the value of the classical Hamiltonian is preserved,

$$H(q, p) = \tilde{H}(\tilde{q}, \tilde{p}) \quad (68)$$

The systematic theory is based on use of one or the other of four possible generating functions, $F_1(q, \tilde{q})$, $F_2(q, \tilde{p})$, $F_3(p, \tilde{q})$, and $F_4(p, \tilde{p})$, each dependent on one old and one new variable in such a way that the remaining variables may be generated by the partial differential relations

$$p = (\partial F_1/\partial q), \qquad \tilde{p} = -(\partial F_1/\partial \tilde{q}) \quad (69a)$$

$$p = (\partial F_2/\partial q), \qquad \tilde{q} = (\partial F_2/\partial \tilde{p}) \quad (69b)$$

$$q = -(\partial F_3/\partial p), \qquad \tilde{p} = -(\partial F_3/\partial \tilde{q}) \quad (69c)$$

$$q = (\partial F_4/\partial p), \qquad \tilde{q} = (\partial F_4/\partial \tilde{p}) \quad (69d)$$

These generators are subject to the interrelations

$$F_2(q, \tilde{p}) = F_1(q, \tilde{q}) + \tilde{p}\tilde{q} \quad (70a)$$

$$F_3(p, \tilde{q}) = F_1(q, \tilde{q}) - pq \quad (70b)$$

$$F_4(p, \tilde{q}) = F_1(q, \tilde{q}) + \tilde{p}\tilde{q} - pq \quad (70c)$$

The origin and significance of Eqs. (69) and (70) are discussed in detail in standard texts,[36,37] but the following example may serve to illustrate their use. This refers to transformation of the Hamiltonian

$$H(q, p) = \frac{1}{2m} p^2 + A e^{-\alpha q} \quad (71)$$

to a new system with the coordinate

$$\tilde{q} = e^{-\alpha q} \tag{72}$$

A suitable generator of type $F_3(p, \tilde{q})$ may be obtained by setting $F_1(q, \tilde{q})$ to zero in Eq. (70b) and using Eq. (72) to obtain

$$F_3(p, \tilde{q}) = (p/\alpha) \ln \tilde{q} \tag{73}$$

This ensures that the first equation of Eqs. (69c) will recover the correct coordinate $q$. The second equation yields

$$\tilde{p} = -(p/\alpha \tilde{q}) \tag{74}$$

Hence on combining Eqs. (71), (72), and (74) the transformed Hamiltonian may be written

$$\tilde{H}(\tilde{q}, \tilde{p}) = \alpha^2 \tilde{q}^2 p^2 / 2m + A\tilde{q} \tag{75}$$

This completes the present purely classical discussion.

The corresponding transformation in quantum mechanics may be expressed in the form

$$\psi(q) = \int U_1(q, \tilde{q}) \tilde{\psi}(\tilde{q}) \, d\tilde{q} \tag{76}$$

where the transform $U_1(q, \tilde{q})$ must be defined in such a way that $\psi(q)$ and $\tilde{\psi}(\tilde{q})$ are eigenfunctions of the quantal Hamiltonians $H(q, -i\hbar\partial/\partial q)$ and $\tilde{H}(\tilde{q}, i\hbar\partial/\partial\tilde{q})$, belonging to the same eigenvalue $E$, for any allowed value of $E$. The implication is that the analogue of Eq. (68) may be written

$$H(q, -i\hbar\partial/\partial q)\psi(q) = \int U_1(q, \tilde{q})\tilde{H}(\tilde{q}, -i\hbar\partial/\partial\tilde{q})\tilde{\psi}(\tilde{q}) \, d\tilde{q} \tag{77}$$

This is readily satisfied, in the semiclassical limit, by use of the generator

$$U_1(q, q) = A_1(q, \tilde{q}) \exp[iF_1(q, \tilde{q})/\hbar] \tag{78}$$

where $F_1(q, \tilde{q})$ is the corresponding classical function, because according to Eq. (76),

$$H(q, -i\hbar\partial/\partial q)\psi(q) = \int H(q, -i\hbar\partial/\partial q)U_1(q, \tilde{q})\psi(\tilde{q}) \, d\tilde{q} \tag{79}$$

Now it follows from Eqs. (68), (69a), and (77) that if only terms of the highest power in $\hbar$ are retained,

$$H(q, -i\hbar\partial/\partial q)U_1(q, \tilde{q}) = H(q, p)U_1(q, \tilde{q})$$

$$= \tilde{H}(\tilde{q}, \tilde{p})U_1(q, \tilde{q}) \tag{80}$$

$$= \tilde{H}(\tilde{q}, i\hbar\partial/\partial\tilde{q})U_1(q, \tilde{q})$$

The final step involves repeated partial integration on the right-hand side of the equation to give

$$H(q, -i\hbar\partial/\partial q)\psi(q) = \int \psi(\tilde{q})\tilde{H}(\tilde{q}, i\hbar\partial/\partial\tilde{q}) U_1(q, \tilde{q}) \, d\tilde{q}$$

$$= \int U_1(q, \tilde{q})\tilde{H}(\tilde{q}, -i\hbar\partial/\partial\tilde{q})\psi(\tilde{q}) \, d\tilde{q} \tag{81}$$

in agreement with Eq. (77), provided that the integrand takes the same value at the two integration limits. A second implicit assumption is that the quantal and classical Hamiltonian functions are identical, which is in general valid only in a semiclassical sense.

The determination of the preexponent $A_1(q, \tilde{q})$ in Eq. (77) also raises points of general interest. The most convenient requirement is that $U_1(q, \tilde{q})$ should be unitary,

$$\int_{-\infty}^{\infty} U_1(q, \tilde{q})U_1^*(q, \tilde{q}') \, dq = \delta(\tilde{q} - \tilde{q}') \tag{82}$$

In other words,

$$\int_{-\infty}^{\infty} A_1(q, \tilde{q})A_1^*(q, \tilde{q}) \exp\left\{\frac{i}{\hbar}\left[F_1(q, \tilde{q}) - F_1(q, \tilde{q}')\right]\right\} dq = \delta(q - q') \tag{83}$$

Now the presence of the factor $\hbar^{-1}$ means that the exponent will be typically large and rapidly varying with $q$; hence the resulting oscillations may be expected to lead to almost complete cancellation unless $\tilde{q} \approx \tilde{q}'$. It is therefore permissible to write

$$F_1(q, \tilde{q}) - F_1(q, \tilde{q}') \approx (\partial F_1/\partial\tilde{q})(\tilde{q} - \tilde{q}') \tag{84}$$

and to introduce the new integration variable

$$\tilde{p} = (\partial F_1/\partial\tilde{q}) \tag{85}$$

provided $\tilde{p}$ varies monotically with $\tilde{q}$. Equation (84) then becomes

$$\int_{-\infty}^{\infty} |A_1(q, \tilde{q})|^2(dq/d\tilde{p}) \exp\left[\frac{i}{\hbar}\tilde{p}(\tilde{q} - \tilde{q}')\right] d\tilde{p} = \delta(\tilde{q} - \tilde{q}') \tag{86}$$

It follows by comparison with the standard form,[1]

$$\int_{-\infty}^{\infty} \exp(ixy) \, dx = \delta(y) \tag{87}$$

that

$$|A_1(q, q)|^2 = \frac{1}{2\pi\hbar}\left(\frac{d\tilde{p}}{dq}\right) = \frac{1}{2\pi\hbar}\left(\frac{\partial^2 F_1}{\partial q \, \partial\tilde{q}}\right) \tag{88}$$

With the conventional choice of phase[34] this implies

$$A_1(q, \tilde{q}) = \left[\frac{1}{2\pi i\hbar}\frac{\partial^2 F_1}{\partial q\, \partial\tilde{q}}\right]^{1/2} = \left[\frac{1}{2\pi i\hbar}\left(\frac{\partial p}{\partial\tilde{q}}\right)\right]^{1/2} = \left[\frac{i}{2\pi\hbar}\left(\frac{\partial\tilde{p}}{\partial q}\right)\right]^{1/2} \qquad (89)$$

where the final form has been derived from Eqs. (69a). However, the normalization is more complicated if $\tilde{p}(q, \tilde{q})$ is not a single-valued function of $q$ at given $\tilde{q}$.

Similar arguments may be used to obtain a unitary transform $U_2(q, \tilde{p})$ in terms of the corresponding classical generator of type $F_2(q, p)$, because $U_2(q, \tilde{p})$ is simply the Fourier transform of $U_1(q, \tilde{q})$,

$$U_2(q, \tilde{p}) = (2\pi i\hbar)^{-1/2}\int_{-\infty}^{\infty} U_1(q, \tilde{q})\exp\left(\frac{i}{\hbar}\tilde{p}\tilde{q}\right)d\tilde{q}$$
$$= (2\pi i\hbar)^{-1/2}\int_{-\infty}^{\infty} A_1(q, \tilde{q})\exp\left\{\frac{i}{\hbar}[F_1(q, \tilde{q}) + \tilde{p}\tilde{q}]\right\}dq \qquad (90)$$

after use of Eq. (78). Again, for any particular values of $q$ and $\tilde{p}$, the integrand will show rapid oscillations unless $\tilde{q}$ lies close to a stationary phase point given by

$$\partial F_1/\partial\tilde{q} = -\tilde{p} \qquad (91)$$

The existence of such a point, which will be denoted $\tilde{q}(q, \tilde{p})$, is guaranteed by the second equation of Eqs. (69a). It follows, on performing the usual quadratic approximation about $\tilde{q}(q, \tilde{p})$, that

$$U_2(q, \tilde{p}) = A_2(q, \tilde{p})\exp[iF_2(q, \tilde{p})/\hbar] \qquad (92)$$

where, in agreement with Eq. (70),

$$F_2(q, \tilde{p}) = F_1[q, \tilde{q}(q, \tilde{p})] + \tilde{p}\tilde{q}(q, \tilde{p}) \qquad (93)$$

and

$$A_2(q, \tilde{p}) = \left[\frac{1}{2\pi i\hbar}\frac{(\partial^2 F_1/\partial q\, \partial\tilde{q})}{(\partial^2 F_1/\partial\tilde{q}^2)}\right]^{1/2} = \left[\frac{1}{2\pi i\hbar}\left(\frac{\partial\tilde{q}}{\partial q}\right)_{\tilde{p}}\right]^{1/2}$$
$$= \left(\frac{1}{2\pi i\hbar}\frac{\partial^2 F_2}{\partial q\, \partial\tilde{p}}\right)^{1/2} \qquad (94)$$

where the final identity follows from Eqs. (69b).

The general semiclassical result is that a canonical transformation generated by $F_\nu(x, \tilde{y})$, where $x$ and $y$ denote $p$ or $q$, has the quantum mechanical analogue

$$U_\nu(x, \tilde{y}) = \left(\frac{1}{2\pi i\hbar}\frac{\partial^2 F_\nu}{\partial x\, \partial\tilde{y}}\right)^{1/2}\exp\left[\frac{i}{\hbar}F_\nu(x, \tilde{y})\right] \qquad (95)$$

and this may be generalized to $n$ dimensions in the form[34,35]

$$U_\nu(\mathbf{x}, \tilde{\mathbf{y}}) = \left[ \left( \frac{1}{2\pi i\hbar} \right)^n \left| \frac{\partial^2 F_\nu}{\partial \mathbf{x}\, \partial \tilde{\mathbf{y}}} \right| \right]^{1/2} \exp\left[ \frac{i}{\hbar} F_\nu(\mathbf{x}, \tilde{\mathbf{y}}) \right] \tag{96}$$

where $|\partial^2 F_\nu/\partial\mathbf{x}\,\partial\tilde{\mathbf{y}}|$ denotes the Van Vleck determinant[34] of the matrix of second partial derivatives. Equations (69) show that $|\partial^2 F_\nu/\partial\mathbf{x}\,\partial\tilde{\mathbf{y}}|$ may be regarded as the Jacobian of the transformation from $\tilde{\mathbf{x}}$ to $\mathbf{x}$ or from $\mathbf{y}$ to $\tilde{\mathbf{y}}$.

Miller[16,38,39,50,52,55] and co-workers make extensive use of the above transformations in deriving primitive semiclassical transition probabilities from the classical limit of the Feynman propagator[22]

$$K_{\text{cl}}(\mathbf{q}_2, \mathbf{q}_1) = \left[ \frac{1}{(2\pi i\hbar)^n} \left| \frac{\partial^2 \phi}{\partial \mathbf{q}_1\, \partial \mathbf{q}_2} \right| \right]^{1/2} \exp\left[ \frac{i}{\hbar} \phi(\mathbf{q}_2, \mathbf{q}_1) \right] \tag{97}$$

where $\phi(\mathbf{q}_2, \mathbf{q}_1)$ is the classical action evaluated along a trajectory from $\mathbf{q}_1$ to $\mathbf{q}_2$,

$$\phi(\mathbf{q}_2, \mathbf{q}_1) = \int_{\mathbf{q}_1}^{\mathbf{q}_2} \mathbf{p} \cdot d\mathbf{q} \tag{98}$$

and $K_{\text{cl}}(\mathbf{q}_2, \mathbf{q}_1)$ is the probability amplitude associated with that event. Because the typical measured quantity is a momentum or quantum number change, this must be transformed to an $S$-matrix element in the appropriate momentum representation

$$S(\mathbf{p}_2, \mathbf{p}_1) = \left( \frac{1}{2\pi h} \right)^n \int\int K_{\text{cl}}(\mathbf{q}_2, \mathbf{q}_1) \exp\left[ \frac{i}{\hbar}(\mathbf{p}_1 \cdot \mathbf{q}_1 - \mathbf{p}_2 \cdot \mathbf{q}_2) \right] d\mathbf{q}_1\, d\mathbf{q}_2$$

$$= \left[ \frac{1}{(2\pi i\hbar)^n} \frac{\partial^2 \psi}{\partial \mathbf{p}_1\, \partial \mathbf{p}_2} \right]^{1/2} \exp\left[ \frac{i}{\hbar} \psi(\mathbf{p}_2, \mathbf{p}_1) \right] \tag{99}$$

where

$$\psi(\mathbf{p}_2, \mathbf{p}_1) = -\int_{\mathbf{p}_1}^{\mathbf{p}_2} \mathbf{q} \cdot d\mathbf{p} \tag{100}$$

The same results are obtained in Section 3.2 by a method that allows extension to uniform as well as primitive semiclassical approximations, and space does not allow full justice to be done to the classical propagator formulation of the theory. The reader is strongly urged to make reference to an existing review[16] and to the original papers.[38,39]

## 2.4. Energy–Time and Action–Angle Representations

Two canonical systems of particular importance for the later discussion are those in which the new coordinates and momenta $(\tilde{q}, \tilde{p})$ are time and energy $(t, E)$ and angle and action $(w, N)$, respectively. The transformations to both these systems may be obtained by a generator of type $F_2(q, \tilde{p})$ derived from the Hamilton–Jacobi equation

$$H[q, (\partial F_2/\partial q)] = E \tag{101}$$

which automatically ensures that

$$(\partial F_2/\partial q) = p \tag{102}$$

as required by Eqs. (69b).

The simpler case is that in which the new momentum is the energy, and the conjugate coordinate $(\partial F_2/\partial E)$ is readily identified with the time, because, according to Eq. (101),

$$\left(\frac{\partial H}{\partial E}\right)_q = \left(\frac{\partial H}{\partial p}\right)_q \left(\frac{\partial^2 F_2}{\partial q\, \partial E}\right) = 1 \tag{103}$$

whereas by Hamilton's equations

$$\left(\frac{\partial H}{\partial p}\right)_q = \frac{dq}{dt} \tag{104}$$

The transformed Hamiltonian is therefore simply

$$\tilde{H}(t, E) = E \tag{105}$$

with a quantum mechanical analogue

$$\hat{H} = -i\hbar\, \partial/\partial t \tag{106}$$

which has eigenfunctions

$$\tilde{\psi}(t) = \exp(iEt/\hbar) \tag{107}$$

Finally it may be noted that the typical Cartesian Hamiltonian

$$H(q, p) = \frac{1}{2m} p^2 + V(q) \tag{108}$$

implies via Eq. (101) a classical generator from $(q, p)$ to $(t, E)$ of the form

$$F_2(q, E) = \int_{q_0}^{q} p(q, E)\, dq \tag{109}$$

where

$$p(q, E) = \{2m[E - V(q)]\}^{1/2} \tag{110}$$

The corresponding unitary transform

$$\begin{aligned} U_2(q, E) &= \left(\frac{1}{2\pi i\hbar} \frac{\partial^2 F}{\partial q\, \partial E}\right)^{1/2} \exp\left[\frac{i}{\hbar} F_2(q, E)\right] \\ &= \left[\frac{m}{2\pi i\hbar p(q, E)}\right]^{1/2} \exp\left[\frac{i}{\hbar} \int_{q_0}^{q} p(q, E)\, dq\right] \end{aligned} \tag{111}$$

is seen by comparison with Eq. (10) to be the JWKB wave function normalized to a delta function of energy. In other words, this wave function may be regarded as the generator of a transformation to a representation in which the new momentum, $E$, is diagonal. Furthermore, the standard semiclassical normalization, as given by Eqs. (96) and (97), is seen to ensure conservation of flux, and this may be confirmed to apply also to non-Cartesian systems.[1]

Transformation to the action–angle representation, which applies to periodic or quasi-periodic[2,3,36,37] systems, is more complicated, particularly in the multidimensional case.[40–43] The present theory is therefore restricted to one degree of freedom. The special feature of this angle–action representation is that the Hamiltonian depends only on the action $N$, defined as

$$N = \frac{1}{2\pi} \oint p\, dq \tag{112}$$

Note that $N$ is used in place of the conventional symbol $J^{(36)}$ in order to emphasize the connection with the quantum number; similarly, the factor $(2\pi)^{-1}$ is introduced so that the conjugate angle $w$ is periodic over $(0, 2\pi)$, rather than $(0, 1)$.

The classical equations of motion in the new representation therefore become

$$\tilde{H}(N) = E \tag{113}$$

$$\dot{N} = -(\partial\tilde{H}/\partial w) = 0$$
$$\dot{w} = (\partial\tilde{H}/\partial N) = \omega \tag{114}$$

so that

$$N = \text{constant}$$
$$w = \omega t + \text{constant} \tag{115}$$

Finally, the connection between the $w$ and the Cartesian coordinate $q$ is given by

$$w = (\partial F_2/\partial N) \tag{116}$$

where the generator $F_2(q, N)$ is determined by combining Eqs. (101) and (113) in the form

$$H[q, (\partial F_2/\partial q)] = \tilde{H}(N) \tag{117}$$

The harmonic oscillator offers a simple example of this procedure. In this case

$$H(q, p) = \frac{1}{2m}p^2 + \frac{1}{2}kq^2 = E \tag{118}$$

so that, according to Eq. (112),

$$N = \frac{1}{\pi} \int_{-q_0}^{q_0} [2m(E - \tfrac{1}{2}kq^2)]^{1/2} \, dq = E/\omega \qquad (119)$$

$$q_0 = (2E/k)^{1/2}, \qquad \omega = (k/m)^{1/2} \qquad (120)$$

In other words,

$$\tilde{H}(N) = E = N\omega \qquad (121)$$

This means that Eq. (117) takes the form

$$\frac{1}{2m}\left(\frac{\partial F_2}{\partial q}\right)^2 + \frac{1}{2}kq^2 = N\omega \qquad (122)$$

from which

$$w = \left(\frac{\partial F_2}{\partial N}\right)_q = \frac{\partial}{\partial N} \int_0^q \left[2m\left(N\omega - \frac{1}{2}kq^2\right)\right]^{1/2} dq$$

$$= \sin^{-1}(q/q_0) \qquad (123)$$

with $q_0$ given in terms of $N$ by Eqs. (120) and (121). This completes the classical picture.

The main significance of the angle–action representation in quantum mechanics is again that the Hamiltonian is independent of the angle, so that the action is again a constant of the motion. There is, however, a complication in defining the action operator $\hat{N}$, because the most general representation consistent with the commutation relation

$$[\hat{N}, w] = -i\hbar \qquad (124)$$

contains an arbitary constant $\delta$,

$$\hat{N} = -i\hbar \frac{\partial}{\partial w} + \delta\hbar \qquad (125)$$

The corresponding eigenfunction

$$\tilde{\psi}_n(w) = e^{inw} \qquad (126)$$

which is single valued for integral values of $n$, is, however, independent of $\delta$; only the eigenvalues

$$N = (n + \delta)\hbar \qquad (127)$$

depend on it. This means, because $\hat{N}$ commutes with $\tilde{H}(\hat{N})$, that the eigenfunctions of $\tilde{H}(\hat{N})$ are known exactly and are extremely simple in form but that the energy levels

$$E_n = \tilde{H}[(n + \delta)\hbar] \qquad (128)$$

depend on knowledge of $\delta$. Fortunately in the scattering context an accurate description of the wave function is of more importance than that of the energy levels.

Exact values for $\delta$ can be obtained by reference to an exact solution of the problem in some other representation, and the values $\delta = \frac{1}{2}$, 0, and $\frac{1}{2}$ for the harmonic oscillator,[44] plane rotor,[45] and orbital motion,[45] respectively, have been obtained in this way. Comparison between the Bohr quantization condition (40) and Eq. (127) shows that in the semiclassical limit $\delta$ may be given the value $\frac{1}{2}$ for any bound motion in one Cartesian degree of freedom.

As a final example of the canonical transformation theory, the asymptotic description of a typical scattering problem will be investigated. It is assumed initially that the internal motion is treated in the angle–action $(w, N)$ system and that Cartesian coordinates $(R, P)$ are employed for the translational motion. In the absence of any interaction potential the classical Hamiltonian depends only on $P$ and $N$,

$$H = \frac{1}{2m}P^2 + H_0(N) = E \tag{129}$$

and the corresponding wave function may be written

$$\Psi(w, R) = v^{-1/2} \exp[inw + iPR/\hbar] \tag{130}$$

where $v$ is the translational velocity $(P/m)$. The intention is to transform this description into one in which $(R, P)$ are replaced by time and total energy $(t, E)$. The appropriate generator $F_2(R, w; E, N)$ may be obtained from the Hamilton–Jacobi equation

$$\frac{1}{2m}\left(\frac{\partial F_2}{\partial R}\right)^2 + H_0\left[\left(\frac{\partial F_2}{\partial w}\right)\right] = E \tag{131}$$

in the form

$$F_2(R, w; E, N) = Nw + P(E, N)R \tag{132}$$

where

$$P(E, N) = \{2m[E - E_0(N)]\}^{1/2} \tag{133}$$

Here the internal energy $E_0(N)$ has been treated as a separation constant in solving Eq. (131). It follows that the new coordinates conjugate to $E$ and $N$ are given by

$$\begin{aligned} t &= \partial F_2/\partial E = mR/P(E, N) \\ \bar{w} &= \partial F_2/\partial N = w - (\partial E_0/\partial N)t = w - \omega t \end{aligned} \tag{134}$$

Note that the introduction of $E$ as one of the momenta has led to a change from $w$ to $\bar{w}$ as the coordinates conjugate to $N$.

A full transformation of the wave function $\Psi(w, R)$ from the $(w, R)$ to the $(\bar{w}, t)$ representation is complicated by the impossibility of eliminating $N$ from an $F_1$ type of generator derived from Eqs. (70) and (132), but clearly if $N$ is to remain a constant of the motion, the transformed wave function must contain a factor $\exp(in\bar{w})$ analogous to the internal part of $\Psi(w, R)$. The more interesting translational part of $\Psi(w, R)$ is, however, amenable to the general method, because a valid partial $F_1$ generator is available in the form

$$F_1(R, t) = F_2(R, E) - Et = \frac{mR^2}{2t} - E_0(N)t \tag{135}$$

after neglecting the term $Nw$ in Eq. (132). This yields the unitary transform

$$U_1(R, t) = \left(\frac{1}{2\pi i\hbar}\frac{\partial^2 F_1}{\partial R\,\partial t}\right)^{1/2} \exp[iF_1(R, t)/\hbar]$$

$$= \left(\frac{mR}{2\pi i\hbar t}\right)^{1/2} \exp\left[\frac{imR^2}{2\hbar t} - \frac{iE_0(N)t}{\hbar}\right] \tag{136}$$

Hence on performing the integration by stationary phase,

$$\tilde{\psi}(t) = \int_{-\infty}^{\infty} U_1^*(R, t)v^{-1/2}\exp\left(\frac{i}{\hbar}PR\right)dR$$

$$= e^{iEt/\hbar} \tag{137}$$

The complete transformation therefore yields

$$\tilde{\Psi}(\bar{w}, t) = \exp(in\bar{w} + iEt/\hbar) \tag{138}$$

This has the advantage over $\Psi(w, R)$ given by Eq. (130) that $E$ remains a constant of the motion even in the scattering region.

## 3. The Semiclassical S Matrix

The title "semiclassical $S$ matrix" or "classical $S$ matrix" as used by Miller[38,39] is reserved in this chapter for quantal solutions based on exact classical trajectories. Cases of elastic and inelastic scattering are discussed in detail below. Other *classical path approximations* based on average trajectories of various types are discussed in Section 4.

### 3.1. Elastic Scattering in a Central Field

The problem of elastic scattering by a central potential, $V(R)$, was one of the first to be tackled by semiclassical methods.[14,15] The standard analysis[8,9,14,15] starts from the exact quantal scattering amplitude

$$f(\theta) = \frac{\hbar}{2ip}\sum_{l=0}^{\infty}(2l+1)(\exp 2i\eta_l - 1)P_l(\cos\theta) \tag{139}$$

but it is more convenient for comparison with the previous discussion to work in terms of the scattered part of the wave function,

$$\Psi^+(R, \theta) \sim R^{-1} f(\theta) \exp(ipR/\hbar), \qquad \text{as } r \to \infty \qquad (140)$$

The semiclassical reduction of Eq. (139) or (140) relies on introduction of the JWKB phase shift $\eta_l$ given by Eq. (46) and replacement of the sum in Eq. (139) by an integral over the classical angular momentum

$$L = (l + \tfrac{1}{2})\hbar \qquad (141)$$

on the grounds that the number of terms in Eq. (139) is large. Alternatively, and more rigorously, the Poisson sum formula may be used to convert Eq. (139) into a combination of integrals.[8,9] This means, because

$$\sum_{l=0}^{\infty} (2l+1)P_l(\cos \theta) = 0 \qquad \text{for } \theta \neq 0 \qquad (142)$$

that for nonforward scattering

$$\Psi^+(R, \theta) \sim \frac{1}{i\hbar pR} \int_0^{\infty} L \exp\left[\frac{i}{\hbar} F(R, L)\right] P_l(\cos \theta) \, dL, \qquad \text{as } R \to \infty \qquad (143)$$

where

$$F(R, L) = \int_{R_1}^{a_L} -p_L(R') \, dR' + \int_{a_L}^{R} p_L(R') \, dR' - pR_1 + L\pi \qquad (144)$$

$$p_L(R) = \{2m[E - V(R) - L^2/2mR^2]\}^{1/2} \qquad (145)$$

The signs in Eq. (144) have been chosen to emphasize the relation to the classical trajectory, with negative and positive radial momentum on its incoming and outgoing parts, respectively; hence $F(R, L)$ is the radial part of the classical action evaluated along this trajectory from some distant point $R_1$. Seen in another way, $F(R, L)$ may also be interpreted by the arguments of Section 2.3 as an $F_2$-type generator for the transformation from the $(R, p)$ system to a new system with momentum $L$ and conjugate angle variable

$$\Theta(R, L) = (\partial F/\partial L) = \pi - 2 \int_{a_L}^{R} \frac{L \, dr}{R^2 p_L(R)} \qquad (146)$$

which is, of course, the classical deflection function. This is of some consequence for the theory because, as will soon become apparent, Eq. (143) is in effect a semiclassical transformation from the $L$ to the $\theta$ representation.

To see this, it is necessary to express Eq. (143) in terms of the unitary transform [see Eq. (95)]

$$U(R, L) = \left[\frac{1}{2\pi i\hbar} \frac{\partial^2 F}{\partial R \, \partial L}\right]^{1/2} \exp\left[\frac{i}{\hbar} F(R, L)\right]$$

$$\sim \left(\frac{L}{2\pi i\hbar R^2 p}\right)^{1/2} \exp\left[\frac{i}{\hbar} F(R, L)\right], \qquad \text{as } R \to \infty \qquad (147)$$

and to replace $P_l(\cos \theta)$ by its asymptotic form,

$$P_l(\cos \theta) = \left(\frac{2\hbar}{\pi L \sin \theta}\right)^{1/2} \cos\left(\frac{L\theta}{\hbar} - \frac{\pi}{4}\right) \tag{148}$$

valid for $L \sin \theta \gg \hbar$. Hence

$$\Psi^+(R, \theta) \sim \frac{1}{(p \sin \theta)^{1/2}}$$

$$\times \int_{-\infty}^{\infty} \left(\frac{L}{2\pi i \hbar R^2 p}\right)^{1/2} \exp\left\{\frac{i}{\hbar}[F(R, L) + L\theta]\right\} dL, \qquad \text{as } R \to \infty \tag{149}$$

where the two components of the cosine have been separated in order to extend the integration range to $(-\infty, \infty)$, and the terms $\pm i\pi/4$ implied by Eq. (148) have been absorbed into the square root to accommodate negative values of $L$. This is the expected Fourier transform of $U(R, L)$ defined by Eq. (147). It shows, as anticipated in Section 2.2.1, that a fully quantum mechanical description requires a superposition of all possible angular momentum states, each of which corresponds in a classical description to a different scattering trajectory.

This full integral in Eq. (149) is, however, readily reduced to a combination of classical terms by the methods of Section 2.2.2. For example, the stationary phase approximation that is appropriate at most scattering angles, $\theta$, shows that the dominant contributions to the integral come from $L$ values close to the stationary phase points given by

$$(\partial F/\partial L) + \theta = 0 \tag{150}$$

or, by virtue of Eq. (146),

$$\theta = -\Theta(\infty, L) \tag{151}$$

In other words, the stationary phase values of $L$ are the angular momenta at which the classical deflection is equal to minus the angle in question (this difference in sign cannot, of course, be detected). Three such $L$ values may be identified in Fig. 2. Hence Eq. (149) reduces to a combination of three terms,

$$\Psi^+(R, \theta) = R^{-1} f(\theta) e^{ipR/\hbar} \tag{152}$$

$$f(\theta) = \sum_{i=a,b,c} P_i^{1/2}(\theta) e^{i\gamma_i(\theta)} \tag{153}$$

where

$$P_i(\theta) = L_i/[P_i^2 \sin \theta (\partial \Theta/\partial L)_i]$$

$$\gamma_i(\theta) = [\eta(L_i) + L_i \theta]/\hbar \tag{154}$$

$$\eta(L_i) = \lim_{R\to\infty} [F(R, L_i) - pR]$$

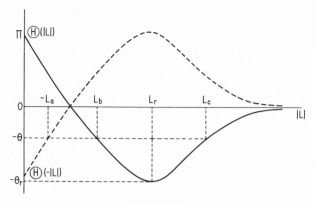

Fig. 2.  Classical deflection function.

with the interpretation that $P_i(\theta)$ is the classical contribution to the differential cross section, whereas the phase $\gamma_i(\theta)$ is the full classical action expressed in units of $\hbar$. This means that the central oscillatory part of the differential cross section, $|f(\theta)|^2$, shown in Fig. 3, may be interpreted as a diffraction pattern, governed by purely "classical" considerations.

Naturally this primitive semiclassical description becomes invalid at the rainbow angle $\theta_r$ in Fig. 2, due to a confluence between the branches $b$ and $c$, but this may be accommodated by using a uniform Airy rather than a stationary phase approximation.[8] It follows from Eq. (58) that in this case

$$f(\theta) = P_a^{1/2}(\theta)e^{i\gamma_a(\theta)} + \pi^{1/2}\exp\left\{\frac{i}{2}[\gamma_b(\theta) + \gamma_c(\theta)]\right\}$$

$$\times\{[|P_b(\theta)|^{1/2} + |P_c(\theta)|^{1/2}]\zeta^{1/4}\,\mathrm{Ai}(-\zeta) - i[|P_b(\theta)|^{1/2}$$

$$+ |P_c(\theta)|^{1/2}]\zeta^{-1/4}\,\mathrm{Ai}'(-\zeta)\} \tag{155}$$

where

$$\tfrac{2}{3}\zeta^{3/2} = -\tfrac{1}{2}[\gamma_b(\theta) - \gamma_c(\theta)] \tag{156}$$

This again depends only on the "classical" quantities $P_i(\theta)$ and $\gamma_i(\theta)$, at least in the classically accessible region.

The extension to nonclassical angles $\theta > \theta_r$, formally covered by $\zeta > 0$ in Eq. (155), merits some discussion, however, because there are by definition no real classical trajectories that terminate in this region. On the other hand, the analysis of the Airy function given in Eqs. (35)–(37) indicates the existence of complex trajectories corresponding to complex angular momenta close to the rainbow value $L_r$, which would cover this region. Similar complex trajectories for nonclassical events play an important part in the theory of inelastic scattering, discussed in Section 3.2.

A similar confluence problem, associated in this case with branches $a$ and $b$, arises at the *glory* angle $\theta = 0$, but the situation is further complicated by the

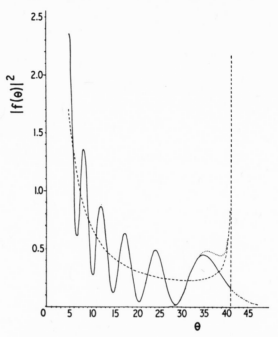

Fig. 3(a). The rainbow contribution to the differential cross section. Dashed curve, classical; dotted curve, primitive semiclassical; solid curve, uniform.

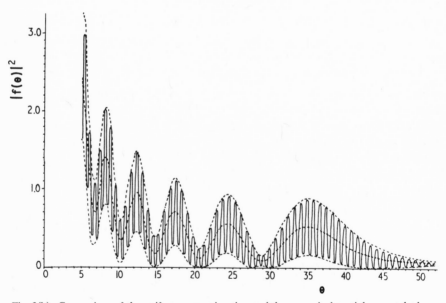

Fig. 3(b). Comparison of the uniform approximation and the numerical partial wave calculation. Solid curve, partial wave; dashed curve, mean and envelope of interface oscillations from the uniform approximation. (After Berry and Mount.[10])

breakdown of Eq. (148). It therefore becomes necessary to replace the asymptotic approximation by

$$P_{(L/\hbar)-1/2}(\cos \theta) \simeq \left(\frac{\theta}{\sin \theta}\right)^{1/2} J_0\left(\frac{L\theta}{\hbar}\right) \tag{157}$$

in terms of which the $a, b$ contribution to the scattering amplitude may be written[9]

$$f_{a,b}(\theta) = \frac{-i}{\pi p\hbar}\left(\frac{\theta}{\sin \theta}\right)^{1/2} \int_0^\pi \int_0^{L*} L \exp\left[\frac{i}{\hbar}(2\eta(L) - L\theta \cos \phi)\right] dL \, d\phi \tag{158}$$

after use of the standard integral representation.[20] Here $L^*$ is assumed to be sufficiently large to include the glory region but not to extend into the high $L$ tail of the deflection function in Fig. 2, which contributes to the $c$ branch of $f(\theta)$. Berry[9] has shown how the integral in Eq. (158) may be approximated by means of a stationary phase integration with respect to $L$, followed by a uniform Bessel approximation to the resulting integral over $\phi$. The final result may be written

$$f_{a,b}(\theta) = [\pi\zeta(\theta)]^{1/2} \exp\left\{\frac{i}{2}[\gamma_a(\theta) + \gamma_b(\theta)]\right\}$$
$$\times \{[|P_a|^{1/2}(\theta) + |P_b|^{1/2}(\theta)]J_0[\zeta(\theta)] - i[|P_a|^{1/2}(\theta) - |P_b|^{1/2}(\theta)]J_0[\zeta(\theta)]\} \tag{159}$$

where

$$\zeta(\theta) = [\gamma_a(\theta) - \gamma_b(\theta)]/2$$

This again depends only on the classical quantities $P_i(\theta)$ and $\gamma_i(\theta)$, but the uniform approximation involves zero-order Bessel functions rather than Airy functions, because the glory caustic is of *axial* rather than *conical* type.[8–10]

Similar methods have been advanced[9] for treatment of the forward diffraction peak and hence of the total cross section,[25] both of which are dominated by the large number of small contributions from high $L$ values. A full description, however, is unavailable due to the absence of a suitable mathematical function for the completely general mapping.

## 3.2. Inelastic Scattering

Pechukas,[46,47] Miller,[38,39] and Marcus[48] have shown how the ideas applied to elastic scattering may be extended to cover the inherently more difficult (because nonseparable) inelastic scattering events. The theory outlined below for the collinear excitation of an oscillator is equally applicable to rotational excitation[49] and to reactive scattering situations.[50]

The first step is to describe the initial internal oscillator motion in terms of angle–action coordinates $(N, w)$ in order to have a uniform description of the unperturbed state; thus, according to Eqs. (126) and (127),

$$\Psi^0(w, R) = v_1^{-1/2} \exp\{i[(N_1 - \delta\hbar)w + P_1 R]/\hbar\} \qquad (160)$$

where $v_1$ is the initial velocity, $P/m$, and $N_1$ the initial action, corresponding to the quantum number

$$n_1 = N_1/\hbar - \delta \qquad (161)$$

The evolution of this initial form is followed by performing the usual semiclassical expansion

$$\Psi(w, R) = A(w, R) \exp\{i[W(w, R) - \hbar\delta w]/\hbar\} \qquad (162)$$

for the solution of the Schrödinger equation

$$\left[ H_0\left(-i\hbar\frac{\partial}{\partial w} + \hbar\delta\right) - \frac{\hbar^2}{2m^2}\frac{\partial^2}{\partial R^2} + V(w, R) \right]\Psi = E\Psi \qquad (163)$$

in which $V(w, R)$ has been converted to the angle representation by the appropriate analogue of Eq. (123).

This means that the exponent $W(w, R)$ must satisfy the Hamilton–Jacobi equation

$$H_0(\partial W/\partial w) + (2m)^{-1}(\partial W/\partial R)^2 + V(w, R) = E \qquad (164)$$

It follows by comparison with the energy conservation condition

$$H_0(N) + (2m)^{-1}P^2 + V(w, R) = E \qquad (165)$$

that $W(w, R)$ may be taken as a line integral,

$$W_{E,N}(w, R) = \int_{w_1}^{w} N\,dw + w_1 N_1 + \int_{R_1}^{R} P\,dR + R_1 P_1 - \frac{\pi}{2} \qquad (166)$$

evaluated along any classical trajectory determined by Hamilton's equation of motion,[36]

$$\dot{N} = -(\partial H/\partial w), \qquad \dot{w} = (\partial H/\partial N)$$
$$\dot{P} = -(\partial H/\partial R), \qquad \dot{R} = (\partial H/\partial P) \qquad (167)$$

Each such trajectory may be defined by the available energy $E$ and initial action $N_1$ and an initial point $(w_1, R_1)$ on the path. This means that the coordinates $(w, R)$ of the representative point at the later time $t$ may be regarded as strictly dependent on $(w_1, R_1)$. Alternatively, $(w, R)$ may be taken to be independent and $(w_1, R_1)$ to be dependent on them. The latter viewpoint is adopted in Eq. (166), and the terms $w_1 N_1 + R_1 P_1$ have been included in order to remove contributions to $(\partial W/\partial w)$ and $(\partial W/\partial R)$ from the lower integration limits. The term $-\pi/2$ is included only on the outgoing part of the trajectory in order to account, by means of Eq. (38), for the semiclassical phase change at the radial turning point. It is readily verified that $W_{E,N_1}(w, R)$

goes over to the exponent in Eq. (161) $(w, R) \rightarrow (w_1, R_1)$, because $N$ and $P$ remain constant in the initial asymptotic region.

The normalization in Eq. (162) is derived either from the flux conservation equation, equivalent to terms of order $\hbar$ in Eq. (26), or from the usual semiclassical unitarity condition given by Eq. (96). Both methods yield

$$A(w, R) = Cv^{-1/2}(\partial w_1/\partial w)^{1/2} \tag{168}$$

provided $w_1$ varies monotonically with $w$. Comparison with Eq. (160) shows that $C = 1$.

The final conclusion upon combining Eqs. (162), (166), and (168) is that the initial function $\Psi^0(w, R)$ given by Eq. (160) evolves to the final form

$$\Psi_{E,N_1}^{(+)}(w, R) \sim v^{-1/2}(\partial w_1/\partial w)_t^{1/2} \exp\{i[W_{E,N}(w, R) - \hbar\delta w]\}, \qquad \text{as } t \rightarrow \infty \tag{169}$$

which may be transformed by the arguments leading to Eq. (135) to the more convenient $(\bar{w}, t)$ representation

$$\tilde{\Psi}_{E,n_1}^{(+)}(w, t) \sim (\partial \bar{w}_1/\partial \bar{w})_R \exp\{i[\chi_E(n_1, \bar{w}) + Et/\hbar]\}, \qquad \text{as } t \rightarrow \infty \tag{170}$$

where

$$n = N - \delta\hbar$$

$$\bar{w} = w - \omega t = w - \omega RP/m \tag{171}$$

$$\omega = \partial H_0/\partial N$$

and

$$\chi_E(n_1, \bar{w}) = n\bar{w} - \int_{n_1}^{n} w \, dn - \frac{1}{\hbar} \int_{P_1}^{P} R \, dP \tag{172}$$

The identity

$$(\partial w_1/\partial w)_t = (\partial \bar{w}_1/\partial \bar{w})_R \tag{173}$$

has also been used in the derivation of Eq. (170). Note that the precise value of $R$ at which this derivative is taken is of no significance because $\bar{w}$, defined by Eqs. (171), is constant in the final asymptotic region.

The first observation about Eq. (170) is that each final reduced phase $\bar{w}$ is accessible at a given initial quantum number $n_1$ by a unique trajectory specified by an initial phase $\bar{w}_1(n_1, \bar{w})$, because $\bar{w}_1$ is assumed to be monotonic in $\bar{w}$. Hence the wave function apparently represented by one term in Eq. (170) contains information about the whole family of trajectories with different end points $\bar{w}$. Second, both the final quantum number $n(n_1, \bar{w})$ and translational momentum $P(n_1, \bar{w})$ are also strictly dependent on $n_1$ and $\bar{w}$, so that when this dependence is taken into account

$$\partial\chi_E/\partial n_1 = \bar{w}_1, \qquad \partial\chi_E/\partial\bar{w} = n \tag{174}$$

Hence $\chi_E(n_1, \bar{w})$ may be regarded as the type $F_2$ generator for transformation from the final $(n, \bar{w})$ to the initial $(n_1, \bar{w}_1)$ internal system.

The elements of the scattering matrix $S$ are readily derived by comparison between Eq. (170) and the $(\bar{w}, t)$ transform of the defining equation

$$\Psi^{(+)}_{E,N_1}(w, R) \sim \sum_{n_2} v_2^{-1/2} S_{n_1 n_2} \exp\left(in_2 w + \frac{i}{\hbar} P_2 R\right) \qquad (175)$$

namely,

$$\Psi^{(+)}_{E,N_1}(\bar{w}, t) \sim \sum_{n_2} S_{n,n_2} \exp\left(in_2 \bar{w} + \frac{i}{\hbar} Et\right) \qquad (176)$$

It follows upon projecting out the $n$th term from $\Psi^{(+)}_{E,n_1,E}(\bar{w}, t)$ and transforming to $\bar{w}_1$ as the independent angle variable that

$$S_{n_1 n_2} = \frac{1}{2\pi} \int_0^{2\pi} (\partial \bar{w}/\partial \bar{w}_1)^{1/2} \exp[i\Delta_{n_1 n_2}(\bar{w})] \, d\bar{w}_1 \qquad (177)$$

where

$$\Delta_{n_1 n_2}(\bar{w}_1) = \chi_E[n_1, \bar{w}(n_1, w_1)] - n_2 \bar{w}(n_1, \bar{w}_1)$$

$$= [n(n_1, \bar{w}_1) - n_2]\bar{w}(n_1, \bar{w}_1) - \int_{n_1}^{n(n_1, \bar{w}_1)} w \, dn - \frac{1}{\hbar} \int_{P_1}^{P(n_1, \bar{w}_1)} R \, dP \qquad (178)$$

Notice that the transition amplitude $S_{n_1 n_2}$ again appears, as in the case of the elastic scattering amplitude, as an integral over all relevant classical trajectories.

Evaluation of the above integral by the now-familiar methods is best understood in terms of the final classical quantum number function $n(n_1, \bar{w}_1)$, plotted in Fig. 4, which shows a division into one classically allowed, $n_{min} < n_2 < n_{max}$, and two classically forbidden, $n_2 < n_{min}$ and $n_2 < n_{max}$, regions, with two real initial phases $\bar{w}_1^{(a)}(n_1, n_2)$ and $\bar{w}_1^{(b)}(n_1, n_2)$ for each classically allowed value of $n_2$. Furthermore, Eqs. (174) and (178) show that $\bar{w}_1^{(a)}$ and $\bar{w}_1^{(b)}$ again correspond to the stationary phase points of $\Delta_{n_1 n_2}(\bar{w}_1)$.

The simplest approximation is therefore obtained by the method of stationary phase, which gives

$$S_{n_1 n_2} = P_a^{1/2} e^{i\Delta_a(n_1, n_2)} + P_b^{1/2} e^{i\Delta_b(n_1, n_2)} \qquad (179)$$

where

$$P_\nu = \left[\frac{1}{2\pi}\left(\frac{\partial \bar{w}}{\partial \bar{w}_1}\right) \Big/ \left(\frac{\partial^2 \Delta}{\partial \bar{w}_1^2}\right)\right] = \left[2\pi\left(\frac{\partial n}{\partial \bar{w}_1}\right)\right]^{-1} \qquad (180)$$

evaluated at $\bar{w}_1^{(\nu)}$. Note that $\partial n/\partial \bar{w}_1$ is positive at $\bar{w}_1^{(a)}$ and negative at $\bar{w}_1^{(b)}$ in Fig. 4. This means, because the initial classical distribution of $\bar{w}_1$ is uniform, that $P_\nu$ has a purely classical interpretation as the density of trajectories from $n_1$ that

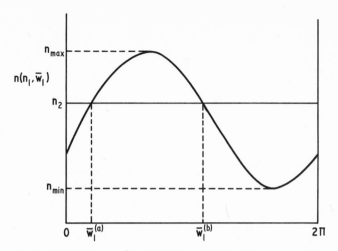

Fig. 4. The final quantum number $n(n_1, \bar{w}_1)$ as a continuous function of the initial phase $\bar{w}_1$. $n_2$ denotes a typical integer value reached by two classical trajectories with initial phases $\bar{w}_1^{(a)}$ and $\bar{w}_1^{(b)}$.

arrive in the neighborhood of $n_2$. The presence of the phase terms $\Delta_\nu(n_1, n_2)$ leads to quantum mechanical interference between the two special $n_1 \to n_2$ trajectories such that the transition probability $|S_{n_1 n_2}|^2$ oscillates around the classical value, as shown in Fig. 5:

$$P_{n_1 n_2} = |S_{n_1 n_2}|^2 = P_a + P_b + 2(P_a P_b)^{1/2} \cos(\Delta_a - \Delta_b) \tag{181}$$

This stationary phase result must as usual be replaced by a suitable uniform approximation for values of $n_2$ in the threshold regions close to $n_{\min}$ and $n_{\max}$, where $\bar{w}_1^{(a)}$ and $\bar{w}_1^{(b)}$ coalesce. Early approaches to this problem[32,39] involved uniform Airy function representations of the type given by Eq. (58) in the immediate neighborhoods of $n_{\max}$ and $n_{\min}$, but these representations have been more recently replaced by the uniform Bessel approximation[33] derived

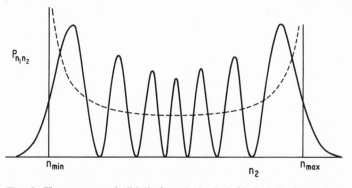

Fig. 5. The quantum (solid line) and classical (dashed line) transition probability $P_{n_1 n_2}$ as a continuous function of the final quantum number $n_2$.

from Eq. (67), which takes into account the periodicity of Fig. 4:

$$S_{n_1 n_2} = (\pi/2)^{1/2} e^{iA} [(|P_a|^{1/2} + |P_b|^{1/2})(\zeta^2 - m^2)^{1/4} J_m(\zeta)$$

$$+ i(|P_a|^{1/2} - |P_b|^{1/2})\zeta(\zeta^2 - m^2)^{-1/4} J'_m(\zeta)] \qquad (182)$$

where

$$m = n_2 - n_1$$

$$A = \tfrac{1}{2}(\Delta_a + \Delta_b - m\pi)$$

$$(\zeta^2 - m^2)^{1/2} - m \arccos(m/\zeta) = \tfrac{1}{2}(\Delta_b - \Delta_a)$$

Computations[33] show that this is markedly superior to the uniform Airy approximation in cases of near elasticity, such that $n_{max} \simeq n_{min} \simeq n_1$.

Figure 5 also indicates that the uniform Bessel approximation may be used to predict transition probabilities for classically forbidden events such that $n_2 > n_{max}$ or $n_2 < n_{min}$ for which $m > \zeta$ and $m < -\zeta$, respectively. Reference to Eqs. (63)–(67) shows that these regions are covered by complex stationary phase points $\bar{w}^{(a)}$ and $\bar{w}^{(b)}$, in other words, by complex classical trajectories. The two possible trajectories are, of course, complex conjugates, but the asymptotic properties of the Bessel function for $|m/\zeta| \gg 1$ show[20] that for strongly forbidden events only one of them contributes to the transition probability, namely, that one associated with the exponentially small transition probability

$$S_{n_1 n_2} \sim |P_a|^{1/2} e^{i\Delta_a}$$

$$\text{Im } \Delta_a > 0 \qquad (183)$$

The prediction and discovery of these hidden complex classical trajectories,[51,52] which are solutions of Hamilton's equations (167) at complex initial phase angles $\bar{w}_1$ and complex times, is one of the most remarkable achievements of the semiclassical theory. A typical complex initial phase $\bar{w}_1$ naturally leads to a physically unacceptable complex final quantum number $n(n_1, \bar{w}_1)$; but there are two lines in the complex $\bar{w}_1$ plane, intersecting the real axis at $\bar{w}_1^{(min)}$ and $\bar{w}_1^{(max)}$, respectively, along which $n(n_1, w_1)$ is real, and integer points along these lines correspond to physically acceptable final quantum numbers. The exponential damping associated with these solutions may be regarded as the analogue in quantum number or action space of quantum mechanical tunneling.

Some practical implications of the above results may now be noted. The first is that the integral representation (177) for $S_{n_1 n_2}$ requires only that $\bar{w}(n_1, \bar{w}_1)$ is monotonic in $\bar{w}_1$. Second, although the functions $(\partial \bar{w}/\partial \bar{w}_1)$ and $\Delta_{n_1 n_2}(\bar{w}_1)$ are specified in the angle–action representation, they may often be more conveniently derived from the Cartesian equations of motion, followed by a transformation based on the appropriate $F_1(q, w)$ generator.[53,54] Direct evaluation of the resulting integral by numerical quadrature may be based on only real trajectories, but the results[25] are found to be reliable only for the classically accessible events. The problem is that all the trajectories sampled start from $n_1$ but only two of them reach $n_2$, and this bias toward $n_1$ may be expected to destroy the required symmetry of $S_{n_1 n_2}$. However, if $n_2$ is classically accessible, the main contributions to the integral came from the real stationary phase regions around $w_1^{(a)}$ and $\bar{w}_1^{(b)}$, and these will be accumulated by integration along the real axis. On the other hand, if $n_2$ is classically inaccessible, the stationary phase regions are complex, and integration over real $\bar{w}_1$ values leads to serious inaccuracy. The various stationary phase and uniform approximations, being based on the symmetrical $n_1 \rightarrow n_2$ paths, do not suffer this disadvantage, but the determination of any particular trajectory becomes a boundary value rather than an initial value problem. Furthermore, considerable care is required to avoid numerical instabilities, particularly when dealing with classically forbidden events.[51,52]

The boundary value problem becomes progressively more serious as the number of internal degrees of freedom is increased, particularly as one may wish to average over several of them. Doll and Miller[55] have therefore introduced a *partial averaging* procedure to deal with these situations.

## 4. Classical Path Approximations

The earliest and simplest semiclassical method is based on a quantal treatment of the internal motion, subject to a time-dependent potential determined by translational motion along an appropriate mean classical trajectory $\mathbf{R}(t)$. This results in replacement of the full Schrödinger equation

$$\left[ -\frac{\hbar^2}{2m} \nabla_R^2 + H_{\text{int}}(\boldsymbol{\rho}) + V(\mathbf{R}, \boldsymbol{\rho}) \right] \Psi = E \Psi \qquad (184)$$

by the time-dependent form

$$\{ H_{\text{int}}(\boldsymbol{\rho}) + V[\mathbf{R}(t), \boldsymbol{\rho}] \} \Phi = i\hbar \, \partial \Phi / \partial t \qquad (185)$$

with the trajectory $\mathbf{R}(t)$ usually taken to be appropriate to elastic scattering by a mean spherical potential $\bar{V}(R)$ at a mean energy $\bar{E} = \frac{1}{2} m \bar{v}^2$. The coordinates of $\mathbf{R}(t)$ are therefore determined by

$$dR/dt = \pm \bar{v} [1 - b^2/R^2 - \bar{V}(R)/\bar{E}]^{1/2}$$

$$d\theta/dt = \bar{v} b / R^2 \qquad (186)$$

$$d\phi/dt = 0$$

where $b$ denotes the impact parameter. See Fig. 6.

Comparison with Eq. (184) shows[25,56,57] that the validity of Eq. (185) requires that the radial motion described by Eqs. (186) should be semiclassical

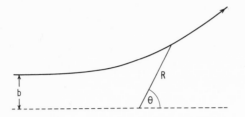

Fig. 6. Mean elastic scattering trajectory,
governed by Eqs. (186).

in the sense of Eq. (13) and that changes in linear and angular momentum due to the collision should be small compared with their absolute values.

Three methods are available for the solution of Eq. (185). The first involves a time-dependent expansion for $\Phi(\boldsymbol{\rho}, t)$ followed by a perturbative or exponential solution of the resulting equations. The second is similar except that Heisenberg's correspondence principle is used[58] to express the internal matrix elements as Fourier components of the interaction potential $V[\mathbf{R}(t), \boldsymbol{\rho}(t)]$ evaluated along a mean internal classical orbit $\boldsymbol{\rho}(t)$. The strong coupling limit of this formulation reduces to the semiclassical $S$-matrix theory of Section 3.2 in the classical perturbation limit. The final method involves reduction of Eq. (185) to a soluble forced oscillator problem.[59]

### 4.1. Perturbation and Sudden Approximations

It is readily verified by means of the following substitutions

$$\Phi(\boldsymbol{\rho}, t) = \sum_n a_n(t)\phi_n(\boldsymbol{\rho}, t)e^{-iE_n t/\hbar} \tag{187}$$

$$\mathbf{a}(t) = \mathbf{U}(t, t_0)\mathbf{a}(t_0) \tag{188}$$

$$\mathbf{U}(t_0, t_0) = \mathbf{U}(t, t) = \mathbf{I} \tag{189}$$

that the time evolution matrix $\mathbf{U}(t, t_0)$ is governed by

$$i\hbar\frac{d\mathbf{U}}{dt} = \mathbf{V}_I(t)\mathbf{U} \tag{190}$$

where $\mathbf{V}_I(t)$ is the matrix of the scattering potential in the interaction picture, with elements

$$[\mathbf{V}_I(t)]_{nm} = V_{nm}(t)e^{i\omega_{nm}t} \tag{191}$$

$$V_{nm}(t) = \int \phi_n^*(\boldsymbol{\rho}) V[\mathbf{R}(t), \boldsymbol{\rho}]\phi_m(\boldsymbol{\rho}) \, d\boldsymbol{\rho} \tag{192}$$

$$\hbar\omega_{nm} = E_n - E_m \tag{193}$$

Perturbation expressions for the $S$ matrix are readily obtained by iterating the integral form of Eq. (191), obtaining

$$\mathbf{S} = \mathbf{U}(\infty, -\infty)$$

$$= \mathbf{I} - \frac{i}{\hbar} \int_{-\infty}^{\infty} \mathbf{V}_I(t_1) \mathbf{U}(t_1, -\infty) \, dt_1$$

$$= \mathbf{I} - \frac{i}{\hbar} \int_{-\infty}^{\infty} \mathbf{V}_I(t_1) \, dt_1 + \frac{i}{\hbar} \int_{-\infty}^{\infty} \int_{-\infty}^{t_1} \mathbf{V}_I(t_1) \mathbf{V}_I(t_2) \, dt_2 \, dt_1 + \cdots \quad (194)$$

Hence to first order in $\mathbf{V}_I(t)$

$$S_{nm} = -\frac{i}{\hbar} \int_{-\infty}^{\infty} V_{nm}(t) e^{i\omega_{nm}t} \, dt \qquad (195)$$

This is the time-dependent equivalent of the first Born approximation. A similar time-dependent analogue of the distorted wave approximation is

$$S_{nm} = -\frac{i}{\hbar} \int_{-\infty}^{\infty} V_{nm}(t) \exp\left\{\frac{i}{\hbar} \int_0^t [W_n(t_1) - W_m(t_1)] \, dt_1\right\} dt \qquad (196)$$

where

$$W_n(t) = E_n + V_{nn}(t) \qquad (197)$$

This is readily obtained by replacing $a_n(t)$ in Eq. (187) by

$$c_n(t) = a_n(t) \exp\left[\frac{i}{\hbar} \int_0^t V_{nn}(t) \, dt\right] \qquad (198)$$

Similarly, the first term in an exponential form of the solution,[60,61]

$$\mathbf{U}(t, t_0) = \exp \mathbf{A}(t, t_0) = \mathbf{I} + \mathbf{A}(t, t_0) + \tfrac{1}{2}\mathbf{A}^2(t, t_0) + \cdots \qquad (199)$$

is found to be

$$\mathbf{A}^{(1)}(t, t_0) = \frac{i}{\hbar} \int_{t_0}^t \mathbf{V}_I(t_1) \, dt_1 \qquad (200)$$

and this may be shown[25,61] to be exact if the elements of $\mathbf{V}(t)$ have a common time constant that is short compared with the internal time period, regardless of the coupling strength. This reduces to the sudden approximation[62,63]

$$\mathbf{S} = \exp\left[-\frac{i}{\hbar} \int_{-\infty}^{\infty} \mathbf{V}(t) \, dt\right] \qquad (201)$$

if all the (small) frequency terms $\omega_{nm}$ are set equal to zero in Eq. (191).

## 4.2. Correspondence Principle Approximations

Methods based on the correspondence principle[62,63] are based on the previous time-dependent equations, with the necessary matrix elements replaced by classical Fourier components. The theory is restricted below to one internal degree of freedom, but extension to the general case introduces no new points of principle.

The underlying philosophy of the method relies on the connection between the quantal level separation and the classical internal frequency, namely,

$$E_n - E_m = \hbar \omega_{nm} \simeq \hbar (n-m) \omega \tag{202}$$

This is a strict equality for harmonic motion, but in the general case $\omega$ depends, according to Eq. (114), on the available action $N$, and the validity of Eq. (202) requires that $n - m \ll n, m$, in which case $N$ may be approximated by the mean

$$N \simeq \tfrac{1}{2}(n+m)\hbar \tag{203}$$

This implies that the classical orbit executes an average motion between those of the initial and final states.

Second, because this mean motion is periodic in time, any function of the internal variable $\rho(\tau)$ may be decomposed into Fourier components,

$$F[\rho(\tau)] = \sum_s F_s^{(c)} \exp(-is\omega t) \tag{204}$$

Heisenberg's correspondence principle[64] states that the quantal matrix element $F_{mn}^{(q)}$ may be identified with the classical Fourier component $F_{n-m}^{(c)}$,

$$F_{mn}^{(q)} = \int_{-\infty}^{\infty} \phi_m^*(\rho) F(\rho) \phi_n(\rho) \, d\rho$$

$$\simeq F_{n-m}^{(c)} = \frac{\omega}{2\pi} \int_0^{2\pi/\omega} F[\rho(\tau)] \exp[i(n-m)\omega\tau] \, d\tau \tag{205}$$

$$\simeq \frac{\omega}{2\pi} \int_0^{2\pi/\omega} F[\rho(\tau)] \exp(i\omega_{nm}\tau) \, d\tau$$

Written in another way, with the angle $w$ in place of $\omega\tau$, this may be recognized as the matrix element $F_{mn}^{(q)}$ in the angle–action representation,

$$F_{mn}^{(q)} = \frac{1}{2\pi} \int_0^{2\pi} F[\rho(w/\omega)] \exp[i(n-m)w] \, dw \tag{206}$$

The correspondence limit of the perturbation formula (195) may now be obtained by identifying $F(\rho)$ with the interaction potential $V[\mathbf{R}(t), \rho]$, so that, upon using Eq. (205),

$$
\begin{aligned}
S_{nm} &\simeq -\frac{i}{\hbar}\frac{\omega}{2\pi}\int_0^{2\pi/\omega}\left\{\int_{-\infty}^{\infty} V[\mathbf{R}(t),\rho(\tau)]\exp[i(n-m)\omega(t-\tau)]\,dt\right\}d\tau \\
&= -\frac{i}{\hbar}\frac{\omega}{2\pi}\int_0^{2\pi/\omega}\left\{\int_{-\infty}^{\infty} V[\mathbf{R}(t),\rho(t+\tau')]\exp[-i(n-m)\omega\tau']\,dt\right\}d\tau'
\end{aligned}
\tag{207}
$$

Here the infinite integral is a classical action function taken over a trajectory such that the external motion follows a mean path $\mathbf{R}(t)$, while the internal motion traces out the orbit $\rho(t+\tau')$, where $\tau'$ governs the position of the bound particle at time $t=0$. A related formula that takes into account changes in the internal frequency with external time $t$ may be derived from Eq. (196).[65]

A similar variant of the sudden approximation (201) may be derived by setting

$$
F(\rho)=\exp\left\{-\frac{i}{\hbar}V[\mathbf{R}(t),\rho]\right\}
\tag{208}
$$

in Eq. (205), so that in the correspondence limit

$$
S_{nm}=\frac{\omega}{2\pi}\int_0^{2\pi/\omega}\exp\left(i\left\{(n-m)\omega\tau-\frac{i}{\hbar}\int_{-\infty}^{\infty} V[\mathbf{R}(t),\rho(\tau)]\,dt\right\}\right)d\tau
\tag{209}
$$

Notice that the integration over $t$ now affects only the external motion, because the interaction time is assumed to be short compared with the period of the internal motion.

Equations (207) and (209) are applicable to very weak and very short-lived interactions, respectively, with the added restriction in both cases that the quantum number change $|n-m|$ be small compared with the initial and final values. The latter corresponds to the conditions for classical perturbation theory. A third approximation, also governed by the validity of classical perturbation theory but unlimited by the strength or the time scale of the coupling, may also be derived directly from Eqs. (188) and (205). This is obtained by expressing the evolution matrix from time $-T$ to $T$ as a product of $N$ elementary steps,

$$
\mathbf{U}(T,-T)=\prod_{k=0}^{N-1}\mathbf{U}(t_{k+1},t_k)
\tag{210}
$$

with

$$
t_0=-T, \qquad t_N=T, \qquad t_k=t_0+k\,\Delta t
\tag{211}
$$

and truncating the expansion (194) for $U(t_{k+1}, t_k)$ after two steps, for sufficiently small values of $t$; thus

$$U_{n_{k+1},n_k}(t_{k+1}, t_k) \simeq 1 - i\frac{\Delta t}{\hbar} V_{n_{k+1},n_k}(t_k) \exp(i\omega_{n_{k+1},n_k}t_k)$$

$$\simeq 1 - i\frac{\Delta t}{\hbar} V^{(c)}_{n_{k+1}-n_k}(t_k) \qquad (212)$$

where Eq. (205) has been used to replace the matrix element $V_{n_k,n_{k+1}}e^{i\omega_{n_{k+1},n_k}t_k}$ by the classical Fourier component,

$$V^{(c)}_{n_{k+1}-n_k}(t_k) \exp(i\omega_{n_{k+1},n_k}t_k)$$

$$= \frac{\omega}{2\pi} \int_0^{2\pi/\omega} V[\mathbf{R}(t_k), \rho(\tau)] \exp[i(n_{k+1} - n_k)\omega(t_k - \tau)] d\tau$$

$$= \frac{\omega}{2\pi} \int_0^{2\pi/\omega} V[\mathbf{R}(t_k), \rho(t_k + \tau')] \exp[-i(n_{k+1} - n_k)\omega\tau'] d\tau' \qquad (213)$$

Hence a typical element of $U(T, -T)$ may be written as

$$U_{nm}(T, -T) = \sum_{n_1} \cdots \sum_{n_{N-1}} \prod_{k=0}^{N-1} [\delta_{n_k n_{k+1}} - i\frac{\Delta t}{\hbar} V^{(c)}_{n_{k+1}-n_k}(t_k)] \qquad (214)$$

Now it is readily verified that the Fourier components of a product function $F(t)G(t)$ satisfy the rules for matrix multiplication,

$$(fg)_{n-m} = \sum_l f_{n-l}g_{l-m} \qquad (215)$$

where

$$f_{n-l} = \frac{\omega}{2\pi} \int_0^{2\pi/\omega} F(\tau)e^{i(n-l)\omega\tau} d\tau \qquad (216)$$

and similarly for $(fg)_{n-m}$ and $g_{l-m}$. Hence the sums in Eq. (214) may be performed directly to give

$$U_{nm}(T, -T) = \frac{\omega}{2\pi} \int_0^{2\pi/\omega} \prod_{k=0}^{N-1} \left\{1 - \frac{i\,\Delta t}{\hbar} V[\mathbf{R}(t_k), \rho(t_k + \tau)]\right\} e^{i(n-m)\omega\tau} d\tau \qquad (217)$$

Finally, each term in Eq. (217) may be replaced to first order in $\Delta t$ by an exponential, and the resulting sum may be converted to an integral by taking the limit $N \to \infty$. The final result for the $S$-matrix element therefore becomes

$$S_{nm} = \lim_{T \to \infty} U_{nm}(T, -T)$$

$$= \frac{\omega}{2\pi} \int_0^{2\pi/\omega} \exp\left\{i(n-m)\omega\tau - \frac{i}{\hbar} \int_{-\infty}^{\infty} V[\mathbf{R}(t), \rho(t+\tau)] dt\right\} d\tau \qquad (218)$$

This has been termed the strong coupling correspondence principle.[58] Its validity again depends on the assumption of a mean internal trajectory $[\mathbf{R}(t), \rho(t+\tau)]$ with the internal coordinate at the point $\rho(\tau)$ at time $t=0$. The validity criterion on the correspondence principle, $|n-m| \ll n, m$, justifies the use of a common internal frequency $\omega$ throughout the calculation. It is readily apparent that Eq. (218) reduces to Eqs. (207) and (209) in the perturbation and sudden approximation limits, respectively.

Equation (218) may also been related to the semiclassical $S$-matrix element in Eq. (177) by use of classical perturbation theory.[66] The first step is to identify the initial phase angle $\bar{w}_1$ in Eq. (177) with $\omega\tau$ in Eq. (218), but the main problem is to relate the two exponents in the integrands. To this end, $\Delta_{n_1 n_2}(2_1)$ in Eq. (177) may be expressed in the form

$$\Delta_{n_1 n_2}(\bar{w}_1) = (n_1 - n_2)\bar{w}_1 + \frac{1}{\hbar}\int_{\bar{w}_1}^{\bar{w}} (N-N_1)\,d\bar{w} + \frac{1}{\hbar}\int_{R_1}^{R} (P-\bar{P})\,dR + \phi \quad (219)$$

where $\phi$ is a radial term independent of $\bar{w}_1$, and $\bar{P}(R)$ is a suitable average radial momentum function. Now the classical path approximation employs a mean trajectory governed by a separable zeroth-order Hamiltonian

$$H^0 = H_0(N) + \frac{1}{2m}P^2 + \bar{V}(R) \quad (220)$$

in place of the exact trajectory required for strict evaluation of Eq. (177), and this can cause no change from the values $N = N_1$ and $P(R) = \bar{P}(R)$ in Eq. (219), so that the only contribution to $\Delta_{n_1 n_2}(\bar{w}_1)$ comes from the phase shift $\phi$. Changes in $N$ and $P(R)$ are, however, induced by the interaction potential $V(\bar{w}, R)$, because by Hamilton's equations

$$\frac{d}{dt}(N-N_1) = -\left(\frac{\partial V}{\partial w}\right), \qquad \frac{d}{dt}(P-\bar{P}) = -\left(\frac{\partial V}{\partial R}\right) \quad (221)$$

Hence by integration along the mean trajectory $[\mathbf{R}(t), w(t+\bar{w}/\omega)]$ and substitution into Eq. (220),

$$\Delta_{n_1 n_2}(\bar{w}) = (n_1 - n_2)\bar{w}_1 - \frac{1}{\hbar}\int_{-\infty}^{\infty}\left[\int_{\bar{w}_1}^{\bar{w}}\left(\frac{\partial V}{\partial w}\right)dw + \int_{R_1}^{R}\left(\frac{\partial V}{\partial R}\right)dR\right]dt + \phi$$

$$= (n_1 - n_2)\bar{w}_1 - \frac{1}{\hbar}\int_{-\infty}^{\infty} V[\mathbf{R}(t), w(t+\bar{w}/\omega)]\,dt + \phi \quad (222)$$

in agreement, apart from the term $\phi$, with the exponent in Eq. (218). Finally, it may be noted that $\bar{w} = \bar{w}_1$ along the mean trajectory, and hence $(\partial\bar{w}/\partial\bar{w}_1) = 1$ in Eq. (177). This confirms that the semiclassical $S$-matrix result goes over to the strong coupling correspondence principle in the limit of classical perturbation theory.

### 4.3. The Impulse Approximation

A final method of practical importance for vibrational excitation problems is the impulse approximation based on the known closed solution of the forced harmonic oscillator problem. The most convenient[67] of the various approaches to this solution involves removal of the forcing term from the fundamental equation

$$\left[ -\frac{\hbar^2}{2\mu}\frac{\partial^2}{\partial\rho^2} + \frac{1}{2}k\rho^2 - \rho F(t) \right]\Psi = i\hbar\left(\frac{\partial\Psi}{\partial t}\right)_\rho \tag{223}$$

by means of the transformation

$$\Psi(\rho, t) = \Phi(\xi, t)\exp[\rho v(t) + w(t)]$$
$$\xi(\rho, t) = \rho - u(t) \tag{224}$$

in terms of which Eqs. (224) may be reduced to

$$\left( -\frac{\hbar^2}{2\mu}\frac{\partial^2}{\partial\xi^2} + \frac{1}{2}k\xi^2 \right)\Phi = i\hbar\left(\frac{\partial\Phi}{\partial t}\right)_\xi \tag{225}$$

by choosing the functions $u(t)$, $v(t)$, and $w(t)$ such that

$$\mu\frac{d^2u}{dt^2} + ku(t) = F(t) \tag{226}$$

$$v(t) = i(\mu/\hbar)(du/dt) \tag{227}$$

$$w(t) = -\frac{i}{\hbar}\int_{-\infty}^{t}\left[\frac{1}{2}\mu\left(\frac{du}{dt}\right)^2 - \frac{1}{2}ku^2\right]dt \tag{228}$$

This means that $\Phi(\xi, t)$ reduces to a standard harmonic oscillator solution,

$$\Phi_n(\xi, t) = \phi_n(\xi)e^{-iE_nt/\hbar}, \qquad E_n = (n+\tfrac{1}{2})\hbar\omega$$
$$\phi_n(\xi) = (\beta/\pi)^{1/4}(2^nn!)^{-1/2}H_n(x)e^{-1/2x^2} \tag{229}$$
$$x = \beta^{1/2}\xi, \qquad \beta = (\mu k/\hbar^2)^{1/4}$$

provided the mean point $u(t)$ of the displaced coordinate $\xi(\rho, t)$ follows the path of the equivalent forced classical oscillator.

The transition probability at time $t \to \infty$ is obtained by taking the projection of $\Psi_n(\rho, t)$ defined by Eqs. (224) and (229) onto the unperturbed solution $\phi_n(\rho)e^{-iE_mt/\hbar}$,

$$P_{nm} = \left| \lim_{t\to\infty}\int_{-\infty}^{\infty}\phi_m(\rho)\Psi_n(\rho, t)\,d\rho \right|^2 \tag{230}$$

It follows on evaluating the integral[67] that

$$P_{nm} = n!m!\epsilon^{m+n}e^{-\varepsilon}\left| \sum_{l=0}^{\min(n,m)}\frac{(-1)^l\varepsilon^{-l}}{l!(m-l)!(n-l)!} \right|^2 \tag{231}$$

where $\varepsilon$ is the mean classical energy transfer expressed in units of $\hbar\omega$. Thus on setting $u(t) = du/dt = 0$ at $t \to \infty$ and solving Eq. (226),

$$\varepsilon = \lim_{t \to \infty} \left[\tfrac{1}{2}ku^2 + \tfrac{1}{2}\mu(du/dt)^2\right]/\hbar\omega$$

$$= (2\mu\hbar\omega)^{-1}\left|\int_{-\infty}^{\infty} F(t)e^{i\omega t}\,dt\right|^2 \tag{232}$$

Hence $\varepsilon$ depends simply on the Fourier component of the force $F(t)$ at the resonant frequency. The attractive feature of the result (231) is that it is quantum mechanically exact for any force $F(t)$.

Some approximations are required, however, in a collision problem in order to convert the combined effect of the collision velocity and the interaction potential into an equivalent force $F(t)$. The assumptions of the impulse approximation are that the interaction is between the incident atom A and a single atom B in the target molecule and that the effective force is that due to an elastic collision between A and B, at an average relative velocity $\bar{v}$ appropriate to the initial and final channels. In other words, the oscillatory component of the B atom motion is ignored in calculating the force. This means that the time dependence of the $r_{ab}$ separation is given by

$$\frac{dr_{ab}}{dt} = \pm\bar{v}(1 - V(r_{ab})/\bar{E})^{1/2} \tag{233}$$

where $\bar{E}$ denotes the initial relative AB kinetic energy

$$\bar{E} = \tfrac{1}{2}(m_a m_b/m_{ab})\bar{v}^2$$

and the sign in Eq. (233) depends on the inward or outward character of the motion. Integration of Eq. (233) determines the time dependence of $V(r_{ab})$ and hence of its derivative with respect to $r_{ab}$, which gives the force on atom B. because motion of B represents only a fraction $(m_c/m_{bc})$ of the oscillator displacement, this corresponds to a net force on the oscillator given by

$$F(t) = -(m_c/m_{bc})(dV/dr_{ab}) \tag{235}$$

with the derivative evaluation at time $t$. Substitution into Eq. (232) then determines $\varepsilon$ and hence via Eq. (231) the transition probability.

The calculation is simply illustrated in the case of exponential potential

$$V(r_{ab}) = Ae^{-\alpha r_{ab}} \tag{236}$$

for which Eq. (233) may be integrated in closed form for $m$ to give

$$V[r_{ab}(t)] = A\,\exp[-\alpha r_{ab}(t)] = \bar{E}\,\mathrm{sech}^2(\alpha\bar{v}t/2) \tag{237}$$

so that

$$F(t) = \tfrac{1}{2}\alpha(m_a m_b m_c/m_{ab}m_{bc})\bar{v}^2\,\mathrm{sech}^2(\alpha\bar{v}t/2) \tag{238}$$

It follows from Eq. (233) that

$$\varepsilon = (2\mu\hbar\omega)^{-1}\left|\int_{\infty}^{\infty} F(t)e^{i\omega t}\,dt\right|^2$$

$$= 2\pi^2(\omega/\hbar\alpha^2)(m_a^2 m_b m_c/m_{ab}^2 m_{bc})\,\text{csch}(\pi\omega/\alpha\bar{v}) \tag{239}$$

because[68]

$$\int_{-\infty}^{\infty} \text{sech}^2\gamma t e^{i\omega t}\,dt = \frac{\pi\omega}{\gamma^2}\,\text{csch}\,\frac{\pi\omega}{2\gamma} \tag{240}$$

Use of this expression in Eq. (232) has been shown to give close agreement with exact numerical results[59] for the same model at energies such that the mean trajectory assumption is applicable.

The same general approach has also been applied to the general anharmonic oscillator problem[69] by use of a mean frequency $\omega$ for the states in question. The justification is similar to that employed in Section 4.2, namely, that the states coupled by the interaction must involve changes in the quantum number or action that are small compared with their absolute values.

ACKNOWLEDGMENT

The author is a fellow of St. Edmund Hall and Lecturer in theoretical chemistry at the University of Oxford.

## References

1. P. A. M. Dirac, *Principles of Quantum Mechanics*, 4th ed., Oxford University Press, London (1958).
2. M. Born, *The Mechanics of the Atom*, G. Bell & Sons Ltd., London (1960).
3. V. P. Maslov, *Théorie des Perturbations et Méthodes Asymptotiques*, Dunod, Gautier Villars, Paris (1972).
4. L. D. Landau and E. M. Lifshitz, *Electrodynamics of Continuous Media*, Addison-Wesley Publishing Company, Inc., Reading, Mass. (1965).
5. M. Born and E. Wolf, *Principles of Optics*, Pergamon Press, Inc., Elmsford, N.Y. (1970).
6. S. C. Miller and R. H. Good, A WKB-type approximation to the Schrödinger equation, *Phys. Rev.* **91**, 174–179 (1953).
7. C. Chester, B. Friedmann, and F. Ursell, An extension of the method of steepest descents, *Proc. Cambridge Philos. Soc.* **53**, 599–611 (1957).
8. M. V. Berry, Uniform approximation for potential scattering involving a rainbow. *Proc. Phys. Soc., London* **89**, 479–490 (1966).
9. M. V. Berry, Uniform approximations for glory scattering and diffraction peaks, *J. Phys. B* **2**, 381–392 (1969).
10. M. V. Berry and K. E. Mount, Semi-classical approximations in wave mechanics, *Rep. Prog. Phys.* **35**, 315–397 (1972).
11. J. N. L. Connor, Evaluation of multi-dimensional canonical integrals in semi-classical collision theory, *Mol. Phys.* **26**, 1371–1378 (1973).

12. P. Pechukas, Analysis of the Miller–Good method for approximating bound states, *J. Chem. Phys.* **54**, 3864–3873 (1971).
13. L. D. Landau and E. M. Lifshitz, *Quantum Mechanics: Non-Relativistic Theory,* 2nd ed., Addison-Wesley Publishing Company, Inc., Reading, Mass. (1965).
14. K. W. Ford and J. A. Wheeler, Semi-classical description of scattering, *Ann. Phys. NY* **7**, 259–286 (1959).
15. K. W. Ford and J. A. Wheeler, Application of semi-classical analysis, *Ann. Phys. NY* **7**, 286–322 (1959).
16. W. H. Miller, Classical limit quantum mechanics and the theory of molecular collisions. *Adv. Chem. Phys.* **25**, 69–177 (1974).
17. N. Fröman and P. O. Fröman, *J. W. K. B. Approximation: Contributions to the Theory,* North-Holland Publishing Company, Amsterdam (1965).
18. J. Heading, *Phase Integral Methods,* Methuen & Co. Ltd., London (1962).
19. H. Jeffreys, *Asymptotic Approximations,* Oxford University Press, London (1962).
20. M. Abramowitz and I. A. Stegun, *Handbook of Mathematical Functions,* Dover Publications, Inc., New York (1965).
21. H. Jeffreys and B. S. Jeffreys, *Methods of Mathematical Physics,* 3rd ed., Cambridge University Press, London (1956).
22. R. P. Feynman and A. R. Hibbs, *Quantum Mechanics and Path Integrals,* McGraw-Hill Book Company, New York (1965).
23. R. E. Langer, On the connection formulas and solutions of the wave equation, *Phys. Rev.* **51**, 669–676 (1937).
24. M. S. Child, Uniform evaluation of one-dimensional matrix elements, *Mol. Phys.* **29**, 1421–1429 (1975).
25. M. S. Child, *Molecular Collision Theory,* Academic Press, Inc., New York (1974).
26. B. R. Johnson, A generalized JWKB approximation for multichannel scattering, *Chem. Phys.* **2**, 381–399 (1973).
27. B. C. Eu, Theory of inelastic collisions: Uniform asymptotic (WKB) solutions and semi-classical *S*-matrix elements for multichannel problems, *J. Chem. Phys.* **56**, 2507–2516 (1972).
28. J. N. L. Connor, On the semi-classical description of molecular orbiting collisions, *Mol. Phys.* **15**, 621–631 (1968).
29. A. S. Dickinson, An approximate treatment of shape resonances in elastic scattering, *Mol. Phys.* **18**, 441–449 (1970).
30. M. S. Child, Semi-classical theory of tunneling and curve crossing problems: A diagrammatic approach, *J. Mol. Spectrosc.* **53**, 280–301 (1974).
31. M. S. Child, in: *Molecular Spectroscopy* (R. F. Barrow, D. A. Long, and D. J. Millen, eds.), Vol. 2, pp. 466–512, Specialist Periodical Report, The Chemical Society, London (1974).
32. J. N. L. Connor and R. A. Marcus, Theory of semi-classical transition probabilities for inelastic and reactive collisions. II. Asymptotic evaluation of the *S* matrix, *J. Chem. Phys.* **55**, 5636–5643 (1971).
33. J. R. Stine and R. A. Marcus, Semi-classical transition probabilities by an asymptotic evaluation of the *S* matrix for elastic and inelastic collisions, *J. Chem. Phys.* **59**, 5145–5150 (1973).
34. J. H. Van Vleck, The correspondence principle in the statistical interpretation of quantum mechanics, *Proc. Natl. Acad. Sci. USA* **14**, 178–188 (1928).
35. V. A. Fock, On the canonic transformation in classical and quantum mechanics, *Vestn. Leningr. Univ. Ser. Mat. Fiz. Khim.* **16** (3), 67–70 (1959) [Tech. Transl. No. 60-17464, **4**, 53–58 (1960)].
36. H. Goldstein, *Classical Mechanics,* Addison-Wesley Publishing Company, Inc., Reading, Mass. (1959).
37. H. C. Corben and P. Stehle, *Classical Mechanics,* John Wiley & Sons, Inc., New York (1960).
38. W. H. Miller, Semi-classical theory of atom–diatom collisions: Path integrals and the classical *S* matrix, *J. Chem. Phys.* **53**, 1949–1959 (1970).
39. W. H. Miller, Classical S matrix: Numerical applications to inelastic collisions, *J. Chem. Phys.* **53**, 3578–3587 (1970).

40. J. Keller, Corrected Bohr–Sommerfeld quantum corrections for non-separable systems, *Ann. Phys. NY* **4**, 180–188 (1958).

41. W. H. Miller, Classical quantization of non-separable system, *J. Chem. Phys.* **56**, 38–45 (1972).

42. W. Eastes and R. A. Marcus, Semi-classical calculation of bound states of a multi-dimensional system, *J. Chem. Phys.* **61**, 4301–4307 (1974).

43. I. C. Percival, Regular and irregular spectra, *J. Phys. B* **6**, L229–232 (1973).

44. B. Leaf, Canonical operators for the simple harmonic oscillator, *J. Math. Phys. (N.Y.)* **10**, 1980–1987 (1969).

45. P. A. M. Dirac, The elimination of the nodes in quantum mechanics, *Proc. R. Soc. London Ser. A* **111**, 281–305 (1926).

46. P. Pechukas, Time-dependent semi-classical theory. I: Potential scattering, *Phys. Rev.* **181**, 166–173 (1969).

47. P. Pechukas, Time-dependent semi-classical theory. II: Atomic collisions, *Phys. Rev.* **181**, 174–185 (1969).

48. R. A. Marcus, Theory of semi-classical transition probabilities (*S* matrix) for inelastic and reactive collisions, *J. Chem. Phys.* **54**, 3965–3979 (1971).

49. H. Kreek and R. A. Marcus, Semi-classical collision theory, multidimensional integral method, *J. Chem. Phys.* **61**, 3308–3312 (1974).

50. W. H. Miller and T. F. George, Classical *S* matrix theory of reactive tunneling: Linear H + $H_2$ collisions, *J. Chem. Phys.* **57**, 2458–2467 (1972).

51. J. R. Stine and R. A. Marcus, Theory of semi-classical transition probabilities for inelastic and reactive collisions. IV. Classically inaccessible transitions calculated by integration along complex valued trajectories, *Chem. Phys. Lett.* **15**, 536–544 (1972).

52. W. H. Miller and T. F. George, Analytical continuation of classical mechanics for classically forbidden collision processes, *J. Chem. Phys.* **56**, 5668–5681 (1972).

53. R. A. Marcus, Semi-classical *S* matrix theory. VI: Integral expression and transformation of conventional coordinates, *J. Chem. Phys.* **59**, 5135–5144 (1973).

54. S. J. Fraser, L. Gottdiener, and J. N. Murnell, The relationship between classical phases in Cartesian and action–angle variables, *Mol. Phys.* **29**, 415–419 (1975).

55. J. D. Doll and W. H. Miller, Classical *S* matrix for vibrational excitation of $H_2$ by collision with He in three dimensions, *J. Chem. Phys.* **57**, 5019–5026 (1972).

56. D. R. Bates and D. S. F. Crothers, Semi-classical treatment of atomic collisions, *Proc. R. Soc. London Ser. A* **315**, 465–478 (1970).

57. J. B. Delos, W. R. Thorson, and S. K. Knudson, Semiclassical theory of inelastic collisions. I. Classical picture and semiclassical formulation, *Phys. Rev. A* **6**, 709–720 (1972).

58. I. C. Percival and D. Richards, A correspondence principle for strongly coupled states, *J. Phys. B* **3**, 1035–1046 (1970).

59. F. E. Heidrich, K. R. Wilson, and D. Rapp, Collinear collisions of an atom and a harmonic oscillator, *J. Chem. Phys.* **54**, 3885–3897 (1971).

60. W. Magnus, On the exponent solution of differential equations for a linear operator, *Commun. Pure Appl. Math.* **7**, 649–673 (1954).

61. P. Pechukas and J. C. Light, On the exponential form of time-displacement operators in quantum mechanics, *J. Chem. Phys.* **44**, 3897–3912 (1966).

62. K. H. Kramer and R. B. Bernstein, Sudden approximation applied to rotational excitation of molecules by atoms. I. Low-angle scattering, *J. Chem. Phys.* **40**, 200–203 (1964).

63. R. B. Bernstein and K. H. Kramer, Sudden approximation applied to rotational excitation of molecules by atoms. II. Scattering of polar diatomics, *J. Chem. Phys.* **44**, 4473–4485 (1966).

64. B. L. Van der Waerden, *Sources of Quantum Mechanics*, North-Holland Publishing Company, Amsterdam (1967).

65. I. L. Beigman, L. A. Vainshtein, and I. L. Sobel'man, Classical approximations in the theory of inelastic collisions, *J. Exp. Theor. Phys.* **30**, 920–923 (1970).

66. D. Richards, private communication, (1975).

67. E. Kerner, Note on the forced and damped oscillator in quantum mechanics, *Can. J. Phys.* **36**, 371–377 (1958).

68. H. B. Dwight, *Tables of Integrals and Other Mathematical Data*, 4th ed., The Macmillan Company, New York (1961).
69. R. I. Morse and R. J. LaBrecque, Collinear collisions of an atom and a Morse oscillator: An approximate semi-classical approach, *J. Chem. Phys.* **55**, 1522–1530 (1971).

# Nonadiabatic Processes in Molecular Collisions

## John C. Tully

### 1. Introduction

Substantial effort has been directed toward developing methods for describing molecular collision processes that are *electronically adiabatic*, i.e., for which it can be assumed that nuclear motion evolves on a single potential energy hypersurface. A number of recent reviews are devoted to this subject.[1-8] Considerably less attention has been paid to processes that are *nonadiabatic*, i.e., that involve electronic transitions between potential energy surfaces. This is in spite of the fact that nonadiabatic behavior is both common and important, even in thermal energy collisions.

Several articles reviewing aspects of the theory of nonadiabatic collisions have appeared recently.[9-19] These articles have been concerned primarily with inelastic atom–atom scattering, usually at relatively high collision energies. The emphasis of this chapter is on *low-energy molecular collisions*, i.e., encounters for which the kinetic energy associated with nuclear motion is less than $\sim 20\,\text{eV}$ and for which at least one of the colliding partners is a molecule. The low-energy regime includes almost all chemically interesting reaction and energy transfer processes occurring in planetary atmospheres, interstellar space, gas lasers, and plasmas as well as those studied in laboratory chemiluminescence and molecular beam experiments.

Electronically nonadiabatic behavior is observed in a great number of fundamental molecular processes.[17,20-30] These include

1. Electronic energy transfer,

$$A^* + B \rightarrow A + B^* \tag{1}$$

where the asterisk denotes electronic excitation.

*John C. Tully* • Bell Laboratories, Murray Hill, New Jersey

2. Charge transfer, e.g.,

$$A^+ + B \rightarrow A + B^+ \tag{2}$$

3. Quenching of electronic excitation,

$$A^* + B \rightarrow A + B^\dagger \tag{3}$$

where the dagger denotes internal (vibrational and rotational) excitation of molecule B.

4. Chemical reaction,

$$A + B \rightarrow C + D \tag{4}$$

Although many chemical reactions can be described within the adiabatic hypothesis, many cannot. Reactions involving electronically excited reactants or products are likely to exhibit nonadiabatic transitions because of the expected proximity of neighboring excited state potential energy surfaces. Similarly, ion–molecule reactions are frequently nonadiabatic due to the possibility of charge transfer. But electronic transitions are common even in reactions of ground state species at room temperature. In some cases the role of nonadiabatic coupling is readily apparent. Examples are spin forbidden reactions and reactions that are accompanied by excitation transfer. There are other more subtle cases where nonadiabatic effects dramatically alter the chemical forces that determine the course of reaction, but do not reveal themselves explicitly in the products.

Accurate theoretical descriptions of low-energy nonadiabatic molecular collision processes can be very difficult to obtain. The standard perturbation techniques of quantum scattering theory, applicable to high-energy collisions between weakly interacting systems, are generally inappropriate here where we are faced with the opposite situation, slow collisions with interactions that are strong and specific (e.g., curve crossings).

Only now is it becoming feasible to obtain accurate numerical solutions to the quantum mechanical equations describing low-energy atom–atom inelastic collisions. For molecular collisions we must resort to approximate procedures. The many types of nonadiabatic processes noted above are accompanied by an even larger selection of approximate theories, none of which are applicable to all cases of interest. We shall make no attempt to describe all these methods but rather shall concentrate on the fundamental aspects of the theory of low-energy nonadiabatic molecular collisions.

Two generally useful qualitative "rules" have emerged. The first of these is the Wigner spin rule, which states that total spin angular momentum is conserved throughout collision. This rule is obeyed in the majority of cases, but dramatic exceptions have been documented.[26] The second is the energy defect rule. This rule can be stated in several slightly different ways, one of which is that nonadiabatic transitions requiring the least transfer of electronic energy to or from nuclear kinetic energy are most probable. Regardless of how the rule is formulated, countless examples of failure can be cited.[15] Therefore these two rules, although useful as guidelines, cannot be considered general laws.

There is, however, one fundamental simplification applicable to virtually all low-energy molecular collision processes: the Born–Oppenheimer separation of electronic and nuclear motion.[31,32] This separation is basic to the theory of both adiabatic and nonadiabatic processes, even though the latter involve breakdown of the Born–Oppenheimer approximation. Its almost universal validity is a consequence of the huge ratio $M/m$ of nuclear to electron mass, which is a minimum of 1800 for the hydrogen atom. This mass discrepancy is manifested in the qualitatively different behavior of electrons and nuclei, the former being fast and highly quantal and the latter slow and nearly classical.

The description of an adiabatic collision process can be divided into two parts. The first, the electronic part, involves construction of a potential energy surface to describe the forces acting upon the nuclei. To do this to the required accuracy can be a very difficult task, ultimately requiring quantum mechanical molecular structure calculations of the highest level. The second part is treatment of the motion of the nuclei subject to the interactions embodied by the potential energy surface. This can be attempted by quantum mechanical, classical, or semiclassical methods.

Following Miller,[33] we shall distinguish among these methods in the following way. Quantum treatments require solving the Schrödinger equation governing nuclear motion to obtain a transition amplitude $S_{12}$. The square modulus of this $S$-matrix element is the observable quantity, the transition probability $P_{12}$:

$$P_{12} = |S_{12}|^2 \tag{5}$$

Classical and semiclassical procedures both employ the solution of the classical equations of nuclear motion but in different ways. In the classical method $P_{12}$ is obtained directly, without reference to a transition amplitude. Semiclassical procedures, on the other hand, employ a classical approximation to $S_{12}$ and obtain the transition probability by the quantum prescription, Eq. (5). Thus quantum interference effects arising from the superposition of transition amplitudes can be incorporated by semiclassical formulations but not by classical methods.

If nonadiabatic effects are important, the problem becomes even more demanding. We still take advantage of the huge nuclear/electron mass ratio to divide the problem into two parts. The electronic part is more difficult, however, requiring construction of two or more potential energy surfaces, each corresponding to a particular electronic state. In addition, calculation of the off-diagonal (nonadiabatic) interactions that promote transitions between potential energy surfaces is required. In Section 2 the electronic part of the problem is discussed, beginning with a presentation of the basic equations effecting separation of electronic and nuclear motion and the definitions of some useful concepts such as adiabatic and diabatic representations.

The treatment of nuclear motion is also more difficult if nonadiabatic transitions can occur. We can again distinguish quantum mechanical, classical, and semiclassical methods, according to the classification procedure described above. Quantum treatments of nuclear motion are discussed in Section 3. Two alternative classical approaches, the *multiple-crossing*[34] and *surface-hopping trajectory*[35] models, are described in Section 4. Some promising recent advances in the development of semiclassical methods, notably the theory of Miller and George,[36,37] are discussed in Section 5. Summarizing remarks are contained in Section 6.

## 2. Formulation of the Problem

### 2.1. Separation of Electronic and Nuclear Motion

Consider two molecules A and B with initial electronic, vibrational, and rotational states denoted, respectively, by $k_a$, $v_a$, $j_a$ and $k_b$, $v_b$, $j_b$. We shall distinguish two general classes of collisions. The first is inelastic (nonreactive) encounters in which the internal states of the molecules may be altered but not their molecular arrangements. The second class is chemically reactive collisions, during which nuclear rearrangement occurs. We can represent both classes by

$$A(k_a, v_a, j_a) + B(k_b, v_b, j_b) \rightarrow C(k_c, v_c, j_c) + D(k_d, v_d, j_d) \tag{6}$$

where the product molecules C and D will be identical to A and B if the collision is nonreactive but different if it is reactive.

To simplify notation, we shall absorb all quantum numbers describing the isolated reactants into two composite indices. The first index, $k$, designates the electronic state, i.e., $k$ encompasses the indices $k_a$ and $k_b$. The second index $\alpha$ is a combination of all the other reactant quantum numbers, i.e., $v_a$, $j_a$, $v_b$, $j_b$. Similar indices $k'$ and $\alpha'$ are defined for the products. If reactions are possible, $\alpha$ and $\alpha'$ include designation of the molecular arrangements.

The entire system is composed of $n$ electrons, whose positions relative to the center of mass of the system are designated collectively by the vector $\mathbf{r} = (\mathbf{r}_1, \ldots, \mathbf{r}_n)$, and $N$ nuclei, whose positions relative to the center of mass are specified by $\mathbf{R} = (\mathbf{R}_1, \ldots, \mathbf{R}_N)$.

We shall employ standard time-independent scattering theory. The collision process can be described completely in terms of a wave function $\Psi(\mathbf{r}, \mathbf{R})$, which is a solution of the time-independent Schrödinger equation and which, in addition, possesses the asymptotic behavior appropriate for the particular process of interest. For most molecular collision problems, in regions of asymptotically large separation of reactants or products, $\Psi$ will behave as an incoming plane wave corresponding to initial state $k\alpha$ and as outgoing spherical waves corresponding to all possible product states (and arrangements) $k'\alpha'$.[16] $\Psi$ describes both electronic and nuclear motion and depends on the total energy $\mathscr{E}$ and the initial state $k\alpha$, i.e.,

$$\Psi = \Psi(\mathbf{r}, \mathbf{R}; \mathscr{E}, k, \alpha) \tag{7}$$

For simplicity of notation we shall suppress the dependences on $\mathscr{E}$, $k$, and $\alpha$. The amplitudes of the various outgoing spherical waves of $\Psi$ are directly related to the elements of the $S$ matrix, Eq. (5), from which all observable information about the collision process can be obtained. In what follows we shall concentrate on methods for obtaining the wave function $\Psi$. Procedures for extracting from $\Psi$ the desired cross sections, angular distributions, etc., can be found in any of the standard scattering theory texts.[16,38–41]

The Hamiltonian operator $\mathscr{H}$ governing the motion of $n$ electrons and $N$ nuclei can be written as follows:

$$\mathscr{H} = \mathscr{T}_R + \mathscr{H}_0 \tag{8}$$

$\mathscr{T}_R$ is the nuclear kinetic energy operator,

$$\mathscr{T}_R = \sum_{M=1}^{N-1} -(\hbar^2/2\mu_M)\nabla_M^2 \tag{9}$$

The reduced masses $\mu_M$ appearing in Eq. (9) arise after separation of the motion of the center of mass of the system*.

$$\mathscr{H}_0 = \mathscr{H}_{el} + \mathscr{H}_{so} \tag{10}$$

where $\mathscr{H}_{el}$ is the usual electronic Hamiltonian for fixed positions of the nuclei:

$$\mathscr{H}_{el} = \sum_{i=1}^{n} -\frac{1}{2}\hbar^2\nabla_i^2 + \sum_{i=1}^{n-1}\sum_{j>1}^{n} \frac{1}{|\mathbf{r}_i - \mathbf{r}_j|} - \sum_{i=1}^{n}\sum_{M=1}^{N} \frac{z_M}{|\mathbf{r}_i - \mathbf{R}_M|} + \sum_{M=1}^{N-1}\sum_{M'>M}^{N} \frac{z_M z_{M'}}{|\mathbf{R}_M - \mathbf{R}_{M'}|} \tag{11}$$

$z_M$ is the charge on nucleus $M$. $\mathscr{H}_{so}$ is the spin–orbit operator, which we include here because even for systems involving light atoms, the effects of spin–orbit interaction are frequently as important as those of nonadiabatic coupling.

We shall now exploit the great disparity between the electron and nuclear masses to separate electronic and nuclear motion. This can be accomplished by selecting some basis set of electronic wave functions $\varphi_k(\mathbf{r}; \mathbf{R})$ which depend parametrically on the nuclear positions $\mathbf{R}$. For any value of $\mathbf{R}$, the $\varphi_k$ are assumed to be orthonormal and complete (i.e., span the subspace defined by the electronic coordinates $\mathbf{r}$). They are also assumed to vary in a continuous manner with $\mathbf{R}$. Otherwise they are, for now, unspecified. The total wave

---

*Equations (8)–(11) neglect "mass polarization" terms that result from transformation to center of mass coordinates.[42,43] The effect of these terms is entirely negligible in low-energy collisions.

function $\Psi$ can be expanded in terms of the electronic basis functions:

$$\Psi(\mathbf{r}, \mathbf{R}) = \sum_k \varphi_k(\mathbf{r}; \mathbf{R})\chi_k(\mathbf{R}) \tag{12}$$

Roughly speaking, the nuclear wave function $\chi_k(\mathbf{R})$ describes the motion of the nuclei on the potential energy surface associated with electronic state $k$. Substitution of Eq. (12) into the Schrödinger equation

$$[\mathcal{H} - \mathcal{E}]\Psi(\mathbf{r}, \mathbf{R}) = 0 \tag{13}$$

results in the usual infinite set of coupled equations for the $\chi_k(\mathbf{R})^{(44)}$:

$$[\mathcal{T}_R + \mathcal{T}''_{kk} + U_{kk} - \mathcal{E}]\chi_k = -\sum_{k' \neq k} [\mathcal{T}'_{kk'} + \mathcal{T}''_{kk'} + U_{kk'}]\chi_{k'} \tag{14}$$

where

$$U_{kk'} = U^{(el)}_{kk'} + U^{(so)}_{kk'} \tag{15}$$

$$U^{(el)}_{kk'} = \langle \varphi_k | \mathcal{H}_{el} | \varphi_{k'} \rangle \tag{16}$$

$$U^{(so)}_{kk'} = \langle \varphi_k | \mathcal{H}_{so} | \varphi_{k'} \rangle \tag{17}$$

$$\mathcal{T}'_{kk'} = \sum_{M=1}^{N-1} \left(\frac{-\hbar^2}{2\mu_M}\right) \mathbf{d}^{(M)}_{kk'} \cdot \nabla_M \tag{18}$$

$$\mathcal{T}''_{kk'} = \sum_{M=1}^{N-1} \left(-\frac{\hbar^2}{2\mu_M}\right) D^{(M)}_{kk'} \tag{19}$$

$$\mathbf{d}^{(M)}_{kk'} = \langle \varphi_k | \nabla_M \varphi_{k'} \rangle \tag{20}$$

$$D^{(M)}_{kk'} = \langle \varphi_k | \nabla_M^2 \varphi_{k'} \rangle \tag{21}$$

Brackets denote integration over electronic coordinates $\mathbf{r}$ only. The diagonal elements $U_{kk}(\mathbf{R})$ are the effective *potential energy surfaces* that govern nuclear motion. The additional diagonal terms $\mathcal{T}''_{kk}$ are nonadiabatic corrections to the potential energy surfaces.[32] These terms are invariably small and can usually be neglected. The off-diagonal terms $\mathcal{T}'_{kk'}$, $\mathcal{T}''_{kk'}$, and $U_{kk'}$ promote transitions between potential energy surfaces. We call $U^{(so)}_{kk'}$ the *spin–orbit interaction* and $\mathcal{T}'_{kk'}$ and $\mathcal{T}''_{kk'}$, or alternatively $\mathbf{d}^{(M)}_{kk'}$ and $D^{(M)}_{kk'}$, the *nonadiabatic interactions*. Note that $\mathcal{T}'_{kk'}$, which provides the dominant nonadiabatic coupling, is velocity dependent because it involves $\nabla_M$ operating on the nuclear wave function $\chi_{k'}$.

We have achieved our goal, a formally exact separation of electronic and nuclear motion. The electronic part consists of first selecting the electronic basis functions $\varphi_k$ and then obtaining the potential energy surfaces $U_{kk'}$ and off-diagonal coupling terms defined in Eqs. (15)–(21). The infinite set of coupled equations, Eq. (14), can then be solved for the motion of the nuclei. Although it may not be readily apparent that achieving this separation is of any benefit, in fact it invariably leads to tremendous simplification. As a consequence of the great disparity between electron and nuclear masses, it is almost always possible to select some set of electronic basis functions $\varphi_k$ that, for low-energy collisions, will allow the infinite set of equations, Eq. (14), to be very accurately approximated by one equation or a very few equations. This fact not only provides the justification of the adiabatic hypothesis in cases where it is applicable but is also central to all the treatments of nonadiabatic collision processes discussed in this chapter. In the remainder of this section we shall briefly discuss ways of selecting the basis functions $\varphi_k$ and calculating the properties defined in Eqs. (15)–(21), i.e., the electronic part of the problem. In Sections 3–5 we shall deal exclusively with methods for obtaining approximate solutions to a finite set of coupled equations of the form of Eq. (14), i.e., the nuclear motion part of problem.

## 2.2. Electronic Representations

### 2.2.1. Adiabatic Representation

If full advantage of the Born–Oppenheimer separation is to be taken, the electronic basis functions $\varphi_k$ must be chosen intelligently. It has long been recognized that it is not adequate to take the $\varphi_k$ to be the undistorted wave functions of the isolated molecules. Explicit allowance must be made for the distortion of the wave functions as the molecules interact at finite separations.[45] This is particularly true for reactive collisions, because it is these distortion effects that are primarily responsible for the weakening and strengthening of chemical bonds that accompany reaction.

We shall describe the two electronic representations that have proved most useful for describing low-energy molecular collisions. Each incorporates accurately the adjustment of the electrons to changing nuclear positions. Consequently, calculation of the electronic basis functions is frequently the most difficult part of the description of a molecular collision process.

The first and generally most useful representation, the *adiabatic representation*, employs for the electronic basis functions the eigenfunctions of $\mathcal{H}_0$:

$$[\mathcal{H}_0 - E_k]\varphi_k = 0 \tag{22}$$

The effective potential energy surfaces governing nuclear motion in this representation are

$$U_{kk}(R) = E_k(R) \tag{23}$$

The only nonzero coupling terms on the right-hand side of Eq. (14) are $\mathcal{T}'_{kk'}$ and $\mathcal{T}''_{kk'}$, the nonadiabatic interactions. It is for this reason that we call this the adiabatic representation.

In cases where spin–orbit coupling is not negligible, it may be useful to distinguish between this representation and one defined in terms of eigenfunctions of the electronic Hamiltonian operator $\mathscr{H}_{\text{el}}$:

$$[\mathscr{H}_{\text{el}} - E_k]\varphi_k = 0 \tag{24}$$

This is also frequently referred to as the adiabatic representation. In the absence of spin–orbit coupling the two representations become identical. To avoid confusion in the discussions that follow, unless it is specifically stated to the contrary, spin–orbit interactions will be neglected, and the term adiabatic will refer to the eigenstates of $\mathscr{H}_0 = \mathscr{H}_{\text{el}}$.

### 2.2.2. Diabatic Representations

In low-energy molecular collisions, nonadiabatic transitions occur with high probability only if two adiabatic potential energy surfaces approach each other closely. Very frequently these regions of close approach arise because the adiabatic states are mixtures predominantly of two simple electronic (molecular orbital or valence bond) structures whose corresponding potential energy surfaces cross. For example, approximate potential surfaces of $LiH_2$ associated with ionic and covalent configurations cross as shown in Fig. 1(b).[44] In one-dimensional (atom–atom) systems, if the approximate states are of the same symmetry, then the "noncrossing rule"[46] requires that the exact adiabatic potential energy curves, formed from linear combinations of the approximate ones, will *not* cross. Instead, they will exhibit an "avoided crossing," similar to those of Fig. 1(a). The generalization of the noncrossing rule to systems with $N$ internal nuclear degrees of freedom is as follows[47–49]: The locus of points defined by the intersection of two $N$-dimensional exact adiabatic potential energy hypersurfaces corresponding to states of the same symmetry forms a hypersurface of at most $N-2$ dimensions. A corresponding $(N-1)$-dimensional *avoided crossing* hypersurface can also be defined, as illustrated in Fig. 2. We frequently refer to the latter as an avoided crossing "seam."

During a molecular collision event, if the nuclei are moving sufficiently slowly, they will tend to follow a single adiabatic potential energy surface, even in the vicinity of an avoided crossing. On the other hand, if the nuclei move very rapidly through an avoided crossing, the probability of a transition from one adiabatic surface to the other will approach unity; there will not be enough time for the electrons to rearrange to a new configuration. In this limit, it is appropriate to choose for the electronic basis functions $\varphi_k$ the approximate electronic functions corresponding to particular electronic configurations, e.g., the ionic and covalent states in Fig. 1(b). Lichten has called this the *diabatic representation*.[50] The noncrossing rule does not apply to diabatic potentials, i.e., the dimension of the locus of points of intersection of two $N$-dimensional diabatic potential energy hypersurfaces can be as large as $N-1$.

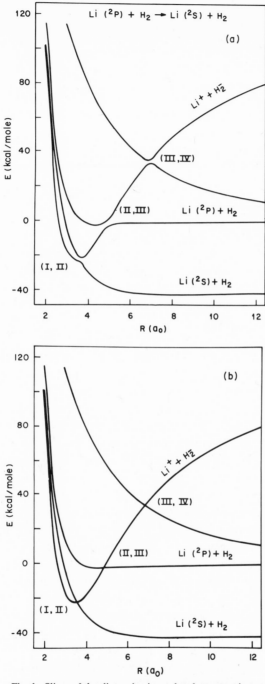

Fig. 1. Slices of the diatomics-in-molecules approximation to the four lowest $^2A'$ potential energy surfaces of
LiH$_2$ as a function of Li–H$_2$ separation. The orientation
of the H$_2$ internuclear axis is 60° relative to the Li–H$_2$
direction, and the H$_2$ internuclear distance is $1.4a_0$. (a)
Adiabatic representation, (b) diabatic representation.

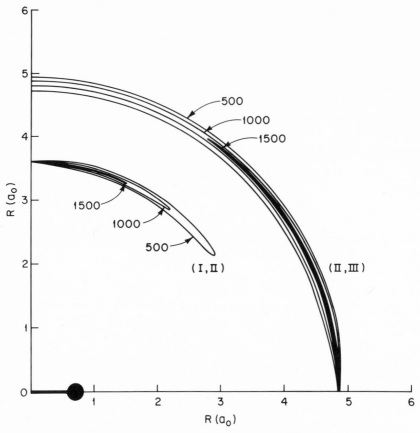

Fig. 2. Contours of the nonadiabatic interaction strength $|d_{kk'}^{(z)}/E_k - E_{k'}|$, in units of $a_0^{-1} \cdot \text{a.u.}^{-1}$, as a function of Li–$H_2$ separation and $H_2$ orientation. $z$ is taken along the Li–$H_2$ direction, and the $H_2$ internuclear distance is $1.4a_0$. (I, II) and (II, III) refer to the avoided crossings indicated in Fig. 1(a).

There is no generally accepted unique definition of the diabatic representation. Several different procedures for constructing diabatic potential energy surfaces have been employed. Some involve minimizing the electronic energy with respect to a subspace of electronic functions. This can be achieved using projection operator techniques[51] or by carefully selecting electronic functions only of the desired character, e.g., only ionic functions.[52] The simplest and possibly most useful method is to construct diabatic surfaces by extrapolation of the adiabatic potential energy surfaces in the vicinity of an avoided crossing.[9] As discussed by Smith,[53] a feature common to all these variously defined diabatic representations is that the electronic basis functions $\varphi_k$ are constructed so that the resulting off-diagonal nonadiabatic coupling $\mathscr{T}'_{kk'}$ is small.[53-54] Therefore, in a diabatic representation transitions between surfaces are induced primarily by off-diagonal elements $U_{kk'}$ of the Hamiltonian $\mathscr{H}_0$. If these off-diagonal elements are sufficiently small, then a diabatic representation may be preferred over the adiabatic representation, even at low collision energies.

It is important to note, however, that the concept of diabatic states will not be as useful in molecular collision problems as it has been for atom–atom processes. In systems with $N$ nuclear degrees of freedom, avoided crossings are not located at a single point but rather define a seam of $N-1$ dimensions. The separation between adiabatic surfaces, and therefore the nonadiabatic coupling, will in general vary at different positions along the seam. In fact, in most systems studied to date, this variation is quite dramatic.[44,55–57] See, for example, Figs. 2 and 3. This suggests that a diabatic representation might be appropriate in some regions and an adiabatic representation in others. In such cases, neither representation will affect a reduction in the number of coupled

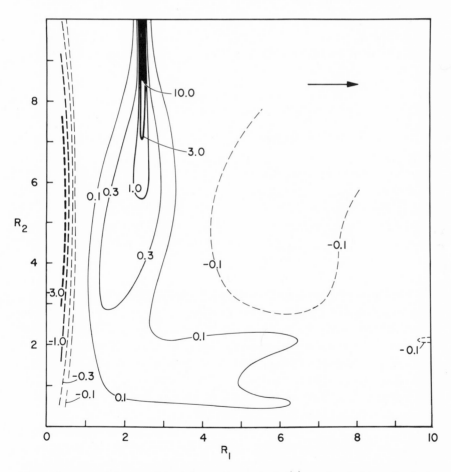

Fig. 3. Contours of the nonadiabatic interaction strength $d_{kk'}^{(\rightarrow)}$ associated with motion in the direction of the arrow, in units of $a_0^{-1}$, for collinear $H_3^+$. The asymmetry of the figure is a reflection of the vastly greater propensity of vibrational motion to induce transitions than of relative motion.

equations (14), so the rigorously founded adiabatic representation is preferable.

## 2.3. Electronic Motion

### 2.3.1. Choice of Representation

The electronic part of the problem requires choosing an optimal representation, deciding which states to include among the coupled equations, Eq. (14), and computing for these states the potential energy surfaces $U_{kk}$ and off-diagonal interactions $U_{kk'}$ and $\mathcal{T}'_{kk'}$ within the chosen representation. Choice of representation is frequently obvious from the outset. Whether it is obvious or not, usually the best procedure is to calculate all the required quantities, Eqs. (15)–(21), in terms of eigenfunctions of $\mathcal{H}_{el}$ (the Born–Oppenheimer wave functions). Transformations to another representation can be performed later, if desired.

### 2.3.2. Number of Coupled Equations

A more critical step is to determine which states must be retained in Eq. (14). There is no foolproof way of doing this. Generally, one must include all states that correlate with the reactants and perhaps several nearby states as well. Any potential energy surface that lies close to the initial surface (e.g., within a few tenths of 1 eV in energetically accessible regions) must be suspected of playing a role in the collision. The following criterion,[9] first proposed by Massey,[58] is sometimes useful as a rough guide:

$$\hbar \mathbf{v} \cdot \mathbf{d}_{kk'} / |E_k - E_{k'}| \ll 1 \qquad (25)$$

is the condition for adiabatic behavior, where $\mathbf{d}_{kk'}$ is the nonadiabatic coupling defined in Eq. (20), $\mathbf{v}$ is the (classical) nuclear velocity, and $E_k$ is the adiabatic energy defined in Eq. (22). In questionable cases, sample calculations of nuclear motion should be performed with the larger set of equations to test whether further reduction is permissible. Unfortunately, all too often practice has been to assume that nuclear motion evolves on a single potential energy surface unless nonadiabatic effects are so blatant they cannot possibly be ignored.

### 2.3.3. Potential Energy Surfaces

Computations are required of diagonal and off-diagonal matrix elements of $\mathcal{H}_{el}$, $\mathcal{H}_{so}$, and $\nabla$, Eqs. (16), (17), and (20). As noted above, it is almost always preferable to calculate these properties with respect to the Born–Oppenheimer wave functions defined in Eq. (24) and later to transform to a different representation if necessary. Off-diagonal elements of $\mathcal{H}_{el}$ are then identically zero, and the diagonal elements are the usual Born–Oppenheimer potential energy surfaces. Enormous effort has been directed toward developing methods for constructing potential energy surfaces suitable for describing molecular collision processes. We shall make no attempt to review these here. We shall mention only that, in addition to the usual constraints imposed for adiabatic collisions (e.g., correct dissociation limits), nonadiabatic processes require that ground and excited potential energy surfaces be of comparable quality and that features such as avoided intersections arising from the interaction between surfaces be included in a realistic manner. Because of these considerations, single-configuration approaches, whether *ab initio* (Hartree–Fock) or semiempirical (CNDO, INDO, MINDO, extended Hückel), are not generally useful. Of much more promise are *ab initio* configuration interaction methods[59] and approximate or semiempirical approaches based on pseudo-potential theory,[60–62] the *atoms-in-molecules* method,[63,64] and the *diatomics-in-molecules* method.[65–67]

### 2.3.4. Spin–Orbit Interactions

Very accurate wave functions are required for direct evaluation of the spin–orbit coupling, Eq. (17), making this a very difficult task for *ab initio* treatment. Because both one- and two-electron contributions are important, semiempirical approaches based on one-electron models are inadequate. It has been shown, however, that the sum of the one- and two-electron contributions from electrons on a single atomic center usually far outweigh two-center terms.[68,69] This fact has led to the development of semiempirical methods that include two-electron effects but ignore two-center terms.[44,69–71] These methods are straightforward to apply and appear to give results of accuracy quite adequate for the description of most low-energy collision processes.

### 2.3.5. Nonadiabatic Interactions

We have defined two nonadiabatic coupling terms, $\mathcal{T}'_{kk'}$ and $\mathcal{T}''_{kk'}$ of Eqs. (18) and (19). $\mathcal{T}'_{kk'}$ is usually the dominant term, and we shall restrict discussion to it, i.e., to calculation of $\mathbf{d}^{(M)}_{kk'}$ of Eq. (20). $\mathcal{T}''_{kk'}$ is usually negligible and can be computed using similar techniques.[44]

If the electronic functions $\varphi_k$ are chosen to be real and orthonormal, then the diagonal elements $\mathbf{d}^{(M)}_{kk}$ are identically zero. Two qualitatively different kinds of off-diagonal terms arise. In the case of atom–atom collisions, the gradient $\nabla_M$ in Eq. (20) has a radial component involving $\partial/\partial\rho$ and an angular component involving $\partial/\partial\theta$, where $\rho$ is the internuclear distance and $\theta$ defines the orientation of the molecular axis with respect to a space-fixed axis.[53] The two resulting types of nonadiabatic coupling are referred to, respectively, as *radial* and *rotational* (or *angular*, or *coriolis*). In systems of more than two atoms, a distinction can still be made between the ordinary coupling that results from changing internuclear separations and rotational coupling that arises from rotation of the entire molecular system.

Calculation of rotational nonadiabatic coupling is straightforward.[72,73] The spurious $\rho^{-2}$ behavior that appears in standard treatments is insignificant in low-energy collisions[73] and will not be discussed here.

Calculation of nonadiabatic coupling arising from relative motion is sometimes more difficult. It is important in molecular systems to recognize the implications of having more than one internuclear separation coordinate. It has already been noted that the strength of nonadiabatic coupling can vary significantly along an avoided crossing seam, so an avoided crossing cannot be characterized by a single coupling strength, as it can in atom–atom systems. More generally, contour plots of the nonadiabatic coupling must be constructed, as in Fig. 2 and 3. In addition, $\mathbf{d}_{kk'}^{(M)}$ is a vector quantity. There will usually be some preferred direction of nuclear motion that is most effective for promoting electronic transitions, e.g., the direction perpendicular to an avoided crossing seam. In molecular problems, avoided crossing seams of all different shapes can be found. In some cases, radial motion is most effective in inducing nonadiabatic behavior, e.g., the quenching of Li ($^2P$) by $H_2$, illustrated in Fig. 2. However, there are many cases where motion in other directions dominates. For example, vibrational motion is primarily responsible for promoting electronic transitions in $H_3^+$, as shown in Fig. 3. These considerations can be of crucial importance in determining the outcome of molecular collisions and are too often overlooked.

Nonadiabatic coupling strengths can be computed by several procedures.[72,74] The most general uses the finite difference expression

$$\langle \varphi_k(\mathbf{r}, \mathbf{R}) | \partial \varphi_{k'}(\mathbf{r}, \mathbf{R}) / \partial R_i \rangle \simeq a^{-1} \langle \varphi_k(\mathbf{r}, \mathbf{R}) | \varphi_{k'}(\mathbf{r}, \mathbf{R} + a\mathbf{i}) \rangle \qquad (26)$$

which follows from the orthogonality of $\varphi_k$ and $\varphi_{k'}$ at any $\mathbf{R}$. There is a potential complication that may arise from failure to properly account for the fact that electrons must be moving with their respective atomic centers. This introduces a spurious contribution to the nonadiabatic coupling that is nonvanishing at large internuclear separation. In atom–atom collisions the problem can be removed by inclusion of a *momentum transfer* factor.[75-78] It is more difficult to deal with in molecular systems. Fortunately, in the low-energy regime considered in this chapter, this effect is not significant.[77] Furthermore, the problem does not arise in most of the approximate methods mentioned below.

Electronic transitions are in most cases unlikely during low-energy collisions unless nonadiabatic interactions are quite large, i.e., unless a relatively abrupt change of electronic configuration occurs. Because only such major effects are important, approximate methods can be very useful in many situations. If the wave function $\varphi_k$ is expressed as a linear combination of basis functions that vary only slowly with **R**, then the **R** dependence of $\varphi_k$ is contained primarily in the linear coefficients. Therefore, if $\varphi_k$ is computed by an *ab initio* configuration interaction procedure, the nonadiabatic coupling can be obtained directly from the secular equation. Furthermore, this provides the justification for a very promising semiempirical method for computing nonadiabatic interactions, based on the diatomics-in-molecules method.[44]

If the $\varphi_k$ can be assumed to be exact solutions to the fixed nuclei problem, then a Hellman–Feynman type of relation might be useful[45,53]:

$$\langle \varphi_k | \partial \varphi_{k'} / \partial R \rangle \simeq (E_{k'} - E_k)^{-1} \langle \varphi_k | \partial \mathcal{H}_0 / \partial R | \varphi_{k'} \rangle \qquad (27)$$

Finally, if a well-defined avoided crossing exists, then the nonadiabatic coupling strength can be estimated directly from the shapes of the adiabatic surfaces. If $x$ is the perpendicular distance from the crossing, $\Delta F$ is the difference in slopes of the diabatic curves drawn through the crossing, and $2H_{12}$ is the minimum separation of the adiabatic curves, then[9]

$$\langle \varphi_k | \partial \varphi_1 / \partial x \rangle \simeq \Delta F \cdot H_{12} / [(\Delta F x)^2 + 4H_{12}^2] \qquad (28)$$

## 3. Quantum Mechanical Methods

### 3.1. Classification of Collision Processes

In Sections 3, 4, and 5 we shall deal exclusively with obtaining approximate solutions to the coupled equations governing nuclear motion, Eq. (14). The methods employed depend to a large extent on the nature of the process to be described. We can distinguish among several different situations.

### 3.1.1. Adiabatic

In the simplest case, Eqs. (14) uncouple in the adiabatic representation, i.e., nonadiabatic coupling is negligible and to a good approximation nuclear motion evolves on a single adiabatic potential energy surface. Methods for treating nuclear motion in such cases are reviewed elsewhere.[1–8]

### 3.1.2. Diabatic

In some cases, Eqs. (14) do not decouple in the adiabatic representation but do in some properly chosen diabatic representation; nuclear motion evolves on a single diabatic potential energy surface. Strictly speaking, such processes are nonadiabatic, but any of the standard techniques for treating adiabatic collisions apply to diabatic ones as well.

### 3.1.3. Symmetric Resonant Transfer[11,17,79,80]

Excitation transfer or charge transfer in symmetric systems where exact resonance applies can be described within the adiabatic approximation. A simple example is the process

$$He^+ + He \rightarrow He + He^+ \qquad (29)$$

At low collision energies this process can be described in terms of the two electronic states, one *gerade* and one *ungerade*, that dissociate to the ground state ion and atom. Before collision the relative phase of the *g* and *u* states must be prescribed so that the charge is initially localized on the proper atom. Nuclear motion evolves independently on the two adiabatic potential energy surfaces; there are no nonadiabatic transitions between surfaces. Therefore standard single potential surface treatments can be used. However, the probability of electron transfer depends on the relative phase of the scattering amplitudes $f_g$ and $f_u$, i.e., the elastic ($+$) and charge exchange ($-$) cross sections take the form

$$\sigma_{\pm} = |f_g \pm f_u|^2 \qquad (30)$$

### 3.1.4. Degenerate Reactant States

In many cases one or both of the isolated reactant species exhibit electronic degeneracies that are removed as the colliding partners begin to interact at finite separations. An example is the reaction of $F\,(^2P_{3/2})$ with $H_2$. As the reactants approach the degenerate $F(^2P_{3/2})$ splits into two states corresponding in collinear configurations to $^2\Sigma^+$ and $^2\Pi$ states of FHH. Because there are obviously regions where these potential energy surfaces lie very close to each other, nonadiabatic transitions between them will be likely. If one is concerned with the polarization of the F atom, as with depolarization or ESR line width experiments, then these transitions must be treated explicitly, and the process must be included among those of Section 3.1.5.[81] If one is not interested in polarization effects, and, in addition, transitions between the surfaces are likely only in the asymptotic region, then nonadiabaticity can be ignored. Standard procedures for treating nuclear motion on a single potential energy surface can be applied, complicated only by the fact that calculations must be performed individually on each potential energy surface and the results weighted by the correct statistical probability.[82–84]

### 3.1.5. Electronic Transitions

There are a great many cases for which Eq. (14) cannot be decoupled in any representation.* Transitions between potential energy surfaces must be

---

*A set of coupled equations like Eq. (14) can always be decoupled formally, for example, by employing effective potentials that are in general nonlocal and complex-valued. However, construction of such an effective potential is equivalent to solving the original set of coupled equations.

accounted for. All the accidentally resonant and nonresonant processes mentioned in Section 1 fall into this category. Most methods designed to treat nuclear motion on a single potential energy surface are not directly applicable to collisions involving electronic transitions, so these methods must be extended or new methods developed.

### 3.1.6. Collisional Ionization

If any of the possible product channels involve a free electron, then the coupled equations (14) become inappropriate, because an infinite number of equations representing the continuum of possible electronic states would have to be retained. Processes such as Penning ionization and associative ionization are examples. The most generally useful procedure for treating collisional ionization involves construction of complex effective potential energy surfaces, the imaginary part determining the rate of electron ejection. We shall not discuss these processes in this chapter, and we refer the reader elsewhere.[14,51,85,86]

## 3.2. Quantum Treatments of Nuclear Motion

We shall focus attention now, and in the next two sections, on the description of processes in Section 3.1.5, those explicitly involving electronic transitions. In this section we shall discuss quantum mechanical methods for describing nuclear motion, postponing classical and semiclassical approaches until Sections 4 and 5.

It is now practical to obtain quite accurate numerical solutions of the coupled equations (14) for low-energy atom–atom collision processes, at least if relatively few equations need to be retained. Results have been reported recently for nonresonant excitation transfer,[87,88] nonresonant charge transfer,[89] fine structure transitions,[90,91] and model problems.[92-95] This development is very significant, both because it is now feasible to describe important atomic collision processes with predictive accuracy and also because such solutions provide standards against which various approximation schemes potentially applicable to molecular systems can be tested. The atom–atom methods themselves usually cannot be carried over directly to molecular processes. In the latter, electronic transitions generally occur concurrently with vibrational and rotational transitions and chemical reaction. Methods capable of describing all these processes must be designed.

Most quantum mechanical approaches begin with a close coupling expansion of the nuclear wave functions $\chi_k(\mathbf{R})$ of Eqs. (12) and (14):

$$\chi_k(\boldsymbol{\rho}, \mathbf{R}') = \sum_{\alpha} u_{k\alpha}(\boldsymbol{\rho})\xi_{k\alpha}(\boldsymbol{\rho}; \mathbf{R}') \tag{31}$$

where $\mathbf{R}'$ encompasses all the internal (rotational–vibrational) coordinates of the reactant molecules, and $\boldsymbol{\rho}$ is the relative position of one molecule with respect to the other. $\xi_{k\alpha}$ is the nuclear motion wave function for the $\alpha$ internal vibrational–rotational state of the $k$ electronic state of the colliding molecules. In some cases $\xi_{k\alpha}$ can be taken to be a simple product of the two internal wave functions of the isolated molecules and is then independent of $\boldsymbol{\rho}$. More frequently, it is desirable to account for distortion of vibrational and rotational motion as the molecules approach by introducing a parametric dependence of $\xi_{k\alpha}$ on $\boldsymbol{\rho}$. The wave functions $u_{k\alpha}$ describing relative motion play the role of linear expansion coefficients of the various possible internal states. Usually the $u_{k\alpha}$ are expanded further in terms of spherical harmonics involving the relative orientation angles. We shall omit this step here.

Substitution of Eq. (31) into Eq. (14) produces the following infinite set of coupled differential equations:

$$\Omega_{k\alpha}u_{k\alpha}(\boldsymbol{\rho}) = {\sum_{k'\alpha'}}' \Lambda_{k\alpha,k'\alpha'}u_{k'\alpha'}(\boldsymbol{\rho}) \tag{32}$$

where

$$\Omega_{k\alpha} = -\frac{\hbar^2}{2\mu}\nabla_\rho^2 + \mathscr{U}_{k\alpha,k\alpha}(\boldsymbol{\rho}) - \mathscr{E} \tag{33}$$

$$\Lambda_{k\alpha,k'\alpha'} = -\mathscr{U}_{k\alpha,k'\alpha'}(\boldsymbol{\rho}) - \frac{\hbar^2}{2\mu}(\gamma_{k\alpha,k'\alpha'}\langle\varphi_k\nabla_\rho\varphi_{k'}\rangle$$

$$+ \delta_{kk'}\int \xi_{k\alpha}\nabla_\rho\xi_{k'\alpha'}d\mathbf{R}')\cdot\nabla_\rho \tag{34}$$

and

$$\mathscr{U}_{k\alpha,k'\alpha'} = \int \xi_{k\alpha}[\mathscr{T}'_{kk'}+U_{kk'}]\xi_{k'\alpha'}\,d\mathbf{R}' \tag{35}$$

$\delta_{kk'}$ is the Kronecker delta function, and $\gamma_{k\alpha,k'\alpha'}$ is the overlap between internal nuclear wave functions on different potential energy surfaces:

$$\gamma_{k\alpha,k'\alpha'} = \int \xi_{k\alpha}\xi_{k'\alpha'}\,d\mathbf{R}' \tag{36}$$

The prime on the summation sign in Eq. (32) indicates that the term with both $k = k'$ and $\alpha = \alpha'$ is omitted. Brackets again indicate integration over electronic coordinates $\mathbf{r}$ only. Equation (32) can be rewritten in matrix form,

$$\boldsymbol{\Omega}\mathbf{u} = \boldsymbol{\Lambda}\mathbf{u} \tag{37}$$

where the matrix operator $\boldsymbol{\Omega}$ is diagonal with elements given by Eq. (33), and $\boldsymbol{\Lambda}$ has zeros along the diagonal and off-diagonal elements given by Eq. (34).

The problem, then, has been reduced to one of obtaining approximate solutions to Eq. (37). Unfortunately, due to the huge number of rotational–vibrational states associated with molecular systems, this can be an enormously difficult task. There are as yet no reported realistic quantum descriptions of nuclear motion for low-energy nonadiabatic molecular collisions, even for triatomic systems. Still, the situation is not hopeless. Some of the available approaches are outlined below.

### 3.2.1. Close Coupling

Several powerful algorithms for obtaining numerical solutions to large systems of coupled equations have recently been developed.[96–102] Some of these algorithms address the differential equations, Eq. (37), directly, whereas others employ the equivalent set of integral equations. These methods have been applied to electronically adiabatic molecular collision processes with modest success. Rotationally inelastic collisions at low energies (small number of open channels) can be described to high accuracy.[103,104] Vibrational inelasticity is more difficult to treat, because if the energy is sufficient to excite vibration, then many rotational levels will be accessible as well. Most studies have used simple models such as the "breathing sphere" approximation or collinear pictures.[1,5] Even for atom–diatom encounters, extremely few results have been reported that explicitly include both vibrational and rotational channels in the solution of the coupled equations.[105]

Rearrangement collisions (chemical reactions) impose the additional difficulty that a set of vibration–rotation basis functions $\xi_{k\alpha}$ suitable for describing reactants is usually not suitable for products. This problem can be surmounted either by defining a set of "reaction coordinates" that vary smoothly from reactants to products[106] or by carrying out a transformation between reactant and product bases. Neither approach is without complication. Partly due to this problem, but primarily because of the huge number of (open and closed) channels that must be included, solutions have been limited, with one notable exception,[107] to collinear and coplanar models.[6,8,108]

There is no fundamental additional difficulty in treating nonadiabatic molecular collisions by close coupling techniques. In practice, the non-orthogonality of vibrational states on different potential energy surfaces and the dominant role of velocity-dependent coupling terms may cause annoyances, but these problems can certainly be overcome. A convenient procedure for handling velocity-dependent nonadiabatic coupling, based on Eq. (26), has been suggested by Johnson and Levine.[109]

Techniques for describing reactive collisions may require more significant modification. In particular, reaction coordinates that are appropriate for one potential energy surface may not be for another. Nakamura[110] has described a method for treating nonadiabatic collinear reactive collisions that assumes that vibrational excitation and electronic transitions occur in distinctly separate regions. Under this assumption, a reaction coordinate can be usefully defined. This may not be the case in general, however, and methods involving transformations between (or matching of) reactant and product basis functions may be required.

Close coupling treatments of electronically nonadiabatic collinear collisions of an atom with a diatom have been reported recently by two groups. Top and Baer[111] have studied a model reactive system in which nonadiabatic transitions are induced by vibrational motion. They conclude that, for their model, it is possible to describe reaction in terms of motion along the initial potential energy surface and to treat nonadiabatic interaction separately later.

Zimmerman and George[112,113] have chosen interactions to model collinear $X + H_2$ reactions, were $X = F$, Cl, Br, or I. They observe that, near threshold, energy resonance between electronic and vibrational levels can be an important factor, whereas at higher collision energies resonance effects are less noticeable.

Model studies such as these can be very instructive, and more should be carried out in the near future. Unfortunately, the computer technology is not yet available to solve the coupled equations for realistic three-dimensional nonadiabatic molecular collisions, even for triatomic systems, except perhaps in the very simplest cases. However, development of computational methods for treating adiabatic collisions is currently a very active field, and we can hope for breakthroughs that can be carried over to nonadiabatic processes.

### 3.2.2. Variational Principles

The application of variational principles[6,16] to molecular collision problems has begun only recently. Methods for treating large systems of coupled equations, including rearrangements, appear to have promise,[114–117] but it would be premature to speculate on their potential applicability to electronically nonadiabatic processes.

### 3.2.3. Perturbation Approaches

The most widely applied perturbation approach in scattering theory is the *Born approximation*.[16] It is essentially a high-energy, weak coupling approximation and is rarely useful in low-energy molecular collision problems, chiefly because it cannot adequately account for the influence of strong chemical forces on nuclear motion. However, the Born approximation has been shown to be useful in certain near-resonant processes dominated by long-range interactions.[118]

Like the Born approximation, the *distorted wave approximation*[16] DWA) is applicable only in weak coupling situations. However, the DWA can take into account chemical forces acting on the nuclei and is therefore not restricted to high collision energies or long-range interactions. In the present context, a DWA treatment consists of initially setting the off-diagonal elements $\Lambda$ of Eq. (37) equal to zero and solving the resulting uncoupled homogeneous equations for the initial and final channels of interest, $u_i^{(0)}$ and $u_f^{(0)}$. The first-order DWA transition probability is then proportional to the integral

$$\int u_i^{(0)} \Lambda_{if} u_f^{(0)} \, d\rho$$

This method is applicable only in cases where transitions from initial to final state are direct, i.e., no intermediate states are involved. The method does not satisfy flux conservation. In fact, deviations from flux conservation can sometimes serve as a useful indication of the accuracy of the method. Like the Born approximation, the DWA can be extended to higher order. However, both are likely to be useful only when accurate to first order.

The internal wave functions $\xi_{k\alpha}$ of Eq. (31) can be defined in a variety of ways. How this is done can be crucial to the success of perturbation approaches such as the DWA. Usually, the most accurate procedure is to employ adiabatic wave functions, i.e., to take the $\xi_{k\alpha}$ to be eigenfunctions of the Hamiltonian describing internal nuclear motion at every relative position $\rho$. This is entirely analogous to the adiabatic representation for electronic motion described in Section 2.2.1 and frequently results in an important reduction of the magnitude of the off-diagonal elements of $\Lambda$ in Eq. (37). This in turn can greatly improve the accuracy of the DWA.

There have been no reported applications of the DWA to nonadiabatic molecular collisions. The method has been applied to nonadiabatic atom–atom collisions at low energy and has proved successful in situations of weak coupling, i.e., low transition probability.[119] The method has also been of value in describing certain nonreactive adiabatic molecular collisions[1,120] and may even be useful for reactive collisions in some cases.[121]

Because there appear to be regimes within which the DWA is valid for all these types of processes, it is likely that there will also exist cases where it can be usefully applied to low-energy nonadiabatic molecular collisions. Unitarization procedures for extending the range of the DWA, such as the exponential approximation,[4,122,123] should also be valuable.

### 3.2.4. Reduction of Number of Channels

Because the basic difficulty associated with the application of close coupling methods to molecular collision processes is the enormous number of channels required, there has naturally been substantial effort directed toward developing approximations that will reduce the number of equations needed. Notable among these are methods based on the vibrationally adiabatic approximation,[124–130] the sudden approximation,[131–135] and the effective Hamiltonian approach.[136–144] These methods have not yet been applied to nonadiabatic processes, but they appear very promising. In fact, in the opinion of the author, continued progress in this area is our greatest hope for developing tractable quantum mechanical procedures for describing low-energy nonadiabatic molecular collisions.

## 4. Classical Mechanical Methods

### 4.1. Classical Path Equations

*4.1.1. Derivation*

The classical trajectory method is the most popular and most successful technique for describing the dynamics of adiabatic molecular collisions.[1,3,7] Extension of the classical approach to electronically nonadiabatic processes is not entirely straightforward. An electronic transition is a quantum mechanical phenomenon and cannot be accurately incorporated into any purely classical method. Nevertheless, the *classical path* approach, a hybrid of classical and quantum mechanics, has become the most useful technique for describing low-energy nonadiabatic molecular collisions. The classical path equations were first suggested by Mott[45] and have since been derived in a variety of ways. We shall obtain the equations in a very simple way that is not restricted to one-dimensional (atom–atom) situations.

We shall separate the electronic and nuclear coordinates into two groups, *internal* coordinates $\mathbf{q}$ and *external* coordinates $\mathbf{Q}$. Usually $\mathbf{q}$ will be taken to be the electronic coordinates $\mathbf{r}$ and $\mathbf{Q}$ the nuclear coordinates $\mathbf{R}$. We shall use different notation here to permit inclusion of some nuclear (vibration, rotation) coordinates among the internal coordinates $\mathbf{q}$, if we so desire.

The total Hamiltonian describing both internal and external motion can be written, analogous to Eq. (8),

$$\mathcal{H} = \mathcal{T}_Q + \mathcal{H}_q \tag{38}$$

where $\mathcal{H}_q$ is the internal motion Hamiltonian and $\mathcal{T}_Q$ is the kinetic energy operator for the external coordinates. Following Section 2, we select an orthonormal set of basis functions $\phi_j(q; Q)$ that describe the internal motion for fixed values of the external coordinates. We are free to choose an adiabatic or diabatic representation, exactly the same as in Section 2.2.

We shall now assume that the motion of the external coordinates can be described by some as yet unspecified classical trajectory:

$$\mathbf{Q} = \mathbf{Q}(t) \tag{39}$$

The internal motion Hamiltonian $\mathcal{H}_q$ is now a time-dependent operator, depending on $t$ through $\mathbf{Q}(t)$. We can define a wave function $\Phi(\mathbf{q}, t)$ describing internal motion and satisfying the time-dependent Schrödinger equation

$$\mathcal{H}_q(\mathbf{q}, \mathbf{Q})\Phi(\mathbf{q}, t) = i\hbar\frac{\partial}{\partial t}\Phi(\mathbf{q}, t) \tag{40}$$

We expand $\Phi$ in terms of the internal wave functions $\phi$:

$$\Phi(\mathbf{q}, t) = \sum_j a_j(t)\phi_j(\mathbf{q}, \mathbf{Q}) \exp\left[-\frac{i}{\hbar}\int^t W_{jj}(\mathbf{Q})\, dt\right] \tag{41}$$

where it is understood that $\mathbf{Q}$ and $t$ are related by Eq. (39). $W_{ij}$ is defined by

$$W_{ij}(\mathbf{Q}) = \langle \phi_i \mathcal{H}_q \phi_j \rangle \tag{42}$$

where the angle brackets now indicate integration over internal coordinates $\mathbf{q}$. In Eq. (41), the expansion coefficients have been written as a product of a function $a_j(t)$ and an exponential factor simply for convenience. Substituting Eq. (41) into Eq. (40) and multiplying from the left by $\phi_k$, we obtain

$$i\hbar \dot{a}_k = \sum_{j \neq k} a_j \left( W_{kj} - i\hbar \left\langle \phi_k \Big| \frac{\partial \phi_j}{\partial t} \right\rangle \right) \exp\left[ -\frac{i}{\hbar} \int^t (W_{jj} - W_{kk})\, dt \right] \tag{43}$$

Using the chain rule

$$\left\langle \phi_k \Big| \frac{\partial \phi_i}{\partial t} \right\rangle = \dot{\mathbf{Q}} \cdot \langle \phi_k | \nabla_Q \phi_j \rangle \tag{44}$$

we obtain the basic equations of the classical path method:

$$i\hbar \dot{a}_k = \sum_{j \neq k} a_j \left( W_{kj} - i\hbar \dot{\mathbf{Q}} \cdot \langle \phi_k | \nabla_Q \phi_j \rangle \right) \exp\left[ -\frac{i}{\hbar} \int^t (W_{jj} - W_{kk})\, dt \right] \tag{45}$$

The probability of transition $|a_k(t \to \infty)|^2$ can be obtained from the coupled equations (45) for any assumed trajectory $\mathbf{Q}(t)$. Analogous to the quantum treatment, Eq. (14), there are two types of terms that contribute to the promotion of transitions among internal states. The first involve $W_{kj}$, the off-diagonal elements of the internal Hamiltonian, and therefore vanish if the $\phi_j$ are defined in an adiabatic representation. The second are velocity dependent and involve the matrix element of $\nabla_Q$, completely analogous to the nonadiabatic coupling of Eq. (20).

### 4.1.2. Validity

The development above leaves two important questions largely unanswered. First, under what conditions are the classical path equations accurate? Second, how should the classical path be selected? These questions are related, and neither is easily resolved. We shall address first the question of justification of the method.

We can state one criterion immediately: The classical path approach can be valid only in situations where the classical trajectory method would be valid in the absence of electronic transitions. Assessment of the accuracy of the classical trajectory method is a difficult task for which we refer the reader elsewhere.[3,7]

If transitions between potential energy surfaces are to be described accurately, additional criteria must be satisfied. Unfortunately, these are equally difficult to quantify. Derivations of the classical path equations that employ semiclassical (WKB) arguments shed some light on the validity of the approach as applied to systems with a single external coordinate.[145–148] Delos et al.[147,148] have given careful consideration to this problem. They conclude that the classical path equations are valid if, in addition to the validity of classical mechanics mentioned above, the following two conditions are met. First, the difference in energy of the interacting potential curves, in the region of significant interaction must be small compared to the kinetic energy of the nuclei. Second, either the off-diagonal interaction must be small near the turning points, *or* the forces must have the same sign near the turning points. There are examples that indicate that the classical path equations may be valid under conditions less restrictive than these.[149] Nevertheless, the analysis of Delos et al., although not strictly applicable in cases involving more than one external degree of freedom, can serve as a useful guide.

In systems involving more than one external coordinate, we can say at least the following: Provided that external motion on a single potential energy surface can be adequately described by classical mechanics, the classical path equations (45) should be valid in two types of regions. The first type is regions where off-diagonal coupling is sufficiently small that transitions between surfaces are unimportant (e.g., regions A and C of Fig. 4). An approximate

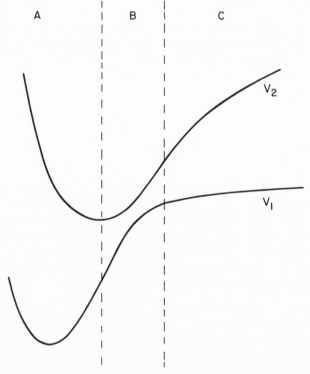

Fig. 4. Schematic illustration of two interacting adiabatic potential energy surfaces. Off-diagonal coupling is assumed weak in regions A and C and strong in region B.

criterion for this was presented earlier in Eq. (25). The classical path equations should also be accurate in regions where nonadiabatic interaction is large enough to promote transitions, but only between potential energy surfaces that are so similar in the interaction regions that nuclei would follow essentially the same trajectory on either surface (e.g., region B of Fig. 4). A very rough criterion is

$$|W_{22} - W_{11}|/T \ll 1 \tag{46}$$

where $T$ is the kinetic energy of (external) nuclear motion. These criteria are not mutually exclusive, so regions of types 1 and 2 might overlap. In fact, there appear to be many situations where this occurs, i.e., for which Eq. (25) is satisfied in regions of large and moderate splittings between surfaces and Eq. (46) is satisfied in regions of moderate and small splittings, so the classical path equations are valid everywhere. Unfortunately, there are also cases that exhibit intermediate regions where neither Eq. (25) nor (46) is satisfied.[84] How far the classical path approach can be extended to treat such cases is unknown at present and awaits future investigation.

### 4.1.3. Prescription of Trajectory

The second unanswered question is how to select the classical trajectory $\mathbf{Q}(t)$. The most commonly used method, the *impact parameter* method,[16,150] assumes that nuclei move in straight lines. The impact parameter method, like the closely related Born approximation,[151] is applicable only in cases where interactions are weak and long range and is therefore rarely useful in low-energy molecular collisions. Related approaches based on the eikonal method have been employed,[12,152–155] but these, too, are generally applicable only for collision energies somewhat higher than those considered here.

In the majority of cases trajectories must be prescribed in a way that accurately reflects the influence of strong chemical forces on the nuclei. For adiabatic processes this can be accomplished by numerical solution of the classical equations of motion associated with the appropriate potential energy surface. It is more difficult for nonadiabatic processes involving more than one potential energy surface. Consider the example illustrated in Fig. 4. Assume that nonadiabatic coupling is sufficiently small in regions A and C that transitions between surfaces do not occur in these regions. Coupling is assumed large and transitions likely in region B. $\mathbf{Q}(t)$ is frequently taken to be the classical trajectory defined by an effective potential, $V_{\text{eff}}(\mathbf{Q})$, that is some sort of average of the two potential surfaces 1 and 2 of Fig. 4. Common choices are

$$V_{\text{eff}} = \tfrac{1}{2}(V_1 + V_2) \tag{47}$$

$$\frac{\partial V_{\text{eff}}}{\partial Q} = \left(\frac{\partial V_1}{\partial Q}\right)^{1/2}\left(\frac{\partial V_2}{\partial Q}\right)^{1/2} \tag{48}$$

and

$$V_{\text{eff}} = |a_1|^2 V_1 + |a_2|^2 V_2 \qquad (49)$$

The third of these equations is a weighted average involving the populations $|a_i|^2$ of the two internal states. There is an obvious shortcoming associated with any prescription of this type. After collision, some fraction of the receding products will be in internal state 1 of Fig. 4, and their motion should be governed by the potential surface $V_1$. The remaining products should be governed by $V_2$. Equations (47)–(49) demand that both types of products follow the same unphysical potential energy surface, $V_{\text{eff}}$, even in regions of weak nonadiabatic coupling. Nevertheless, when applied to atom–atom collisions, single-trajectory* procedures such as those of Eqs. (47)–(49) often provide accurate transition probabilities, particularly for relatively high collision energies where the differences between potential surfaces are less significant. Single-trajectory methods may also be adequate for describing certain molecular collisions, namely, large impact parameter, weakly interacting encounters, especially at high collision energies. In general, however, explicit account must be taken of the fact that trajectories on different potential energy surfaces may follow substantially different paths.[156] There are a great many examples in which some potential surfaces are attractive and can lead to chemical reaction, whereas others are repulsive and cannot.[84] It would be meaningless to attempt a single-trajectory description in such cases. Even in nonreactive systems, forces governing nuclear motion frequently differ so greatly from one potential surface to another that a single trajectory is clearly inadequate. In the vast majority of low-energy nonadiabatic molecular collisions, multiple (branching) trajectories must be employed.

Using the Feynman path integral approach to quantum mechanics,[157] Pechukas and Pechukas and Davis have developed a formal prescription that has all the desired properties,[158,159] i.e., a different trajectory is defined for each final internal state, and the effective potential that determines one of these trajectories changes from that of the initial state to that of the final state, regardless of the magnitude of the transition probability. Unfortunately, the effective potentials are nonlocal, and in general the method is very difficult to apply. We shall discuss this approach and the related method of Miller and George[36] in the next section.

In many (perhaps most) cases, we may not need sophisticated techniques to define the branching trajectories. We have argued above that for the classical path approach to be valid it must be possible to divide configuration space into two types of region, weakly and strongly interacting. In weak interaction

---

*A classical description of a molecular collision process will always require sampling of a great many trajectories. The question we are addressing here is whether a trajectory evolving from a particular specification of initial conditions should be allowed to split into several branches, or whether it should remain intact as a single trajectory.

regions there are no transitions, so trajectories will follow a single potential energy surface. In strong interaction regions where trajectories split into branches, the potential energy surfaces must be sufficiently similar that any reasonable choice of path through these regions will be acceptable. All that remains is to match trajectories at the region borders. This is the basis of the surface-hopping trajectory method to be discussed later in this section.

## 4.2. One-Dimensional Model Solutions

Analytic solutions to Eq. (45) have been derived for one-dimensional, two-state systems with several different simplified interaction potentials. In the past these solutions have proved very valuable in analyzing atom–atom scattering experiments. However, it is now feasible to obtain exact numerical solutions of Eq. (45), or even the quantum equations (14), for one-dimensional problems with arbitrary interactions. Consequently, the value of these approximate solutions is reduced, and it is no longer profitable to examine them in detail. We shall mention briefly a few of the more important approximate expressions for estimating nonadiabatic transition probabilities. They remain important not only for making quick estimates but also because they may be of help in simplifying the description of complicated molecular processes perhaps requiring computation of transition probabilities along thousands of trajectories.

### 4.2.1. Landau–Zener–Stueckelberg Approximation

The Landau–Zener–Stueckelberg (LZS) approximation, the most famous and most important of the simple model solutions, applies to systems that exhibit avoided crossings. The assumptions of the model are as follows:

1. There is only one external coordinate $Q$.
2. Only two internal states, labeled 1 and 2, interact.
3. The velocity $\dot{Q}$ is constant in the coupling region.
4. In terms of diabatic internal wave functions $\phi_j$, defined so that the nonadiabatic coupling of Eq. (44) is zero, the internal Hamiltonian matrix $\mathbf{W}$ of Eq. (42) is given by

$$\mathbf{W} = \begin{pmatrix} W_0 + \frac{1}{2}B(Q - Q_0) & A \\ A & W_0 - \frac{1}{2}B(Q - Q_0) \end{pmatrix} \tag{50}$$

where $A$ and $B$ are constants and $Q_0$ is the position of the crossing of the diabatic potential energy curves. Thus, the splitting between the diabatic curves is taken to be linear in $Q$, and the off-diagonal coupling constant.

For a system initially prepared in internal state 1, the probability of transition to state 2 can be obtained from Eq. (45). Landau,[160] Zener,[161] and

Stueckelberg[162] showed independently that, with assumptions 1–4 above, the transition probability is given by

$$P_{12} = |a_2(t \to \infty)|^2 = \exp\left(\frac{-2\pi A^2}{B\dot{Q}\hbar}\right) \tag{51}$$

In atomic collisions for which nuclei traverse the coupling region twice, the final probability of transition to state 2 is

$$P = 2P_{12}(1 - P_{12}) \tag{52}$$

In many real systems the assumptions of the LZS model are invalid, and Eq. (51) is correspondingly inaccurate. At low collision energies or when the avoided crossing is near a turning point, the constant velocity assumption is poor. At high collison velocities the assumption of constant off-diagonal coupling becomes unacceptable.[163] Nonetheless, the LZS approximation is useful in many situations, its range of validity is well understood,[164,165] and improvements can be made in some cases if necessary.[9,166,167] We shall see below that shortcomings of the approximation that would be unacceptable for atomic processes may lead to only minor errors when applied to certain molecular collision processes, so the emergence of the computer has not made the LZS approximation obsolete.

### 4.2.2. Near-Resonant Transitions

Near-resonant charge transfer or excitation transfer processes frequently involve transitions that occur at relatively large internuclear separation between potential curves that are nearly parallel in this region. Although the mechanism is thus quite different from the avoided crossing situation for which the LZS approximation is appropriate, the process can be described by the same two coupled equations (45) with suitably chosen interactions. One of the more useful approximate expressions for near-resonant transitions was obtained by Demkov, who assumed an exponential interaction.[168,169] Alternative expressions have been discussed, including some that are appropriate for inverse power interactions.[153,170–174] In cases where the transition is optically allowed, expressions involving the energy splitting $\Delta$ and the appropriate transition dipole moment can be derived.[10,175] All these expressions predict transition probabilities that decrease rapidly with increasing resonance defect $\Delta$, a trend frequently observed experimentally.[176]

### 4.2.3. Fine Structure Transitions

Analytic expressions have been derived for the probability of transition between fine structure states of a $^2P$ atom upon collision with a rare gas. These expressions again involve approximate solution of the two coupled equations (45) and for this case have been worked out for exponential[177,178] and van der Waals' interactions.[179]

## 4.3. Surface-Hopping Approach

There are several ways one can define the internal and external coordinates of Eqs. (38)–(45). The most obvious is to choose all electronic coordinates to be internal and all nuclear coordinates to be external:

$$\mathbf{q} = \mathbf{r}, \qquad \mathbf{Q} = \mathbf{R} \tag{53}$$

This is the choice of the *surface-hopping trajectory* (SHT) model.[35] It is the most nearly consistent choice with respect to mixing classical and quantum descriptions of motion. There is more intuitive justification for making a distinction between quantal electron motion and near-classical nuclear motion than there would be for making a similar distinction between different nuclear degrees of freedom. The resulting equality of all nuclear degrees of freedom permits description of both weakly and strongly interacting systems, including nuclear rearrangement (chemical reaction). The effects of strong chemical forces will be directly reflected in the complicated paths traced out by trajectories over multidimensional potential energy hypersurfaces.

A price has to be paid for this generality. Quantization of rotational and vibrational levels must be sacrificed. In addition, calculations may be lengthy, perhaps requiring computation of many thousands of trajectories to adequately describe a molecular collision process. The same drawbacks, of course, have not prevented the ordinary classical trajectory approach from becoming an extremely powerful tool for studying adiabatic molecular collisions.

To completely define the method, a prescription for obtaining the classical trajectory $\mathbf{Q}(t)$ must be specified. We shall describe here the simple approach that was employed in the original application of the SHT method.[35] There are obvious improvements that can be made in cases where the simple version proves inadequate.

The method requires specification of two or more $N$-dimensional (adiabatic or diabatic) potential energy hypersurfaces and the off-diagonal interactions responsible for promoting transitions between these surfaces, i.e., all the information required in Eq. (45). In addition, a relation of the form

$$S(\mathbf{R}) = 0 \tag{54}$$

must be prescribed to define the $(N-1)$-dimensional hypersurface(s) at which "hops" between potential energy surfaces can occur. One hopping hypersurface is defined for each region of strong nonadiabatic interaction. Selection of this "hopping seam" is straightforward in cases where nonadiabatic coupling is localized to relatively narrow regions such as the vicinity of an avoided intersection. It could be problematical if coupling is spread over a large area.

A surface-hopping trajectory begins with the selection of initial conditions by any of the procedures employed in standard classical trajectory studies.[3]

Specification of the initial electronic state must be included. Trajectories are followed by numerical integration of the classical equations of motion along the initial potential energy surface. If and when a hopping seam is encountered [i.e., when $S(\mathbf{R}) = 0$], the trajectory is split into $n$ branches, where $n$ is the number of strongly interacting states in the region. Each branch is assigned a weight $P_i(\mathbf{R}, \dot{\mathbf{R}})$ computed by numerical integration of the coupled equations (45):

$$P_i(\mathbf{R}, \dot{\mathbf{R}}) = |a_i|^2, \qquad i = 1, \ldots, n \tag{55}$$

The coupled equations (45) must be integrated through the entire strong coupling region, not just up to the hopping seam. Any trajectory branch may be selected for this purpose, because by hypothesis the choice of effective potential within a strong coupling region makes little difference.

After leaving the hopping seam, each branch moves along its own potential energy surface as if it were an individual trajectory. This splitting procedure is repeated whenever a branch reaches another hopping location. Thus a trajectory can develop many branches, each corresponding to a particular product in a particular electronic state. A large number of these branched trajectories are sampled, and the results are analyzed analogous to ordinary trajectory studies, taking proper account of the final weighting of each branch.

In cases where interaction regions are narrow and trajectories develop relatively few branches, an SHT calculation is not much more difficult to carry out than a standard classical trajectory calculation. The coupled equations (45) need to be solved only in regions of strong interaction, so numerical integration of the classical equations of motion is often the most lengthy part of the computation. Furthermore, in some cases it may be possible to avoid solving Eqs. (45) by using an approximate expression such as the LZS formula. If trajectories develop a great many branches, it may not be feasible to compute them all. In such cases, branches can be sampled at random, taking proper account of weightings.[35]

Note that probabilities $|a_i|^2$, not amplitudes, are assigned to each trajectory branch. Thus SHT is properly characterized as a classical mechanical method, according to the classification procedure in Section 1. It is possible to keep track of the amplitude $a_i$ along each trajectory branch and thereby to incorporate interference effects into the description. As discussed in Section 5, the semiclassical formulation of Miller and George[36] accomplishes this and, in addition, provides justification for the classical SHT method.

When a diabatic representation is employed, the potential surface intersections provide natural hopping seams. Furthermore, trajectories that branch at such locations automatically conserve energy and angular momentum. On the other hand, if adiabatic potential surfaces are used, hops will occur at places where surfaces are not degenerate. This problem can be rectified simply by adjusting the component of velocity directed along the non-adiabatic coupling vector to conserve energy. This procedure is essentially equivalent to transforming locally to a diabatic representation, performing the branching, and then transforming back to the adiabatic representation. If the basic hypothesis of the method is correct, velocity adjustments should be minor whenever $P_i(\mathbf{R}, \dot{\mathbf{R}})$ is significant; in low-energy collisions, transitions are likely only where potential energy surfaces approach each other closely.

The SHT method has been applied to several triatomic systems,[35,57,180–185] so far with excellent success. The most complete study was performed on the $H_3^+$ system. $H_3^+$ is a two-electron molecule, so the electronic part of the problem can be carried out to high accuracy. This makes it an ideal testing ground for dynamics theories. In spite of its apparent simplicity, $H_3^+$ is an interesting reactive system. Consider collision of $D^+$ with HD:

$$D^+ + HD \begin{cases} \rightarrow D + HD^+ & \Delta E = 1.83 \text{ eV} \\ \rightarrow H + D_2^+ & \Delta E = 1.81 \text{ eV} \\ \rightarrow H^+ + D_2 & \Delta E = -0.04 \text{ eV} \end{cases} \tag{56}$$

The possibility of forming two different kinds of products, atomic ions and molecular ions, is a consequence of the existence of an avoided crossing between the two lowest singlet potential energy surfaces of $H_3^+$. Because each of the four ionic species above are of different mass, the products can be easily identified. Accurate and detailed experimental studies have been carried out on this system and isotopic variants, providing a wealth of information against which to test theory.

The SHT calculations were carried out using diatomics-in-molecules potential energy surfaces, the accuracy of which was documented by comparison with accurate *ab initio* results.[56] The adiabatic representation was employed, and complete three-dimensional calculations were carried out precisely as outlined above, with one additional simplification. After numerical integration of Eqs. (45) through the strong interaction regions was performed for an initial sampling of trajectories, it was found that the hopping probability $P_i(\mathbf{R}, \dot{\mathbf{R}})$ could be accurately approximated by a function of the form of Eq. (51).

A detailed comparison of experimental and SHT cross sections, angular and velocity distributions, and velocity contour plots is given in Reference 183 for collision energies between 3 and 7 eV. The SHT results are in quantitative agreement within experimental uncertainty for almost every measured property. The largest discrepancies appeared in the absolute cross sections. These cross sections were recently remeasured very accurately by Ochs and Teloy.[186] The results, shown in Fig. 5, are in remarkable agreement with the previously reported SHT cross sections.[183]

In summary, although additional verification is required, the success to date provides strong indication that the SHT approach is capable of providing not only reliable qualitative descriptions of the dynamics but in many cases

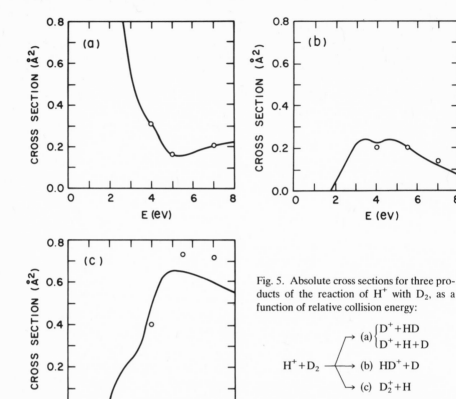

Fig. 5. Absolute cross sections for three products of the reaction of $H^+$ with $D_2$, as a function of relative collision energy:

$$H^+ + D_2 \begin{cases} \text{(a)} \begin{cases} D^+ + HD \\ D^+ + H + D \end{cases} \\ \text{(b)} \ \ HD^+ + D \\ \text{(c)} \ \ D_2^+ + H \end{cases}$$

The solid lines are experimental results of Reference 186. The circles are SHT theory of Reference 183.

quantitative predictions of the outcome of low-energy nonadiabatic molecular collisions. It is important to emphasize that, in at least one respect, nonadiabatic effects may often be somewhat easier to treat in molecular collisions than in atom–atom collisions. The description of a molecular collision requires a large number of trajectories. Some of these presumably will pass through regions where nonadiabatic interactions are large, and some will not. Some trajectories may make many passes through strong coupling regions. Final electronic transition probabilities may frequently depend more on the relative number of trajectories that enter certain regions where coupling is strong than on precise probabilities assigned to individual hops. Hence, in contrast to atom–atom collisions, cross sections may be relatively insensitive to the magnitude of nonadiabatic coupling, and approximate solutions to Eqs. (45) may often suffice.* This was demonstrated conclusively for the $H_3^+$ system, where a drastic

---

*One could imagine defining regions along the avoided crossing seam such that $P(\mathbf{R}, \dot{\mathbf{R}})$ is taken to be unity within these regions and zero elsewhere. A model as crude as this would probably be useless in atom–atom collisions, but there is strong indication that it might be very reasonable for certain molecular processes. Note that this model would eliminate any difficulties associated with branching trajectories.

change in the SHT hopping probability $P(\mathbf{R}, \dot{\mathbf{R}})$ was found to produce almost no change in the computed cross sections for any of the product channels, including charge transfer.[183] If such behavior proves general, the SHT method should enjoy wide applicability.

### 4.4. Multiple-Crossing Approach

An obvious alternative to Eq. (53) is to choose the internal coordinates $\mathbf{q}$ to be the electronic coordinates $\mathbf{r}$ plus all internal (vibration–rotation) nuclear coordinates $\mathbf{R}'$, and the external coordinate $\mathbf{Q}$ to be the relative coordinate $\boldsymbol{\rho}$:

$$\mathbf{q} = (\mathbf{r}, \mathbf{R}'), \qquad \mathbf{Q} = \boldsymbol{\rho} \tag{57}$$

Intermediate choices can be imagined, such as treating relative and rotational motion classically and vibration quantally, but we shall discuss only the prescription of Eq. (57).

We shall employ for the internal basis functions $\phi$ of Eq. (41) a product of an electronic function $\varphi_k$ as defined in Section 2.1, and an internal nuclear motion wave function $\xi_{k\alpha}$ as defined in Section 3.2:

$$\phi_{k\alpha}(\mathbf{r}, \mathbf{R}'; \boldsymbol{\rho}) = \varphi_k(\mathbf{r}; \mathbf{R}', \boldsymbol{\rho}) \xi_{k\alpha}(\mathbf{R}'; \boldsymbol{\rho}) \tag{58}$$

where $\phi_{k\alpha}$ is now labeled by a double index, $k$ denoting the electronic state and $\alpha$ the internal (vibration–rotation) state. Substituting into Eqs. (45), we obtain

$$i\hbar \dot{a}_{k\alpha} = \sum_{k'\alpha'}{}' a_{k'\alpha'} \left\{ W_{k\alpha,k'\alpha'} - i\hbar \dot{\boldsymbol{\rho}} \cdot \left[ \delta_{kk'} \int \xi_{k\alpha} (\nabla_\rho \xi_{k'\alpha'}) \, d\mathbf{R}' \right. \right.$$
$$\left. \left. + \gamma_{k\alpha,k'\alpha'} \langle \varphi_k \nabla_\rho \varphi_{k'} \rangle \right] \right\} \exp\left[ -\frac{i}{\hbar} \int^t (W_{k'\alpha',k'\alpha'} - W_{k\alpha,k\alpha}) \, dt \right] \tag{59}$$

where the angle brackets indicate integration over electronic coordinates only; the prime on the summation omits the term $k = k'$ and $\alpha = \alpha'$; $\gamma_{k\alpha,k'\alpha'}$ is the Franck–Condon overlap, Eq. (36); and $W_{k\alpha,k'\alpha'}$ is given by

$$W_{k\alpha,k'\alpha'} = \int \xi_{k\alpha} \left( \mathcal{T}_I \delta_{kk'} - \sum_i \frac{\hbar^2}{2\mu_i} \langle \varphi_k | \nabla_i \varphi_{k'} \rangle \cdot \nabla_i + \langle \varphi_k | \mathcal{T}_I + \mathcal{H}_0 | \varphi_{k'} \rangle \right) \xi_{k'\alpha'} \, d\mathbf{R}' \tag{60}$$

$\mathcal{T}_I$ is the kinetic energy operator for internal nuclear motion, $\mathcal{H}_0$ is the electronic Hamiltonian defined in Eqs. (10) and (11), and the summation is over internal nuclear degrees of freedom.

The diagonal elements of $W$ behave as effective potential energy curves. Each of these curves corresponds to a particular electronic–vibrational–rotational state, so there can be an enormous number of them in close proximity. Figure 6 is a schematic illustration of these multiple-crossing curves for a case involving interaction of an ionic and covalent state.

Several types of off-diagonal coupling terms appear in Eqs. (59) and (60), corresponding to qualitatively different transition mechanisms. Some promote

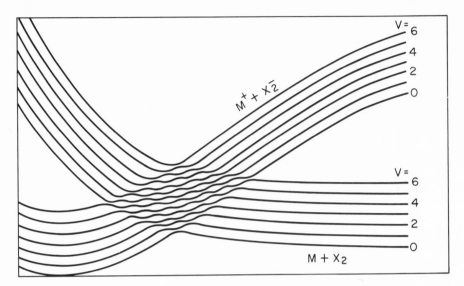

Fig. 6. Schematic illustration of two interacting manifolds of vibrational states, one corresponding to an ionic configuration and the other a covalent configuration. Multiple avoided crossings are exhibited. The interaction region would be much more complex if rotational states were included.

transitions between different vibrational and rotational states belonging to the same electronic state, and others produce simultaneous electronic and vibrational transitions. In cases where nonadiabatic coupling is largest for motion in the radial direction, the term in Eq. (59) involving $\gamma_{k\alpha,k'\alpha'}$ is primarily responsible for inducing electronic transitions. This is consistent with the experimental observation that at high collision energies vibrational states of charge transfer products are frequently populated roughly according to Franck–Condon overlaps. In fact, Lipeles has suggested that deviations from this behavior can sometimes be explained by replacing the Franck–Condon factors by overlaps of vibrational wave functions that have been distorted by the presence of the colliding atom.[187] This would correspond to using vibrationally adiabatic functions for the $\xi_{k\alpha}$ in Eq. (36) and would thus appear quite reasonable.

Because of the enormous number of coupled equations required, there have been no attempts to obtain numerical solutions of Eqs. (59) without further approximations. Distorted wave or exponential approximations[123] should be applicable in some cases, but they have not been tried. Four additional approximations have been invoked in all treatments to date.[34,188–193] The first, resulting in major simplification, is achieved by neglect of all terms in Eq. (59) for which $k = k'$. This corresponds to neglect of ordinary rotational and vibrational excitation occurring on a single potential energy surface. The necessity of invoking this assumption limits the applicability of the method to weakly interacting, large impact parameter (nonreactive) processes. Because rotational states of different quantum number $j$ are orthogonal even when associated with different potential surfaces, this assumption decouples the rotational states; the summation in Eq. (59) includes electronic and vibrational states only. However, it is still necessary to solve this set of equations separately for each rotational state initially populated (or, equivalently, for each initial molecular orientation).

A second simplification occurs if all quantities in Eqs. (59) and (60) are replaced by their averages over rotational states, i.e., by their orientation averages. This reduces the problem to one set of coupled equations involving electronic and vibrational states only. It is frequently assumed that quantities such as nonadiabatic coupling are nearly independent of orientation. In fact, as illustrated in Fig. 2, these quantities usually depend strongly on orientation, and to replace them by some average value may be a poor approximation.

The third approximation is neglect of terms in Eq. (60) involving $\langle \varphi_k | \nabla_i \varphi_{k'} \rangle$. This is equivalent to assuming that nonadiabatic transitions can be induced by relative motion only. Neglect of nonadiabatic coupling associated with internal (vibrational–rotational) motion severely limits the range of applicability of the method. Nevertheless, this assumption has been invoked in all applications of the method to date, including cases where nonadiabatic transitions are known to be promoted almost exclusively by vibrational motion.

The fourth approximation is that nonadiabatic transitions occur only in the vicinity of avoided crossings (Fig. 6) and, furthermore, that the transition probability at each of the crossings can be computed using a two-state approximation, e.g., the LZS formula. The two-state approximation has been criticized recently by Child on the grounds that, except at very low velocities, neighboring crossing points usually lie too close to be treated independently.[194] See, however, the work of Demkov and Osherov.[195]

In spite of these criticisms, with the four approximations above the method becomes tractable and may provide accurate descriptions of certain nonadiabatic molecular processes dominated by large impact parameter trajectories. The assumptions above are essentially those of the Bauer–Fisher–Gilmore (BFG) model.[34] The BFG model employs straight line trajectories, an approximation that is probably accurate whenever the first assumption above is justified. In addition, the BFG model employs for the basis functions $\xi_{k\alpha}$ the undistorted vibrational functions of the isolated molecule. This eliminates any possibility of describing the Lipeles effect mentioned above.[187] Bauer et al. have applied this method to quenching of excited alkali atoms by simple molecules.[34,190] Their results are very reasonable, even though they ignored the strong orientation dependence of nonadiabatic coupling that is exhibited by all these systems, e.g., Fig. 2.

In comparison with the SHT method, the multiple-crossing approach has several important disadvantages. It is usually extremely complicated to implement except when the additional approximations listed above are invoked, and these approximations severely restrict its applicability. Perhaps more important is the fundamental inconsistency associated with the prescription of Eq. (57). It is permissible to treat (classical) relative motion different from (quantal) vibrational–rotational motion only in cases where the relative coordinate remains clearly distinguishable throughout the collision. Therefore the method cannot be applied to chemically reactive systems. Furthermore, it cannot be applied to any low-energy process involving significant transfer of energy between relative and vibrational–rotational motion. This point is well illustrated by a simple example, quenching of the O($^1D$) atom by $N_2$. Potential energy surfaces governing this process are shown in Fig. 7. Reactants approaching on the attractive $^1A_1$ surface can make a transition to one of the three triplet surfaces in the vicinity of the surface crossings, with the aid of spin–orbit coupling. The BFG model[188] and a related straight line trajectory study[196] predict unacceptably small quenching cross sections and extremely little vibrational excitation. These treatments are unable to account for the fact that at low collision energies the $N_2O$ molecule will be temporarily trapped in the deep well, energy will be apportioned more or less statistically among $N_2O$

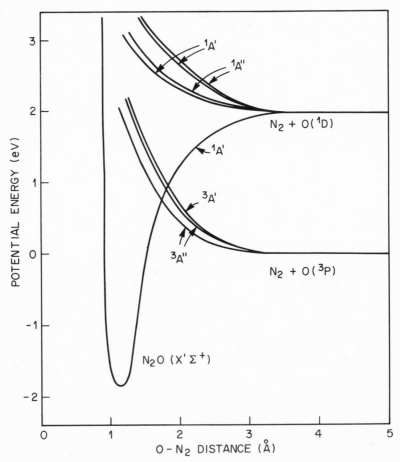

Fig. 7. Schematic illustration of potential energy surfaces of $N_2O$ as a function of O–$N_2$ separation.

vibrational modes, and many passes through the surface crossing region will be accomplished before the collision complex breaks up. The large cross section and large vibrational excitation observed for this process have been quantitatively reproduced by a simple statistical model based on this collision complex picture.[197] These features have also been accurately predicted by SHT calculations.[185] Thus the multiple-crossing approach can be applied only in situations where trajectories describing relative motion remain simple throughout the collision.

In cases where it is applicable, the multiple-crossing approach has two advantages over the SHT method. First, transition probabilities to particular final vibrational quantum levels can be obtained. Second, if the additional assumptions of the BFG model are justified, the method may be somewhat easier to apply than SHT. The two approaches are therefore to some extent complementary.

## 4.5. Approximate Models

Implementation of the SHT and multiple-crossing approaches described above can be fairly tedious. In addition, they generally require accurate potential energy surfaces and nonadiabatic interactions. Not surprisingly, several simplified models for describing nonadiabatic molecular collision processes have been proposed. None of them are of general applicability, but they may be very useful in specific situations. The phase space theory assumes that the probability of formation of any rotation–vibration–electronic state is proportional to its statistical weight.[198,199] The theory suffers from not being able to take into account the fact that electronic transitions occur in much more specific ways than vibrational or rotational transitions, i.e., it does not recognize the Born–Oppenheimer principle. More satisfactory are models that treat nuclear motion by statistical methods but treat electronic transitions explicitly.[185,197,200,201] Models depending on simplified pictures of nuclear motion have also been proposed.[202–206]

## 5. Semiclassical Methods

### 5.1. One-Dimensional WKB Approach

Semiclassical theories take maximum advantage of the near-classical behavior of nuclei and yet, by retaining the quantum superposition principle, can incorporate interference phenomena and classically forbidden (tunneling) processes. They offer the possibility of eliminating the deficiencies of classical methods without the necessity of performing very lengthy quantum mechanical calculations. In addition, they can provide justification for classical path methods and suggest prescriptions for defining optimal classical trajectories. Of particular note in this regard are the contributions of Delos et al.,[147,148] discussed in Section 4.1, and of Pechukas and Davis[158,159] and Miller and George,[36] outlined below.

Semiclassical methods based on the WKB approximation[207] have been very successful in describing low-energy atom–atom inelastic collisions.[94,119,146,167,208–210] As with quantum mechanical methods, approximate solutions to Eq. (14) are sought. The nuclear motion wave functions $\chi_k(\rho)$, depending in the atom–atom case on the internuclear separation $\rho$ only, are written in the form

$$\chi_k(\rho) = A_k^+(\rho) \exp\left[\frac{i}{\hbar}S_k(\rho)\right] + A_k^-(\rho) \exp\left[-\frac{i}{\hbar}S_k(\rho)\right] \tag{61}$$

where $S_k(\rho)$ is the classical action integral

$$S_k(\rho) = \int^\rho p_k(\rho)\, d\rho \tag{62}$$

$p_k$ is the classical nuclear momentum at position $\rho$:

$$p_k(\rho) = \{2\mu[\mathscr{E} - U_{kk}(\rho)]\}^{1/2} \tag{63}$$

where $U_{kk}(\rho)$ is the potential energy defined in Eq. (15). The two terms in Eq. (61) correspond to the two possible directions of nuclear motion, incoming and outgoing. The functions $A_k^+$ and $A_k^-$ are not determined uniquely by Eq. (61), so additional relations between the $A_k$ must be prescribed. Substitution of Eq. (61) into Eq. (14) then leads to a set of coupled equations for the amplitudes $A_k^\pm$. The WKB approximation is obtained by neglect of terms of order equal to or greater than $\hbar^2$ in the equations for $A_k^\pm$ and specification of certain continuity relations.

Most semiclassical treatments of atom–atom collisions can be considered variations of the approach outlined above.[119,145–148,208,209] A basic feature common to all of them is that the phase of the wave function $\chi_k$ is the purely classical mechanical quantity $S_k(\rho)$. Because the transition amplitude $S_{12}$ of Eq. (5) may be a sum of several terms, each with different phase, the possibility of incorporating interference effects is apparent. The striking oscillatory behavior frequently observed in atom–atom inelastic scattering experiments can be accurately reproduced by the semiclassical approach outlined above.[119,167,209]

Semiclassical theories of nonadiabatic behavior share many of the problems of classical mechanical treatments. Underlying these problems is the fundamental inconsistency of treating some motions classically and others quantum mechanically. One way in which this is manifested is nonuniqueness of the trajectory along which the classical action integral $S_k(\rho)$ is computed. In regions where two or more electronic states interact strongly, nuclear motion is governed by an unknown effective potential that is some sort of average of the potentials for each state. Most semiclassical methods employ a single trajectory throughout the collision. Various procedures for defining the effective potential have been proposed.[146,208] In many cases these methods provide accurate descriptions of atom–atom processes. However, there are situations where a single trajectory is inadequate, particularly for reproducing detailed properties such as differential cross sections. In such cases, methods that allow trajectories to branch in strong coupling regions can provide significant improvement.[119]

Analogous to the classical path approach, there are two obvious ways to extend semiclassical techniques to molecular processes. One is to treat electron motion quantally and nuclear motion classically, i.e., use the prescription of the SHT method, Eq. (53). This requires development of many-dimensional semiclassical theories, including allowance for branching trajectories. This is the goal of the theories of Pechukas[158] and Miller and George,[36] discussed below.

The second obvious choice is to treat relative motion classically and electronic, vibrational, and rotational motion quantum mechanically, i.e., to use the prescription of the multiple-crossing approach, Eq. (57). Standard one-dimensional WKB methods can then be employed. However, they must be applicable to a very large number of coupled equations of the form of Eq. (37). In a series of recent papers,[210–215] Eu has derived a uniform WKB formalism capable of treating many-level systems. The method can be simplified by invoking the first-order Magnus (exponential) approximation.[215] Even so, the number of coupled channels required in Eq. (37) is usually so enormous that semiclassical methods are not likely to be of general value unless simplifications similar to those of the BFG model can be invoked. Thus the approach will be applicable only in large impact parameter, weak coupling situations, and even then only when special requirements are satisfied (see Section 4.4).

## 5.2. Path Integral Formulation

In most cases it is clearly preferable to treat only electronic motion quantum mechanically and to treat all nuclear motion semiclassically. A multidimensional semiclassical theory is required for this. None of the WKB methods based on arguments such as those outlined above can be easily extended beyond one dimension because they depend either explicitly or implicitly on Eq. (63), i.e., on the fact that in one dimension, specification of position and energy is sufficient to restrict to two the possible values of the classical momentum.

Using Feynman's path integral approach,[157] Pechukas[158] has formulated a semiclassical description of inelastic collisions that does not depend on relations like Eq. (63) and that can be directly extended to treat many nuclear degrees of freedom. We shall distinguish between internal (electron) coordinates $\mathbf{r}$ and external (nuclear) coordinates $\mathbf{R}$, as in Eq. (53). Electronic wave functions $\varphi_k(\mathbf{r}; \mathbf{R})$ are defined as in Section 2. Pechukas shows that the transition amplitude $K_{21}$ between initial and final electronic states 1 and 2 and initial and final nuclear positions $\mathbf{R}_1$ and $\mathbf{R}_2$ can be expressed in the form

$$K_{21}(\mathbf{R}_2 t_2 | \mathbf{R}_1 t_1) = \int_{\mathbf{R}_1}^{\mathbf{R}_2} D\mathbf{R} \, T_{21}[\mathbf{R}(t)] \exp\left[\frac{i}{\hbar} \int_{t_1}^{t_2} \frac{\mu}{2} \dot{\mathbf{R}}(t)^2 \, dt\right] \qquad (64)$$

Equation (64) is an exact quantum mechanical expression. The symbol $D\mathbf{R}$ denotes "path integral," i.e., Eq. (64) is a sum of contributions from all continuous paths that connect $\mathbf{R}_1(t_1)$ and $\mathbf{R}_2(t_2)$. $T_{21}[\mathbf{R}(t)]$ is the amplitude for the electronic transition $1 \rightarrow 2$ with the nuclei constrained to move along the path $\mathbf{R}(t)$. The phase factor is simply the action integral, Eq. (62), evaluated along the (nonclassical) path $\mathbf{R}(t)$.

There is a direct relationship between the path integral approach and the classical path method of Section 4.1. Both procedures involve selecting a trajectory $\mathbf{R}(t)$ and solving for the time-dependent response of the electrons along this trajectory to obtain $T_{21}$, using, for example, Eq. (45). In the classical path method, usually only a single near-classical trajectory is selected. In the exact quantum mechanical procedure, path integration is carried out over all possible trajectories joining $\mathbf{R}_1(t_1)$ and $\mathbf{R}_2(t_2)$, with the contribution of each trajectory weighted according to the action integral of Eq. (64).

The semiclassical theories of Pechukas and Davis[158,159] and Miller and George[36] result from picking out one or a few most important trajectories from the infinite number that contribute to the path integral. This is accomplished by evaluation of Eq. (64) using the method of steepest descent. The theories differ in the way that the electronic transition amplitude $T_{21}$ is computed. $T_{21}$ poses a problem here because its evaluation requires knowledge of the trajectory $\mathbf{R}(t)$ at all times previous to and past $t$. The equation of motion determining $\mathbf{R}(t)$ depends, in turn, on $T_{21}[\mathbf{R}(t)]$, so in general the trajectory must be obtained iteratively.

In cases where electronic transition probabilities are small, it may be accurate to compute $T_{21}$ by first-order time-dependent perturbation theory. Pechukas and Davis[159] show that, with this approximation, calculation of the effective potential governing $\mathbf{R}(t)$ may be significantly easier, although still requiring iteration. Furthermore, their effective potential has all the desired properties, i.e., it changes continuously from the potential of the initial state to that of the final state, irrespective of the magnitude of the transition probability. This is a very important improvement over ad hoc prescriptions such as Eqs. (47)–(49) and provides a well-defined procedure for splitting trajectories into branches as required by the SHT method without requiring electronic transitions to be localized in space and time.

## 5.3. Complex-Valued Trajectory Approach

The fact that the effective potential must be determined iteratively makes the Pechukas approach difficult to apply to realistic molecular collision processes. It is desirable to find alternative ways to approximate the electronic transition amplitude $T_{21}$ that do not require iteration. Miller and George[36,37] have proposed one such method based on Stueckelberg's semiclassical procedure for evaluating matrix elements[162,216]:

$$T_{21}[\mathbf{R}(t)] \simeq \exp\left\{ -\frac{i}{\hbar} \int_{t_1}^{t^*} U_{11}[\mathbf{R}(t)]\, dt - \frac{i}{\hbar} \int_{t^*}^{t_2} U_{22}[\mathbf{R}(t)]\, dt \right\} \qquad (65)$$

where $U_{11}$ and $U_{22}$ are the *adiabatic* potential energy surfaces for electronic states 1 and 2, defined in Eq. (22). $t^*$ is the complex time at which the surfaces intersect:

$$U_{22}[\mathbf{R}(t^*)] = U_{11}[\mathbf{R}(t^*)] \qquad (66)$$

Equation (65) reproduces most of the simple one-dimensional solutions discussed in Section 4.2.[37] Substituting Eq. (65) into Eq. (64) and evaluating by steepest descent, we define a complex-valued "classical" trajectory that switches from adiabatic potential surface 1 to 2 at the complex position $\mathbf{R}(t^*)$ at which Eq. (66) is satisfied.

The method has been applied to low-energy collisions involving fine structure transitions,

$$F\,(^2P_{3/2}) + X \to F\,(^2P_{1/2}) + X \qquad (67)$$

where $X = H^+$ and Xe.[217] These processes provide a difficult test of the theory because there are no obvious avoided crossings to define intersections in the complex plane. Furthermore, the 404 cm$^{-1}$ splitting between the fluorine atom fine structure levels is comparable to the nuclear kinetic energy in thermal energy encounters, so trajectories on different potential energy curves may differ substantially. Nevertheless, Preston et al.[217] obtain very satisfactory descriptions of these processes using the spin–orbit interaction to define the complex points of intersection. Lin et al. have applied the method to collinear[218] and three-dimensional[219] collisions of $H^+$ and $D_2$. These calculations, although very lengthy, particularly when they involve multiple transition points,[220] demonstrate that the method can be applied to systems involving several nuclear coordinates. Hopefully, simplifications will be discovered that will allow it to become a practical technique for describing realistic nonadiabatic molecular collision processes.

A simplified version of the theory has been suggested that employs only real-valued trajectories.[36] When interference effects are discarded and sums over vibrational–rotational states are performed by a Monte Carlo procedure the method becomes equivalent to the SHT procedure[35] described in Section

4.3. Thus the semiclassical Miller–George theory provides additional justification for the more easily implemented classical SHT approach.

The complex-valued trajectory approach requires construction of adiabatic potential energy surfaces for complex nuclear coordinates **R**. Methods for doing this based on analytic continuation and on explicit solution of the electronic Schrödinger equation for complex coordinates have been investigated.[37,217,221,222] Note that nonadiabatic coupling, Eq. (20), does not occur explicitly in the Miller–George theory but rather is reflected in the location of the complex intersection.

The Miller–George theory is likely to be most useful in situations involving isolated intersections located near the real axis, i.e., when nonadiabatic coupling is localized in relatively narrow regions. Miller[37] has suggested that alternative approximations to the electronic transitions amplitude $T_{12}$ may be required in other cases but that steepest descent evaluation of the path integral, Eq. (64), should be reliable in almost all systems of physical interest.


## 6. Conclusions

The theory of nonadiabatic collisions is an active field. Some recent advances and avenues for future study have been outlined in the preceding sections. Due to space limitations, methods for treating collisional ionization, radiationless transitions, and "half-collision" events such as predissociation and autoionization have not been discussed. These also are nonadiabatic processes, and techniques used to describe them are related and frequently complementary to those described above.

There is conclusive evidence that realistic treatments of low-energy nonadiabatic molecular collisions must take accurate account of the distortion of electron clouds that accompanies nuclear motion, i.e., chemical interactions must be considered. This is true not only for reactive systems but for most inelastic collisions as well. It is also difficult to accomplish. As a result, the present status of the theory closely parallels that of adiabatic molecular collisions. For both adiabatic and nonadiabatic processes, accurate quantum mechanical descriptions of nuclear motion are extremely difficult to obtain, even for the simplest triatomic systems, and it is not apparent that any drastic improvements are on the horizon. Classical mechanical treatments are practical, but in nonadiabatic situations they suffer from the fundamental inconsistency of explicit mixing of quantum and classical descriptions of motion. Semiclassical procedures minimize the inaccuracies of classical descriptions, but they do not remove this basic inconsistency, and their implementation can become unwieldy.

In spite of these difficulties, great progress has been made. The classical SHT approach has been applied to several systems with remarkable success. By treating all nuclear motion classically and only electron motion by quantum mechanics, problems associated with mixing classical and quantal descriptions are minimized as a result of the huge electron/nuclear mass discrepancy. In addition, the effects of strong chemical forces on nuclear motion can be described accurately; trajectories can squeeze around corners and over barriers, or get temporarily trapped in wells, as needed to reflect realistically the course of a collision event. Nonadiabatic transitions (surface hops) are described by this method, usually with high quantitative accuracy. The many-dimensional semiclassical formulations of Pechukas and Miller–George make it possible to incorporate quantum mechanical interference and tunneling effects while retaining all the desired features of the SHT method.

Thus, as for adiabatic processes, the major barrier to obtaining quantitative a priori descriptions of realistic low-energy nonadiabatic collisions is the current scarcity of accurate potential energy surfaces. Even in the absence of accurate potential surfaces, it is now feasible to obtain the kind of reliable qualitative or semiquantitative information that is so desperately needed for many collision processes. Although theories of nonadiabatic molecular processes still lag behind adiabatic theories, the gap is closing. In fact, it should not go unnoticed that, to date, the most detailed, complete, and successful comparison between an a priori theory of chemical reaction and experiment may well be for the reaction of $H^+$ with $D_2$,[183,186] a nonadiabatic process.

# References

1. D. Rapp and T. Kassal, The theory of vibrational energy transfer between simple molecules in nonreactive collisions, *Chem. Rev.* **69**, 61–102 (1969).
2. R. J. Cross, Jr., in: *Molecular Beams and Reaction Kinetics* (Ch. Shlier, ed.), pp. 50–61, Academic Press, Inc., New York (1970).
3. D. L. Bunker, Classical trajectory methods, *Methods Comput. Phys.* **10**, 287–325 (1971).
4. R. D. Levine, in: *MTP International Review of Science, Physical Chemistry* (W. Byers-Brown, ed.), Vol. I, pp. 229–266, University Park Press, Baltimore (1972).
5. D. Secrest, Theory of rotational and vibrational energy transfer in molecules, *Annu. Rev. Phys. Chem.* **24**, 379–406 (1973).
6. T. F. George and J. Ross, Quantum dynamical theory of molecular collisions, *Annu. Rev. Phys. Chem.* **24**, 263–300 (1973).
7. J. C. Polanyi and J. L. Schreiber, in: *Kinetics of Gas Reactions* (H. Eyring, W. Jost, and D. Henderson, eds.), Chap. 9, Academic Press, Inc., New York (1974).
8. D. A. Micha, Quantum theory of reactive molecular collisions, *Adv. Chem. Phys.* **30**, 221–260 (1975).
9. E. E. Nikitin, in: *Chemische Elementarprozesse* (H. Hartmann, ed.), pp. 43–77, Springer-Verlag, Berlin (1968).
10. T. Watanabe, in: *Advances in Chemistry. Radiation Chemistry II* (E. J. Hart, ed.), pp. 176–193, American Chemical Society, Wahington, D.C. (1968).
11. F. T. Smith, Elastic and inelastic atom–atom scattering, *Lect. Theor. Phys.* **XIC**, 95–117 (1969).
12. J. Callaway, Inelastic atom–atom collisions, *Lect. Theor. Phys.* **XIC**, 119–137 (1969).

13. E. Bauer, *in: Kinetic Processes in Gases and Plasmas* (R. A. Hochstim, ed.), pp. 381–429, Academic Press, Inc., New York (1969).

14. R. S. Berry, *in: Molecular Beams and Reaction Kinetics* (Ch. Schlier, ed.), pp. 193–228, Academic Press, Inc., New York, (1970).

15. R. S. Berry, *in: Molecular Beams and Reaction Kinetics* (Ch. Schlier, ed.), pp. 229–248, Academic Press, Inc., New York, (1970).

16. N. F. Mott and H. S. W. Massey, *The Theory of Atomic Collisions,* Oxford University Press, London (1965).

17. H. S. W. Massey, *Electronic and Ionic Impact Phenomena,* Vol. III, Oxford University Press, London (1971), Chap. 18.

18. R. A. Mapelton, *The Theory of Charge Exchange,* John Wiley & Sons, Inc. (Interscience Division), New York (1972).

19. E. W. Thomas, *Excitation in Heavy Particle Collisions,* John Wiley & Sons, Inc. (Interscience Division), New York (1972).

20. E. E. Muschlitz, Collisions of electronically excited atoms and molecules, *Adv. Chem. Phys.* **10**, 171–194 (1966).

21. B. A. Thrush, Gas reactions yielding electronically excited species, *Annu. Rev. Phys. Chem.* **19**, 371–388 (1968).

22. F. R. Gilmore, E. Bauer, and J. W. McGowan, A review of atomic and molecular excitation mechanics in non-equilibrium gases up to 20,000°K, *J. Quant. Spectrosc. Radiat. Transfer* **9**, 157–183 (1969).

23. R. B. Cundall, *in: Transfer and Storage of Energy by Molecules* (G. M. Burnett and A. M. North, eds.), Vol. I, pp. 1–63, John Wiley & Sons, Inc. (Interscience Division), New York (1969).

24. A. B. Callear and J. D. Lambert, *in: Comprehensive Chemical Kinetics* (C. H. Bamford and C. F. H. Tipper, eds.), Vol. 3, pp. 182–273, Elsevier Publishing Company, Amsterdam (1969).

25. I. W. M. Smith, *in: The Role of the Excited State in Chemical Physics* (J. W. McGowan, ed.), John Wiley & Sons, Inc. (Interscience Division), New York (1973).

26. R. J. Donovan and D. Husain, Recent advances in the chemistry of electronically excited atoms, *Chem. Rev.* **70**, 489–516 (1970).

27. J. I. Steinfeld, Quenching of fluorescence in small molecules, *Acc. Chem. Res.* **3**, 313–320 (1970).

28. R. F. Vasil'ev, Chemiluminescence excitation mechanisms, *Russ. Chem. Rev.* **39**, 529–544 (1970).

29. T. Carrington and J. C. Polanyi, *in: MTP International Review of Science, Physical Chemistry* (J. C. Polanyi, ed.), Vol. 9, pp. 135–171, University Park Press, Baltimore (1972).

30. D. H. Stedman and D. W. Setzer, Chemical applications of metastable rare gas atoms, *Prog. React. Kinet.* **6**, 193–238 (1972).

31. M. Born and J. R. Oppenheimer, Zur Quantentheorie der Molekeln, *Ann. der Phys.* **84**, 457–484 (1927).

32. A. Messiah, *Quantum Mechanics,* Vol. II, Chap. 18, John Wiley & Sons, Inc., New York (1962).

33. W. H. Miller, The semiclassical nature of atomic and molecular collisions, *Acc. Chem. Res.* **4**, 161–167 (1971).

34. E. Bauer, E. R. Fisher, and F. R. Gilmore, De-excitation of electronically excited sodium by nitrogen, *J. Chem. Phys.* **51**, 4173–4181 (1969).

35. J. C. Tully and R. K. Preston, Trajectory surface hopping approach to nonadiabatic molecular collisions, *J. Chem. Phys.* **55**, 562–572 (1971).

36. W. H. Miller and T. F. George, Semiclassical theory of electronic transitions in low energy atomic and molecular collisions involving several nuclear degrees of freedom, *J. Chem. Phys.* **56**, 5637–5652 (1972).

37. W. H. Miller, Classical limit quantum mechanics and the theory of molecular collisions, *Adv. Chem. Phys.* **25**, 69–177 (1974).

38. R. G. Newton, *Scattering Theory of Waves and Particles,* McGraw-Hill Book Company, New York (1966).

39. M. L. Goldberger and K. M. Watson, *Collision Theory,* John Wiley & Sons, Inc., New York (1964).

40. J. R. Taylor, *Scattering Theory. The Quantum Theory of Nonrelativistic Collisions,* John Wiley & Sons, Inc., New York, (1972).

41. L. S. Rodberg and R. M. Thaler, *Introduction to the Quantum Theory of Scattering,* Academic Press, Inc., New York (1967).

42. H. Laue, Coupling between nuclear and electronic motion in diatomic molecules, *J. Chem. Phys.* **46**, 3034–3040 (1967).

43. W. Kolos, Adiabatic approximation and its accuracy, *Adv. Quantum. Chem.* **5**, 99–133 (1970).

44. J. C. Tully, Diatomics-in-Molecules potential energy surfaces. II. Nonadiabatic and spin–orbit interactions, *J. Chem. Phys.* **59**, 5122–5134 (1973).

45. N. F. Mott, On the theory of excitation by collisions with heavy particles, *Proc. Cambridge Philos. Soc.* **27**, 553–560 (1931).

46. J. von Neumann and E. P. Wigner, Über das Verhalten von Eigenwerten bei adiabatischen Prozessen, *Phys. Z.* **30**, 467–470 (1929).

47. G. Herzberg and H. C. Longuet-Higgins, Intersection of potential energy surfaces in polyatomic molecules, *Discuss. Faraday Soc.* **35**, 77–82 (1963).

48. E. Teller, The crossing of potential surfaces, *J. Phys. Chem.* **41**, 109–115 (1937).

49. T. F. George, K. Morokuma, and Y.-W. Lin, Real and complex intersections between potential energy surfaces of the same symmetry in polyatomic systems, *Chem. Phys. Lett.* **30**, 54–57 (1975).

50. W. Lichten, Resonant charge exchange in atomic collisions, *Phys. Rev.,* **131**, 229–238 (1963).

51. T. F. O'Malley, Diabetic state of molecules—Quasistationary electronic states, *Adv. At. Mol. Phys.* **7**, 223–249 (1971).

52. R. W. Numrich and D. G. Truhlar, Mixing of ionic and covalent configurations for NaH, KH and MgH$^+$, *J. Phys. Chem.* **79**, 2745–2766 (1975).

53. F. T. Smith, Diabatic and adiabatic representations for atomic collision problems, *Phys. Rev.* **179**, 111–123 (1969).

54. M. Baer, Adiabatic and diabatic representations for atom–molecule collisions, *Chem. Phys. Lett.* **35**, 112–118 (1975).

55. T. Carrington, The geometry of intersecting potential surfaces, *Acc. Chem. Res.* **7**, 20–25 (1974).

56. R. K. Preston and J. C. Tully, Effects of surface crossing in chemical reactions: The H$_3^+$ system, *J. Chem. Phys.* **54**, 4297–4304 (1971).

57. S. Chapman and R. K. Preston, Nonadiabatic molecular collisions: Charge exchange and chemical reaction in the Ar$^+$–H$_2$ system, *J. Chem. Phys.* **60**, 650–659 (1974).

58. H. S. W. Massey, Collisions between atoms and molecules at ordinary temperatures, *Rep. Prog. Phys.* **12**, 248–269 (1949).

59. H. F. Schaefer III, *The Electronic Structure of Atoms and Molecules,* Addison-Wesley Publishing Company, Inc., Reading, Mass. (1972).

60. J. N. Bardsley, Pseudopotentials in atomic and molecular physics, *Case Stud. At. Phys.* **4**, 299–368 (1974).

61. A. C. Roach and M. S. Child, Electronic potential energy surfaces for the reaction K + NaCl → KCl + Na, *Mol. Phys.* **14**, 1–15 (1968).

62. C. F. Melius, W. A. Goddard III, and L. R. Kahn, Use of *ab initio* G1 effective potentials for calculations of molecular excited states, *J. Chem. Phys.* **56**, 3342–3348 (1972).

63. W. Moffitt, Atoms in molecules and crystals, *Proc. R. Soc. London Ser. A* **210**, 245–268 (1951).

64. G. G. Balint-Kurti and M. Karplus, Potential energy surfaces for simple chemical reactions: Li + F$_2$ → LiF + F, *Chem. Phys. Lett.* **11**, 203–207 (1971).

65. F. O. Ellison, A method of diatomics in molecules. I. General theory and application to H$_2$O, *J. Am. Chem. Soc.* **85**, 3540–3544 (1963).

66. P. J. Kuntz and A. C. Roach, Ion–molecule reactions of rare gases with hydrogen, *J. Chem. Soc. Faraday Trans. 2,* **68**, 259–280 (1972).

67. J. C. Tully, Diatomics-in-molecules potential energy surfaces. I. First-row triatomic hydrides, *J. Chem. Phys.* **58**, 1396–1410 (1973).

68. T. E. H. Walker and W. G. Richards, Molecular spin–orbit coupling constants. The role of core polarization, *J. Chem. Phys.* **52**, 1311–1314 (1970).

69. W. H. Moores and R. McWeeney, The calculation of spin–orbit splitting and $g$ tensors for small molecules and radicals, *Proc. R. Soc. London Ser.* A **332**, 365–384 (1973).

70. F. H. Mies, Molecular theory of atomic collisions: Fine-structure transitions, *Phys. Rev. A* **7**, 942–957 (1973).

71. J. S. Cohen and B. Schneider, Ground and excited states of $Ne_2$ and $Ne_2^+$. I. Potential curves with and without spin–orbit coupling, *J. Chem. Phys.* **61**, 3230–3239 (1974).

72. J. C. Browne, Molecular wave functions: Calculation and use in atomic and molecular processes, *Adv. At. Mol. Phy.* **7**, 47–95 (1971).

73. W. R. Thorson, Asymptotic coriolis interactions in slow atomic collisions, *J. Chem. Phys.* **50**, 1702–1704 (1969).

74. V. Sidis, Simple expression for the off-diagonal matrix elements of the $d/dR$ operator between exact electronic states of a diatomic molecule, *J. Chem. Phys.* **55**, 5838–5839 (1971).

75. W. R. Thorson, Theory of slow atomic collisions. I. $H_2^+$, *J. Chem. Phys.* **42**, 3878–3891 (1965).

76. S. B. Schneiderman and A. Russek, Velocity-dependent orbitals in proton-on-hydrogen-atom collisions, *Phys. Rev.* **181**, 311–321 (1969).

77. D. R. Bates and D. Sprevak, Translation factor in basis functions used in perturbed stationary state approximation and capture in $H^+$–H ($1S$) collisions, *J. Phys. B* **4**, L47–51 (1971).

78. C. F. Melius and W. A. Goddard III, The theoretical description of an asymmetric, nonresonant charge transfer process, *Chem. Phys. Lett.* **15**, 524–529 (1972).

79. H. S. W. Massey and R. A. Smith, The passage of positive ions through gases, *Proc. R. Soc. London A* **142**, 142–172 (1933).

80. R. P. Marchi and F. T. Smith, Theory of elastic differential scattering in low-energy $He^+$–He collisions, *Phys. Rev.* **139**, A1025–1038 (1965).

81. E. E. Nikitin, Remarks on different theoretical approaches to the collisionally induced depolarization of atomic states, *Comments At. Mol. Phys.* **3**, 7–14 (1971).

82. D. G. Truhlar, Multiple potential energy surfaces for reactions of species in degenerate electronic states, *J. Chem. Phys.* **56**, 3189–3190 (1972).

83. J. T. Muckerman and M. D. Newton, Comment on "multiple potential energy surfaces for reactions of species in degenerate electronic states" by D. G. Truhlar, *J. Chem. Phys.* **56**, 3191–3192 (1972).

84. J. C. Tully, Collisions of F ($^2P_{1/2}$) with $H_2$, *J. Chem. Phys.* **60**, 3042–3050 (1974).

85. W. H. Miller, Theory of Penning ionization. I. Atoms, *J. Chem. Phys.* **52**, 3563–3572 (1970).

86. H. Nakamura, Theoretical considerations on Penning ionization processes, *J. Phys. Soc. Jpn* **26**, 1473–1479 (1969).

87. S. A. Evans, J. S. Cohen, and N. F. Lane, Quantum-mechanical calculation of cross sections for inelastic atom–atom collisions. I, *Phys. Rev. A* **4**, 2235–2248 (1971).

88. L. Lenamon, J. C. Browne, and R. E. Olson, Theoretical low-energy inelastic-scattering cross sections for He ($2^3S$) + He ($1^1S$) → He ($2^3P$) + He ($1^1S$), *Phys. Rev. A* **8**, 2380–2386 (1973).

89. R. E. Olson, Low-energy theoretical inelastic-scattering differential cross sections for the process $H^+$ + He → $H^+$ + He ($2^3S$), *Phys. Rev. A* **5**, 2094–2103 (1972).

90. F. H. Mies, Molecular theory of atomic collisions: Calculated cross sections for $H^+$ + F ($^2P$), *Phys. Rev. A.* **7**, 957–967 (1973).

91. R. H. Reid, Transitions among the $3p^2$ $P$ states of sodium induced by collisions with helium, *J. Phys. B* **6**, 2018–2039 (1973).

92. J. S. Cohen, S. A. Evans, and N. F. Lane, Quantum-mechanical calculation of cross sections for inelastic atom–atom collisions. II, *Phys. Rev. A* **4**, 2248–2253 (1971).

93. H. Nakamura, Theoretical studies of inelastic atomic collisions in a two-state model problem. *Mol. Phys.* **25**, 577–602 (1973).

94. H. G. Guerin, T. P. Tsien, B. C. Eu, and R. E. Olson, Comment on the accuracy of the uniform WKB theory of inelastic collisions, *Phys. Rev. A* **9**, 995–998 (1974).
95. J. B. Delos, Studies of the potential-curve-crossing problem. III, *Phys. Rev. A* **9**, 1626–1634 (1974).
96. A. C. Allison, The numerical solution of coupled differential equations arising from the Schrödinger equation, *J. Comput. Phys.* **6**, 378–391 (1970).
97. W. A. Lester, Jr., and R. B. Bernstein, Computational procedure for the close-coupled rotational excitation problem, *J. Chem. Phys.* **48**, 4896–4904 (1968).
98. R. G. Gordon, Quantum scattering using piecewise analytic solutions, *Methods Comput. Phys*, **10**, 81–110 (1971).
99. W. N. Sams and D. J. Kouri, Noniterative solutions of integral equations of scattering. II. Coupled channels, *J. Chem. Phys.* **51**, 4815–4819 (1969).
100. B. R. Johnson and D. Secrest, The solution of the nonrelativistic quantum scattering problem without exchange, *J. Math. Phys. (N.Y.)* **7**, 2187 (1966).
101. B. R. Johnson, The multichannel log-derivative method for scattering calculations, *J. Comput. Phys.* **13**, 445–449 (1973).
102. J. C. Light, Quantum theories of chemical kinetics, *Adv. Chem. Phys.* **19**, 1–31 (1971).
103. W. A. Lester, Jr., Calculation of cross sections for rotational excitation of diatomic molecules by heavy particle impact: Solution of close-coupled equations, *Methods Comput. Phys.* **10**, 211–242 (1971).
104. W. A. Lester, Jr. and J. Schaefer, Rotational transitions in $H_2$ by $Li^+$ collisions; comparison with experiment, *J. Chem. Phys.* **60**, 1672–1674 (1974).
105. P. McGuire, Coupled-states approach for elastic and for rotationally and vibrationally inelastic atom–molecule collisions, *J. Chem. Phys.* **62**, 525–534 (1975).
106. R. A. Marcus, Analytical mechanics of chemical reactions. III. Natural collision coordinates, *J. Chem. Phys.* **49**, 2610–2616 (1968).
107. G. Wolken, Jr. and M. Karplus, Theoretical studies of $H + H_2$ reactive scattering, *J. Chem. Phys.* **60**, 351–367 (1974).
108. J. C. Light, Quantum calculations in chemically reactive systems, *Methods Comput. Phys.* **10**, 111–143 (1971).
109. B. R. Johnson and R. D. Levine, A new approach to non-adiabatic transitions in collision theory, *Chem. Phys. Lett.* **13**, 168–171 (1972).
110. H. Nakamura, Theory of electronically non-adiabatic chemical reactions: Quantum formulation of collinear reactions, *Mol. Phys.* **26**, 673–685 (1973).
111. Z. Top and M. Baer, Non-adiabatic transitions in chemical reaction. A quantum mechanical study, *Chem. Phys.* **10**, 95–106 (1975).
112. I. H. Zimmerman and T. F. George, Quantum resonance effects in electronic-to-vibrational energy transfer in molecular collisions, *J. Chem. Phys.* **61**, 2468–2470 (1974).
113. I. H. Zimmerman and T. F. George, Quantum mechanical study of electronic transitions in collinear atom–diatom collisions, *Chem. Phys.* **7**, 323–335 (1975).
114. O. H. Crawford, Calculation of chemical reaction rates by variational methods, *J. Chem. Phys.* **55**, 2571–2574 (1971).
115. W. H. Miller, Coupled equations and the minimum principle for collision of an atom and a diatomic molecule, including rearrangements, *J. Chem. Phys.* **50**, 407–418·(1969).
116. R. Conn and H. Rabitz, Decomposition of $K$ and $T$ matrices for inelastic scattering using variational principles, *J. Chem. Phys.* **61**, 600–608 (1974).
117. J. H. Weare and E. Thiele, Variation procedure for multichannel scattering processes, *Phys. Rev.* **167**, 11–13 (1968).
118. R. G. Gordon and T.-N. Chiu, On a first-order electronic dipole–dipole mechanism for energy transfer in molecular collisions, *J. Chem. Phys.* **55**, 1469–1471 (1971).
119. R. E. Olson and F. T. Smith, Collision spectroscopy. IV. Semiclassical theory of inelastic scattering with applications to $He^+ + Ne$, *Phys. Rev. A* **3**, 1607–1617 (1971).
120. M. Karplus, *in: Molecular Beams and Reaction Kinetics* (Ch. Schlier, ed.), pp. 407–426, Academic Press, Inc., New York (1970).

121. B. H. Choi and K. T. Tang, Theory of distorted-wave Born approximation for reactive scattering of an atom and a diatomic molecule, *J. Chem. Phys.* **61**, 5147–5157 (1974).

122. P. Pechukas and J. C. Light, On the exponential form of time-displacement operators in quantum mechanics, *J. Chem. Phys.* **44**, 3897–3912 (1966).

123. R. D. Levine, Exponential approximations in collision theory, *Mol. Phys.* **22**, 497–523 (1971).

124. R. A. Marcus, On the analytical mechanics of chemical reactions. Quantum mechanics of linear collisions, *J. Chem. Phys.* **45**, 4493–4499 (1966).

125. B. C. Eu and J. Ross, Optical potential for a chemically reactive system, *Discuss. Faraday Soc.* **44**, 39–45 (1967).

126. C. A. Coulson and B. R. Gerber, A lower-bound property of adiabatic phase shifts, *Mol. Phys.* **14**, 117–131 (1968).

127. R. D. Levine, Variational corrections to decoupling approximations in molecular collision theory, *J. Chem. Phys.* **50**, 1–6 (1969).

128. D. A. Micha, Optical potentials in molecular collisions, *J. Chem. Phys.* **50**, 722–726 (1969).

129. R. D. Levine, B. R. Johnson, and R. B. Bernstein, Role of potential curve crossing in subexcitation molecular collisions, *J. Chem. Phys.* **50**, 1694–1701 (1969).

130. R. E. Roberts, Improved perturbation theory for inelastic encounters, *J. Chem. Phys.* **55**, 100–104 (1971).

131. R. B. Bernstein and K. H. Kramer, Sudden approximation applied to rotational excitation of molecules in atoms. II, *J. Chem. Phys.* **44**, 4473–4485 (1966).

132. R. J. Cross, Jr., Semiclassical theory of inelastic scattering: Diagonalization of the phase shift matrix, *J. Chem. Phys.* **49**, 1753 (1968).

133. M. D. Pattengill, C. F. Curtiss, and R. B. Bernstein, Molecular collisions. XIV. First-order approximation of the generalized phase shift treatment of rotational excitation: Atom-rigid rotor, *J. Chem. Phys.* **54**, 2197–2207 (1971).

134. M. Wartell and R. J. Cross, Jr., Semiclassical theory of vibrationally inelastic scattering in three dimensions, *J. Chem. Phys.* **55**, 4983–4991 (1971).

135. R. T. Pack, Relations between some exponential approximations in rotationally inelastic molecular collisions, *Chem. Phys. Lett.* **14**, 393–395 (1972).

136. D. A. Micha and M. Rotenberg, Collision energy dependence of angular distributions for vibrational excitation of $H_2$ by He, *Chem. Phys. Lett.* **13**, 289–291 (1972).

137. H. Rabitz, Effective potentials in molecular collisions, *J. Chem. Phys.* **57**, 1718–1725 (1972).

138. G. Zarur and H. Rabitz, Rotationally inelastic scattering with effective potentials, *J. Chem. Phys.* **59**, 943–951 (1973).

139. R. A. White, A. Altenberger-Siczek, and J. C. Light, Optical potentials in time-dependent quantum theory, *J. Chem. Phys.* **59**, 200–205 (1973).

140. G. Zarur and H. Rabitz, Effective potential formulation of molecule–molecule collisions with application to $H_2$–$H_2$, *J. Chem. Phys.* **60**, 2057–2078 (1974).

141. P. McGuire and D. J. Kouri, Quantum mechanical close-coupling approach to molecular collisions. $J_z$-conserving coupled-states approximation, *J. Chem. Phys.* **60**, 2488–2499 (1974).

142. M. Tamir and M. Shapiro, The approximate conservation of $P$-helicity in rotational excitation: A new decoupling scheme, *Chem. Phys. Lett.* **31**, 166–171 (1975).

143. D. Secrest, Theory of angular momentum decoupling approximations for rotational transitions in scattering, *J. Chem. Phys.* **62**, 710–719 (1975).

144. D. A. Micha, Effective Hamiltonian methods for molecular collision, *Adv. Quantum Chem.* **8**, 231–287 (1974).

145. R. J. Cross, Jr., Semiclassical methods in inelastic scattering, *J. Chem. Phys.* **51**, 5163–5170 (1969).

146. D. R. Bates and D. S. F. Crothers, Semiclassical treatment of atomic collisions, *Proc. R. Soc. London Ser. A* **315**, 465–478 (1970).

147. J. B. Delos, W. R. Thorson, and S. Knudson, Semiclassical theory of inelastic collisions. I. Classical picture and semiclassical formulation, *Phys. Rev. A* **6**, 709–720 (1972).

148. J. B. Delos and W. R. Thorson, Semiclassical theory of inelastic collisions. II. Momentum-space formulation, *Phys. Rev. A* **6**, 720–727 (1972).

149. A. M. Wooley and S. E. Nielsen, On the limits of applicability of the classical trajectory equations in the two-state approximation, *Chem. Phys. Lett.* **21**, 491–494 (1973).

150. D. R. Bates and A. R. Holt, Impact parameter and semi-classical treatments of atomic collisions, *Proc. R. Soc. London Ser. A* **292**, 168–179 (1966).

151. A. M. Arthurs, The mathematical equivalence of the Born approximation and the method of impact parameters, *Proc. Cambridge Philos. Soc.* **57**, 904–905 (1961).

152. J. C. Y. Chen, C. J. Joachain, and K. M. Watson, Electronic transitions in slow collisions of atoms and molecules. IV, *Phys. Rev. A* **5**, 1268–1285 (1972).

153. J. Callaway and E. Bauer, Inelastic collisions of slow atoms, *Phys. Rev.* **140**, A1072–1084 (1965).

154. L. Wilets and S. J. Wallace, Eikonal method in atomic collisions. I, *Phys. Rev.* **169**, 84–91 (1968).

155. J. C. Y. Chen. T. Ishihara, V. H. Ponce, and K. M. Watson, Electronic transitions in slow collisions of atoms and molecules. V, *Phys. Rev. A* **8**, 1334–1344 (1973).

156. A. P. Penner and R. Wallace, Semiclassical normalization of a path integral for a multichannel scattering problem, *Phys. Rev. A* **11**, 149–153 (1975).

157. R. P. Feynman and A. R. Hibbs, *Quantum Mechanics and Path Integrals*, McGraw-Hill Book Company, New York (1965).

158. P. Pechukas, Time dependent semiclassical scattering theory. II. Atomic collisions, *Phys. Rev.* **181**, 174–184 (1969).

159. P. Pechukas and J. P. Davis, Semiclassical theory of weak vibrational excitation, *J. Chem. Phys.* **56**, 4970–4975 (1972).

160. L. D. Landau, Zur Theorie der Energieübertragung. II, *Phys. Z. Sowjetunion* **2**, 46–51 (1932).

161. C. Zener, Non-adiabatic crossing of energy levels, *Proc. R. Soc. London Ser. A* **137**, 696–702 (1932).

162. E. C. G. Stueckelberg, Theorie der unelastischen Stösse zwischen Atomen, *Helv. Phys. Acta* **5**, 369–422 (1932).

163. D. R. Bates, Collisions involving the crossing of potential energy curves, *Proc. R. Soc. London Ser. A* **257**, 22–31 (1960).

164. E. E. Nikitin, The Landau–Zener model and its region of applicability, *Comments At. Mol. Phys.* **1**, 166–172 (1970).

165. M. S. Child, On the Stueckelberg formula for non-adiabatic transitions, *Mol. Phys.* **28**, 495–501 (1974).

166. E. E. Nikitin, The theory of nonadiabatic transitions: Recent development with exponential models, *Adv. Quantum Chem.* **5**, 135–184 (1970).

167. J. B. Delos and W. R. Thorson, Studies of the potential-curve-crossing problem. II, *Phys. Rev. A* **6**, 728–745 (1972).

168. Yu. N. Demkov, Charge transfer at small resonance defects, *Zh. Eksp. Teor. Fiz.* **45**, 195–201 (1963) [*Sov. Phys. JETP* **18**, 138–142 (1964)].

169. R. E. Olson, Charge transfer at large internuclear distances, *Phys. Rev. A* **6**, 1822–1830 (1972).

170. L. Vainshtein, L. Presnyakov, and I. Sobel'man, Excitation of atoms by heavy particles, *Zh. Eksp. Teor. Fiz.* **43**, 518–524 (1962) [*Sov. Phys. JETP* **16**, 370–374 (1963)].

171. D. R. Bates, Collision processes not involving chemical reactions, *Discuss. Faraday Soc.* **33**, 7–13 (1962).

172. E. F. Gurnee and J. L. Magee, Interchange of charge between gaseous molecules in resonant and near-resonant processes, *J. Chem. Phys.* **26**, 1237–1248 (1957).

173. N. Rosen and C. Zener, Double Stern–Gerlach experiment and related collision phenomena, *Phys. Rev.* **40**, 502–507 (1932).

174. D. Rapp and W. E. Francis, Charge exchange between gaseous ions and atoms, *J. Chem. Phys.* **37**, 2631–2645 (1962).

175. H. Nakamura, Collisional excitation transfer between atoms in near-resonant processes, *J. Phys. Soc. Jpn* **20**, 2272–2278 (1965).

176. K. Birkinshaw and J. B. Hasted, Inelastic collisions between atomic ions and diatomic molecules, *J. Phys. B* **4**, 1711–1725 (1971).

177. E. I. Dashevskaya, E. E. Nikitin, and A. I. Reznikov, Theory of collisionally induced intramultiplet mixing in excited alkali atoms, *J. Chem. Phys.* **53**, 1175–1180 (1970).

178. E. E. Nikitin, Nonadiabatic transitions between fine-structure components of alkali metal atoms during atomic collisions, *Opt. Spectros. USSR.* **19**, 19–95 (1965).

179. C. H. Wang and W. J. Tomlinson, Collision-induced anisotropic relaxation in gases, *Phys. Rev.* **181**, 115–124 (1969).

180. J. R. Krenos, R. K. Preston, R. Wolfgang, and J. C. Tully, Reaction of $H^+$ with $H_2$: Experiment, *ab initio* theory and a conceptual model, *Chem. Phys. Lett.* **10**, 17–21 (1971).

181. J. C. Tully, Trajectories in ion–molecule reactions, *Ber. Bunsenges. Phys. Chem.* **77**, 557–565 (1973).

182. R. Düren, Differential cross sections for alkali–halogen collisions from trajectory calculations on intersecting surfaces, *J. Phys. B* **6**, 1801–1813 (1973).

183. J. R. Krenos, R. K. Preston, R. Wolfgang, and J. C. Tully, Molecular beam and trajectory studies of reactions of $H^+$ with $H_2$, *J. Chem. Phys.* **60**, 1634–1659 (1974).

184. R. K. Preston and R. J. Cross, Jr., Competition between charge exchange and chemical reaction: The $D^+ + H_2$ system, *J. Chem. Phys.* **59**, 3616–3622 (1973).

185. G. E. Zahr, R. K. Preston, and W. H. Miller, Theoretical treatment of quenching in $O\,(^1D) + N_2$ collisions, *J. Chem. Phys.* **62**, 1127–1135 (1975).

186. G. Ochs and E. Teloy, Integral cross sections for reactions of $H^+$ with $D_2$, new measurements, *J. Chem. Phys.* **61**, 4930–4931 (1974).

187. M. Lipeles, Simple model for vibrational transfer in ion–molecule charge-exchange excitation, *J. Chem. Phys.* **5**, 1252–1253 (1969).

188. E. R. Fisher and E. Bauer, On the quenching of $O\,(^1D)$ by $N_2$ and related reactions, *J. Chem. Phys.* **57**, 1966–1974 (1972).

189. A. Bjerre and E. E. Nikitin, Energy transfer in collisions of an excited sodium atom with a nitrogen molecule, *Chem. Phys. Lett.* **1**, 179–181 (1967).

190. E. R. Fisher and G. K. Smith, Vibration–electronic coupling in the quenching of electronically excited alkali atoms by diatomics, *App. Opt.* **10**, 1803–1813 (1971).

191. A. M. Wooley, Semiclassical scattering theory and total cross sections for systems with many crossing points, *Mol. Phys.* **22**, 607–618 (1971).

192. G. M. Kendall and R. Grice, Vibrational coordinates in the electron jump model, *Mol. Phys.* **24**, 1373–1382 (1972).

193. E. A. Gislason, Surface crossing model for ion–molecule reactions, *J. Chem. Phys.* **57**, 3396–3400 (1972).

194. M. S. Child, Franck–Condon transitions in multi-curve crossing processes, *Faraday Discuss. Chem. Soc.* **55**, 30–33 (1973).

195. Yu. N. Demkov and V. I. Osherov, Stationary and nonstationary problems in quantum mechanics that can be solved by means of contour integration, *Zh. Eksp. Teor. Fiz.* **53**, 1589–1599 (1967) [*Sov. Phys. JETP* **26**, 916–921 (1968)].

196. J. B. Delos, On the reactions of $N_2$ with O, *J. Chem. Phys.* **59**, 2365–2369 (1973).

197. J. C. Tully, Collision complex model for spin forbidden reactions: Quenching of $O\,(^1D)$ by $N_2$, *J. Chem. Phys.* **61**, 61–68 (1974).

198. P. Pechukas, J. C. Light, and C. Rankin, Statistical theory of chemical kinetics, *J. Chem. Phys.* **44**, 794–804 (1966).

199. J. R. Krenos and J. C. Tully, Statistical partitioning of electronic energy: Reactions of alkali dimers with halogen atoms, *J. Chem. Phys.* **62**, 420–424 (1975).

200. M. Yen Chu and J. S. Dahler, A theory of the collision-induced singlet to triplet transition of methylene, *Mol. Phys.* **27**, 1045–1069 (1974).

201. J. C. Tully, Reactions of $O\,(^1D)$ with atmospheric molecules, *J. Chem. Phys.* **62**, 1893–1898 (1975).

202. R. D. Levine and R. B. Bernstein, Dynamical theory of vibrational state population distribution in electronic-to-vibrational energy transfer, *Chem. Phys. Lett.* **15**, 1–6 (1972).

203. M. A. Gonzalez, G. Karl, and P. J. S. Watson, Electronic–vibrational energy transfer: $Hg^* + CO$, *J. Chem. Phys.* **57**, 4054–4055 (1972).

204. Y. Haas, R. D. Levine, and G. Stein, Electronic excitation induced by reactive molecular collisions: A theoretical model, *Chem. Phys. Lett.* **15**, 7–11 (1972).

205. A. D. Wilson and R. D. Levine, Simple models of vibrational excitation in energy transfer molecular collisions, *Mol. Phys.* **27**, 1197–1216 (1974).
206. R. E. Olson, Absorbing-sphere model for calculating ion–ion recombination total cross sections, *J. Chem. Phys.* **56**, 2979–2984 (1972).
207. A. Messiah, *Quantum Mechanics*, Vol. I, John Wiley & Sons, Inc., New York (1961), Chap. 10.
208. T. A. Green and M. E. Riley, Strong-coupling semiclassical methods: Phase corrected average approximation for atom–atom collisions, *Phys. Rev. A* **8**, 2938–2945 (1973).
209. G. A. L. Delvigne and J. Los, Rainbow, Stueckelberg oscillations and rotational coupling on the differential cross sections of $Na + I \rightarrow Na^+ + I^-$, *Physica (Utrecht)* **67**, 166–196 (1973).
210. B. C. Eu and T. P. Tsien, Uniform WKB theory of inelastic collisions: Application to $He^+$–Ne inelastic collisions, *Phys. Rev. A* **7**, 648–657 (1973).
211. B. C. Eu, Theory of inelastic collisions: Uniform asymptotic (WKB) solutions and semiclassical *S*-matrix elements for two-channel problems, *J. Chem. Phys.* **55**, 5600–5609 (1971).
212. B. C. Eu, Theory of inelastic collisions: Uniform asymptotic (WKB) solutions and semiclassical scattering matrix elements for multichannel problems, *J. Chem. Phys.* **56**, 2507–2516, 5202 (1972).
213. B. C. Eu, Semiclassical theory of rearrangement and exchange collisions, *J. Chem. Phys.* **58**, 472–478 (1973).
214. B. C. Eu, Theory of inelastic collisions: Extension to multiple turning point problems of uniform WKB theory, *J. Chem. Phys.* **59**, 4705–4713 (1973).
215. U.-I. Cho and B. C. Eu, Improved solutions to the equation of motion in the uniform WKB theory for two-channel problems, *J. Chem. Phys.* **61**, 1172–1179 (1974).
216. L. D. Landau and E. M. Lifshitz, *Quantum Mechanics*, Addison-Wesley Publishing Company. Inc., Reading, Mass. (1958), p. 178.
217. R. K. Preston, C. Sloan, and W. H. Miller, Semiclassical theory of collisionally induced fine-structure transitions in fluorine atoms, *J. Chem. Phys.* **60**, 4961–4969 (1974).
218. Y.-W. Lin, T. F. George, and K. Morokuma, Semiclassical treatment of electronic transitions in molecular collisions: $H^+ + D_2 \rightarrow HD^+ + D$, *J. Chem. Phys.* **60**, 4311–4322 (1972).
219. Y.-W. Lin, T. F. George, and K. Morokuma, Semiclassical treatment of electronic transitions in molecular collisions: Three-dimensional $H^+ + D_2 \rightarrow HD^+ + D_2$, *Chem. Phys. Lett.* **30**, 49–53 (1975).
220. T. F. George and Y.-W. Lin, Multiple transition points in a semiclassical treament of electronic transitions in atom (ion)–diatom collisions, *J. Chem. Phys.* **60**, 2340–2349 (1974).
221. K. Morokuma and T. F. George, *Ab initio* calculations of potential energy surfaces in the complex plane. I. General theory and one-electron example. *J. Chem. Phys.* **59**, 1959–1973 (1973).
222. R. L. Jaffe, T. F. George, and K. Morokuma, Calculations of potential energy surfaces in the complex plane. III. Branch-point structure and rational fractions, *Mol. Phys.* **28**, 1489 (1974).

# Statistical Approximations in Collision Theory

*Philip Pechukas*

## 1. Introduction

This chapter is an introduction to statistical approximation in the theory of reactive collisions. The theme of the chapter is the transition state,* and the statistics in statistical theory is in essence just the counting of the various ways a system can pass through a transition state.

The aim of statistical theory is to bypass collision dynamics en route to rate constants and cross sections. To bypass, but not to ignore: One of the attractions of statistical theory is that at heart it is founded in dynamics, not in the trivial sense that every scheme of approximation in collision theory is ultimately founded in dynamics but in the sense that statistics and dynamics are inseparable and complement each other at the base of the theory. To define a transition state properly is to solve a dynamical problem: A transition state is a condition of dynamical instability, motion to one side of the state differing qualitatively in character from motion to the other side.

We shall discuss both the transition state theory of direct reactions (one transition state, separating reactants from products) and the transition state theory of complex reactions (two transition states, one separating reactants from complex and another separating complex from products). Nothing here is new,† and much is very old. Nevertheless, the ideas are definitely part of the

---

*Many writers prefer the term *activated complex*. It is a matter of personal taste.
†With the possible exception of some observations on nuclear symmetry along reaction paths that the author has not been able to find in the literature.

---

*Philip Pechukas* • Department of Chemistry, Columbia University, New York, New York

core of modern theoretical chemistry and are discussed in detail in many recent texts, such as the excellent books by Bunker,[1] Johnston,[2] Laidler,[3] Levine and Bernstein,[4] and Nikitin.[5] The reader should consult these works for other points of view, for information on the history of transition state theory, and for additional bibliography.

A short discussion of a long topic must of necessity be limited in scope, limited by the interests and even more by the competence of the writer. This chapter is seriously deficient in at least two respects. First, we shall not present any numerical calculations. Numerical work is, of course, essential to assess the validity of an approximation and to suggest modifications in the theory. Classical transition state theory has reached a definitive and natural form unlikely to change, and the same is true of the quantum theory of "loose" transition states. Quantum theory of "tight" transition states is another matter, and as accurate quantum calculations in three dimensions become available we may look forward to numerical studies of the sort mentioned in Section 3.1 for guidance in setting the final form of the theory. Second, we shall not discuss the random phase approximations or average $S$-matrix methods applicable both to reactive and nonreactive scattering problems. Here there are excellent presentations by Levine[6] and others[7,8] that the reader may consult with profit.

## 2. Classical Transition State Theory of Direct Reactions

### 2.1. Once Over Lightly

Consider a bimolecular reaction in the gas phase,

$$A+B \underset{k_b}{\overset{k_f}{\rightleftharpoons}} C+D \tag{1}$$

and suppose that the reaction proceeds through an intermediate, designated by $*$:

$$A+B \rightleftharpoons * \rightleftharpoons C+D \tag{2}$$

At chemical equilibrium the concentrations of A, B, C, D, and $*$ are fixed, and if the gas is sufficiently dilute these concentrations may be expressed in terms of partition functions of the individual species; in particular,

$$\frac{[*]}{[A][B]} = \frac{Q^*}{Q_A Q_B} \tag{3}$$

where [ ] indicates number per unit volume and $Q_A$, $Q_B$, and $Q^*$ are the partition functions per unit volume of A, B, and $*$. That is,

$$Q_A = h^{-3N_A} \int d\Gamma_A \exp(-H_A/kT) \tag{4}$$

where $N_A$ is the number of atoms in A, $H_A$ is the Hamiltonian of isolated molecule A, and the integration in the phase space of A is restricted to points $\Gamma_A$ where the internal energy of A lies below the first dissociation limit and the center of mass of A lies within a spatial region of unit volume. [The factor $h^{-3N_A}$ is customarily included in the classical partition function to give an approximately correct counting of quantum states; for the same reason we should have appropriate factorials in the denominator if the atoms of A are not all distinguishable (see Section 2.4.)]

The partition function $Q_B$ is similarly defined; $Q^*$ is a little more difficult to define. The Hamiltonian appropriate to the intermediate $*$ is the full Hamiltonian of the system of $N = N_A + N_B = N_C + N_D$ atoms; the intermediate is not stable but rather a region of phase space through which reactants have to pass to become products, and $Q^*$ is the partition function associated with this region, again per unit volume swept out by the center of mass of the entire $N$-atom system.

In this section we have in mind direct reactions with a more or less abrupt passage from reactant to product. We suppose that the intermediate is a thin region of phase space, of narrow width $\delta$ along a "reaction coordinate" $q$ measuring progress from reactants to products; we suppose further that in the vicinity of the intermediate the full Hamiltonian takes the form

$$H = H^{\ddagger} + p^2/2m^* \qquad (5)$$

where $p$ is the momentum conjugate to $q$, $m^*$ is the associated mass, and $H^{\ddagger}$ is independent of $p$ and $q$. Then

$$Q^* = Q^{\ddagger}(\delta/h)(2\pi m^* kT)^{1/2} \qquad (6)$$

where $Q^{\ddagger}$ is the integral of $\exp(-H^{\ddagger}/kT)$ over the remaining $3N-1$ coordinates and $3N-1$ momentum perpendicular to $p$ and $q$ (subject again to the restriction that the center of mass of the system be confined to a unit volume), divided by $h^{3N-1}$.

Chemical equilibrium is dynamic, and a given intermediate has a rather brief life. In a time $dt$ all intermediates with velocity $p/m^*$ pointing in the product direction along $q$, and lying within $p\,dt/m^*$ of the product "edge," will disappear to become product. The number of complexes that disappear in this manner, per unit volume, is the concentration of complexes times the probability that a given complex suffers this fate, that is,

$$\frac{[*]\int_0^\infty dp(p\,dt/m^*)\exp(-p^2/2m^*kT)}{\delta\int_{-\infty}^{+\infty} dp\,\exp(-p^2/2m^*kT)} = \frac{[*]kT\,dt}{\delta(2\pi m^*kT)^{1/2}} \qquad (7)$$

We identify the rate of disappearance of intermediate over the product edge as the forward rate of the reaction, $k_f[A][B]$; comparison with Eqs. (3) and (6) gives

$$k_f = \frac{kT}{h} \frac{Q^\ddagger}{Q_A Q_B} \tag{8}$$

Similarly, the rate of disappearance of intermediate over the reactant edge is the backward rate $k_b[C][D]$, and we find

$$k_b = \frac{kT}{h} \frac{Q^\ddagger}{Q_C Q_D} \tag{9}$$

The ratio $k_f/k_b$ is of course just the statistical expression for the equilibrium constant, as it should be.

These formulas are purely classical. The $h$ that appears in Eqs. (8) and (9) is Planck's constant, the quantum of action, but it is not really there: $Q^\ddagger$ contains one less $h$ in its denominator than the products $Q_A Q_B$ or $Q_C Q_D$.

This is classical transition state theory, and it is remarkably simple, but there are questions and a paradox. The paradox: Kinetics is supposed to be an exercise in dynamics, but all the dynamics has disappeared—literally, over the edge in the argument above—and we are left only with the task of calculating partition functions. The questions: First, how to define the "transition state" $\ddagger$? And second, what relation do the rates calculated above—which are rates for a macroscopically unobservable process (passage through a surface in phase space) occurring in a macroscopically static situation (complete chemical equilibrium)—bear to the thermal reaction rates measured in the laboratory or calculated by the methods of molecular dynamics?

It is most helpful to discuss complicated questions with the aid of simple models, and in the next two sections we shall look at transition state theory as it applies to one- and two-dimensional models of chemical reactions.

## 2.2. Dynamical Foundations: A One-Dimensional Model

The classical problem of barrier passage in one dimension is trivial: Particles with energy above the top of the barrier can get across, particles with energy below the top of the barrier are reflected, and particles with energy just equal to the barrier height spend their life trying to reach the top. The problem can be used as a crude model for chemical kinetics by imagining a gas of noninteracting particles moving in one dimension in the presence of the barrier, particles to the left labeled reactant and particles to the right labeled product (Fig. 1). We assume an initial equilibrium distribution of reactants, of density one particle per unit length on the potential-free region to the left of the barrier, and no product. We calculate the rate of formation of product assuming that the reactant and product regions extend to infinity—no walls—so that once across the barrier a particle continues to move to the right. In this

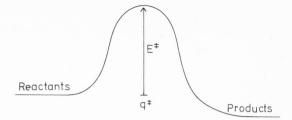

Fig. 1. One-dimensional model for transition state theory of direct reactions.

model there is no back reaction, equilibrium is never reached, and after a period of confusion in which the rate of product formation depends on details of the initial distribution right around the barrier, the rate of product formation is constant and equal to the number of particles per unit time that cross a fixed point $q$ to the left of the barrier with energy sufficient to cross the barrier. Because

$$\rho(p) \, dp = \exp(-p^2/2mkT) \, dp/(2\pi mkT)^{1/2} \tag{10}$$

is the probability that a reactant particle has momentum between $p$ and $p + dp$, and because all particles with positive momentum $p$ and lying in the interval $(q - p/m, q)$ will cross $q$ in a unit of time, the rate of product formation is

$$k = \int_{(2mE^{\ddagger})^{1/2}}^{\infty} \frac{p}{m} \rho(p) \, dp = kT \frac{\exp(-E^{\ddagger}/kT)}{(2\pi mkT)^{1/2}} \tag{11}$$

In this model $k$ is in fact the forward rate constant for reaction, because the density of reactants is unity; we have dropped the subscript $f$ on $k$ and hope the reader will not confuse rate constants $k$ and Boltzmann's constant $k$.

Now for transition state theory. Years of looking in textbooks at diagrams similar to Fig. 1 lead us automatically to choose the point $q^{\ddagger}$—the location of the barrier maximum—as the transition state. The Hamiltonian is $H = p^2/2m + V(q)$. At $q^{\ddagger}$, $H = p^2/2m + E^{\ddagger}$, and because the partition function of reactants per unit "volume" is $(2\pi mkT)^{1/2}/h$, the transition state rate constant [Eq. (8)] is

$$k_{TS} = \frac{kT}{h} \frac{\exp(-E^{\ddagger}/kT)}{(2\pi mkT)^{1/2}/h} = k \tag{12}$$

For this model problem transition state theory is exact.

We can make the entire calculation look slightly less juvenile by introducing an obscure notation. Consider the phase space of a single one-dimensional particle and define on it a function $\chi_r(p, q)$, the characteristic function of reactive phase points, by $\chi_r(p, q) = 1$ if $(p, q)$ lies on a reactive trajectory, and $\chi_r(p, q) = 0$ otherwise. That is, $\chi_r(p, q) = 1$ if in the indefinite past a particle following the classical trajectory through $(p, q)$ lies in the reactant region, and in the indefinite future in the product region. For our model problem $\chi_r(p, q) = 1$ if $p > 0$ and $H(p, q) > E^{\ddagger}$, and 0 otherwise. The function

$\chi_r(p, q) \exp[-H(p, q)/kT]$ is a phase space density (i.e., a nonnegative function of phase point), identical to the equilibrium density $\exp(-H/kT)$ on reactive phase points and zero elsewhere. The equilibrium rate constant $k$, Eq. (11), is proportional to the classical flux of this density through the surface in phase space defined by the equation $q = $ constant, where $q$ is any point in the reactant region:

$$k = (2\pi mkT)^{-1/2} \int dp(p/m)\chi_r(p, q) \exp[-H(p, q)/kT] \qquad (13)$$

Equation (13) is just Eq. (11) in disguise.

Two remarks about the phase space density $\chi_r(p, q) \exp[-H(p, q)/kT]$ are essential to an understanding of equilibrium rate theory and the transition state approximation to it. First, calculating $\chi_r(p, q)$ is a *dynamical* problem: we must follow the classical trajectory through $(p, q)$ and check whether or not we have reactant in the indefinite past and product in the indefinite future. For the present model, of course, this is a trivial dynamical problem, but dynamical nonetheless. Second, $\chi_r(p, q)$ is constant along a given classical trajectory—either zero or one at all points on the trajectory—and therefore the density $\chi_r(p, q) \exp[-H(p, q)/kT]$ is an "invariant density" in phase space, constant along every trajectory (the numerical value of the Hamiltonian of course does not change along a given trajectory). This implies (remember Liouville's equation) that the classical flux associated with this phase space density, through any surface in phase space, is a constant provided only that all reactive trajectories pass through the surface.

For our model problem all reactive trajectories pass every point $q$ on the axis, so that flux integral of Eq. (13) must be completely independent of $q$, a fact that the reader can verify by direct computation. In particular we may just as well evaluate the correct equilibrium rate constant at the transition state $q\ddagger$:

$$k = (2\pi mkT)^{-1/2} \int dp(p/m)\chi_r(p, q^{\ddagger}) \exp[-H(p, q^{\ddagger})/kT] \qquad (14)$$

Transition state theory calculates the rate of reaction as the rate at which phase points leave the transition state *to the product side*. This amounts to replacing $\chi_r(p, q^{\ddagger})$ by $\chi_+(p, q^{\ddagger})$, the characteristic function of phase points with positive momentum:

$$\chi_+(p, q) = \begin{cases} 1, & p > 0 \\ 0, & \text{otherwise} \end{cases} \qquad (15a)$$

$$k_{TS} = (2\pi mkT)^{-1/2} \int dp(p/m)\chi_+(p, q^{\ddagger}) \exp[-H(p, q^{\ddagger})/kT] \qquad (15b)$$

Dynamics disappears in the replacement $\chi_r \rightarrow \chi_+$. To evaluate $\chi_r$ we must examine the entire trajectory through $(p, q)$; to evaluate $\chi_+$ we need only look at the sign of the momentum at $(p, q)$. Transition state theory is exact for this model problem because, *at $q^{\ddagger}$, $\chi_r = \chi_+$* for all $p$.

What governs the choice of transition state? In this model problem *any q* is, in a sense, a point of transition between reactant and product, because every reactive trajectory must at some time cross any $q$. At any $q$ but $q^{\ddagger}$, however, $\chi_+(p, q)$ does *not* equal $\chi_r(p, q)$ for all $p$, and then the transition state rate constant cannot—even by accident—equal the correct equilibrium rate constant: It is not only sufficient but also necessary that $\chi_+(p, q) = \chi_r(p, q)$ at the transition state if transition state theory is to be exact. The reason, in this one-dimensional model, is simple: $\chi_r$ and $\chi_+$ both vanish if the sign of the momentum $p$ is negative, whereas if $p$ is positive, $\chi_+$ by definition equals 1 whereas $\chi_r$ may be zero. Transition state theory, then, gives a rate constant that is never less than the correct equilibrium rate constant, and the best choice of transition state is that state that minimizes the transition state rate constant, i.e., the state that minimizes the partition function $Q^{\ddagger}$. In this one-dimensional model problem the criterion is purely energetic. There is only one coordinate, which is naturally the reaction coordinate, and we should choose the transition state at the point of maximum potential energy along that coordinate. In more than one dimension entropy is a consideration as well, and the best choice of transition state is not automatically the region of highest energy along the reaction coordinate (see Section 2.3).

The correct transition state $q^{\ddagger}$ is a point of instability for dynamics in one dimension. There is a phase point that remains on the transition state for all time, the point with $p = 0$. A little shove to the right and the point inevitably becomes product; a little shove to the left and the point inevitably becomes reactant. At the transition state we have a qualitative change, at $p = 0$, in the character of the classical motion.

It is instructive to compare the discussion of this section with the derivation of transition state theory given in Section 2.1. In that derivation we calculated rates that are macroscopically unobservable, rates of reaction forward and back at complete chemical equilibrium. In this section we posed a problem that in its initial conditions was a step closer to experimental gas kinetics: an initial equilibrium distribution of reactants and, initially, no products. Shortly—within the time it takes reactants to climb the potential barrier—a steady state distribution is reached in the neighborhood of the transition state, and we have *partial* equilibrium in the interaction region. At $q^{\ddagger}$ itself we have an equilibrium distribution in the single-particle phase space *insofar as the positive momenta are concerned*. At other phase points in the neighborhood of the transition state the distribution is the equilibrium distribution if the point can be reached classically by a particle that in the indefinite past was reactant, and zero otherwise.

In general, if isolated reactants are in thermal equilibrium, distributed à la Maxwell–Boltzmann over their possible states, then rapidly—in roughly a single collision—dynamics ensures a partial equilibrium between isolated reactants and reactants in collision: States of interaction that evolve from isolated reactant states are populated according to the Maxwell–Boltzmann distribution. This is an important point,* occasionally missed in discussions of the so-called "equilibrium assumption" in transition state theory.

The equilibrium rate constant, defined by Eq. (13) in terms of the *reactive* phase space density $\chi_r \exp(-H/kT)$, may be expressed as well in terms of the partial equilibrium density of interacting states described above, and we shall do so in the next paragraph because the quantum analogue of the expression is central to recent discussions of quantum transition state theory (see Section 3.2). First, however, we emphasize that rapid equilibrium between isolated reactants and colliding reactants does not in itself guarantee that the equilibrium rate constant is what one measures in a gas–kinetic experiment. A reacting gas mixture is under no obligation to maintain the reactants in thermal equilibrium.

Describing the evolution of a reacting mixture is a difficult statistical mechanical problem that has attracted considerable attention over the years, particularly in the last decade and a half.[10] We shall make only two obvious comments. First, if one enforces thermal equilibrium on the reactants at all times, then one can be sure that the observed rate is the equilibrium rate. In the one-dimensional model of this section we enforced thermal equilibrium on reactants approaching the barrier by providing an infinite sink of thermally equilibrated reactant to the left—remember, no walls. This is a ruse not available in real life, but running the reaction at great dilution in nonreactive "buffer" gas accomplishes the same trick, by ensuring that the rate of equilibration among states of the isolated reactants—by collision with the "buffer"—is much greater than the rate of reaction. Second, we expect nonequilibrium effects to lower the instantaneous rate of reaction below that given by equilibrium theory, for the population of reactant molecules in highly reactive states will be depleted by reaction faster than the population of less reactive states, so the bias in the instantaneous distribution of reactants—relative to an equilibrium distribution—should lie toward the sluggards.

Consider now the partial equilibrium established in phase space by an equilibrium distribution of isolated reactants. The phase points populated are those lying on trajectories that in the indefinite past were reactant. Let $\chi(p, q)$ be the characteristic function of these phase points. For the one-dimensional model of Fig. 1 we have $\chi(p, q) = 1$ if (1) $H(p, q) > E^{\ddagger}$ and $p > 0$; or (2) $H(p, q) = E^{\ddagger}$, $p > 0$, and $q < q^{\ddagger}$; or (3) $H(p, q) < E^{\ddagger}$ and $q < q^{\ddagger}$. Otherwise, $\chi(p, q) = 0$.

We can write the equilibrium rate constant equally well in terms of the characteristic function $\chi$:

$$k = (2\pi mkT)^{-1/2} \int dp(p/m)\chi_r(p, q) \exp[-H(p, q)/kT]$$

$$= (2\pi mkT)^{-1/2} \int dp(p/m)\chi(p, q) \exp[-H(p, q)/kT] \tag{16}$$

*A forceful discussion of this point was given recently by Anderson.[9]

an equality that may be verified by direct calculation. The reason for it is that $\chi$ combines $\chi_r$, the characteristic function of reactive trajectories, with $\chi_{nr}$, the characteristic function of nonreactive trajectories (trajectories that start in the indefinite past as reactant but never make it to product). $\chi_{nr} \exp(-H/kT)$ is an invariant phase space density with no flux through any phase space surface in the product region—because $\chi_{nr}$ is zero in the product region—and therefore no flux through *any* surface separating reactants from products. The nonreactive trajectories are reflected from the barrier (with the exception of the single trajectory at $H = E^{\ddagger}$), and positive flux passing through $q$ on the way in is cancelled by negative flux through $q$ on the way out.

In classical theory there is no good reason to complicate matters by using $\chi$ rather than $\chi_r$. In quantum theory, however, $\chi$ has an analogue (a projection operator onto states that in the indefinite past were reactant), but $\chi_r$ does not.

### 2.3. Dynamical Foundations: A Two-Dimensional Model

As a two-dimensional model for chemical kinetics we shall consider particles of unit mass moving on the plane under the influence of a potential $V(q_1, q_2)$, particles in one region of the plane labeled reactant and particles in another region labeled product. Classical mechanics in two dimensions is a good deal more complicated than in one, and with this model we even make some small contact with reality: If the exchange reaction $A + BC \rightarrow AB + C$ is forced to take place with all three atoms at all times collinear, the potential energy $V$ of the three-atom system is a function of two variables (for instance, the AB and BC distances $d_{AB}$ and $d_{BC}$), and in the center of mass frame and after a suitable linear change of coordinates from $(d_{AB}, d_{BC})$ to $(q_1, q_2)$ the motion of the system is that of a unit mass in the potential $V$ expressed as a function of $(q_1, q_2)$. Details of the coordinate transformation may be found in the standard texts, and considerable discussion of collinear atom–diatom collisions is given in Chapter 2 of Part B. Here we shall concentrate on the dynamical problem of motion of a unit mass on the plane and on the use and validity of transition state theory for reaction rates in such a model.

The potential energy surface $V(\mathbf{q})$ is conveniently represented by a map of the equipotentials $V = $ constant. For a reaction with activation energy the map might look like Fig. 2, where we also indicate some lines of constant $d_{AB}$ and $d_{BC}$ to remind the reader of the collinear atom–diatom problem lurking in the background. In the lower right we have reactant ($d_{AB}$ large, $d_{BC}$ small); in the upper left we have product ($d_{AB}$ small, $d_{BC}$ large); out in "center field" we have the region of three free atoms ($d_{AB}$ and $d_{BC}$ large). At any point the force on a unit mass is $-\nabla V$, where $\nabla$ is the two-dimensional gradient. The force is perpendicular to the equipotential through that point and in the direction of decreasing potential.

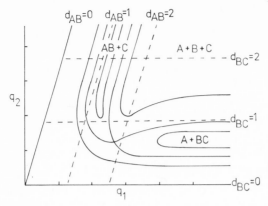

Fig. 2. Two-dimensional model for transition state theory of direct reactions. Map of the equipotentials.

Note in Fig. 2 that two equipotentials cross in the region of interaction ($d_{AB}$ and $d_{BC}$ small). The point of crossing is the so-called saddle point of the surface; $\nabla V = 0$ there, and in the immediate neighborhood the potential is parabolic downward in one direction through the saddle point and parabolic upward in the perpendicular direction through the saddle point. For a potential shaped as in Fig. 2 the potential energy at the saddle point is the classical threshold for reaction: A reactant particle must have total energy at least this great to have a chance of becoming product, for otherwise the motion of the particle is restricted to the U-shaped portion of the reactant region defined by the appropriate equipotential $V = E$.

We are interested in reactive trajectories for the process $A + BC \rightarrow AB + C$, that is, trajectories on which in the indefinite past $d_{BC}$ is bounded but $d_{AB}$ is not, whereas in the indefinite future $d_{AB}$ is bounded but $d_{BC}$ is not. Figure 3 illustrates some of the possibilities for classical motion in two dimensions: a reactive trajectory giving vibrationally excited product, a nonreactive trajectory leading from reactant back to reactant with vibrational deexcitation, and a nonreactive trajectory along which a speedy A smashes into BC and dissociates it. Nonreactive trajectories do not contribute to the reaction rate. We exclude them by defining the characteristic function of reactive trajectories

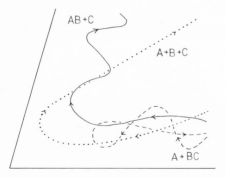

Fig. 3. Classical trajectories in the plane: reactive (——), nonreactive (– –), and nonreactive dissociative (· · ·).

$\chi_r(\mathbf{p}, \mathbf{q}) : \chi_r = 1$ if $(\mathbf{p}, \mathbf{q})$ lies on a reactive trajectory, and zero otherwise. As in one dimension, we need only $\chi_r$ to calculate the reaction rate, and as in one dimension, evaluating $\chi_r$ at any point $(\mathbf{p}, \mathbf{q})$ is a dynamical problem that involves looking at the entire history of the trajectory through $(\mathbf{p}, \mathbf{q})$.

We assume an initial equilibrium distribution of reactants and no products. The phase space density $\chi_r(\mathbf{p}, \mathbf{q}) \exp[-H(\mathbf{p}, \mathbf{q})/kT]$ is invariant, and in the steady state the rate of product formation will be proportional to the flux of this density across any surface in phase space separating reactants from products, that is, any surface through which every reactive trajectory must pass. We can construct such a surface in a particularly simple manner by laying a curve on the plane in such a way that the asymptotic reactant region ($d_{BC}$ bounded, $d_{AB} \to \infty$) lies to one side of the curve and the asymptotic product region ($d_{AB}$ bounded, $d_{BC} \to \infty$) to the other side (see Fig. 4). Any such curve will do; parametrize it by arc length $s$, so that $\mathbf{q}_s$ is the point on the curve a distance $s$ along from the start, and let $\mathbf{n}_s$ be the unit normal to the curve at $\mathbf{q}_s$ pointing to the product side (Fig. 4). The surface in phase space that we have in mind is the three-dimensional set $\{\mathbf{q}_s, \text{all } \mathbf{p}\}$, and the thermal rate constant is proportional to the thermal flux of reactive trajectories through this surface,

$$k = Q^{-1} \int d\mathbf{p} \, ds (\mathbf{p} \cdot \mathbf{n}_s) \chi_r(\mathbf{p}, \mathbf{q}_s) \exp[-H(\mathbf{p}, \mathbf{q}_s)/kT] \qquad (17)$$

where $Q$ is the partition function of isolated reactants per unit distance along the $q_1$ axis, with Planck's constant omitted for simplicity of notation.

The elementary interpretation of Eq. (17) is that all reactive particles with momentum $\mathbf{p}$ and lying in the skewed rectangle with corners at $\mathbf{q}_s$, $\mathbf{q}_{s+ds}$, $\mathbf{q}_s - \mathbf{p} \, dt$, and $\mathbf{q}_{s+ds} - \mathbf{p} \, dt$ will cross the curve in time $dt$. The area of this skewed rectangle is $ds|\mathbf{p} \cdot \mathbf{n}_s| \, dt$. If $\mathbf{p} \cdot \mathbf{n}_s$ is positive, these particles add to the instantaneous population on the product side of the curve; if $\mathbf{p} \cdot \mathbf{n}_s$ is negative, these particles subtract from the instantaneous product population. On balance, of course, we have net addition to the product region, and in fact if we draw the curve $\mathbf{q}_s$ far out in either the reactant or product region, all reactive trajectories will cross with $\mathbf{p} \cdot \mathbf{n} > 0$. If $\mathbf{q}_s$ lies in the region of interaction, however, we may find reactive trajectories that in the course of their meandering through the

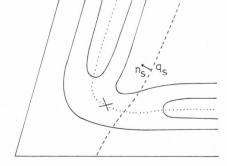

Fig. 4. Equipotentials (——), transition state curve $\mathbf{q}_s$ (– –), unit normal $\mathbf{n}_s$ in the product direction, and reaction path ($\cdots$).

plane cross the curve in the "wrong" direction. These crossings must still be counted, with appropriate sign, in the flux integral.

Transition state theory sets the rate of reaction equal to the total rate at which particles, in complete equilibrium, cross the transition state to the product side. Here the transition state is defined by the entire curve $\mathbf{q}_s$, and

$$k_{TS} = Q^{-1} \int d\mathbf{p} \, ds (\mathbf{p} \cdot \mathbf{n}_s) \chi_+(\mathbf{p}, \mathbf{q}_s, \mathbf{n}_s) \exp[-H(\mathbf{p}, \mathbf{q}_s)/kT] \tag{18}$$

where

$$\chi_+(\mathbf{p}, \mathbf{q}, \mathbf{n}) = \begin{cases} 1, & \mathbf{p} \cdot \mathbf{n} > 0 \\ 0, & \text{otherwise} \end{cases} \tag{19}$$

Dynamics disappears in the replacement $\chi_r \rightarrow \chi_+$: To evaluate $\chi_+$ we need merely examine the sign of $\mathbf{p} \cdot \mathbf{n}$ at the point $(\mathbf{p}, \mathbf{q})$.

The momentum integral in Eq. (18) can be done immediately. Because

$$H(\mathbf{p}, \mathbf{q}_s) = \mathbf{p}^2/2 + V(\mathbf{q}_s) = (\mathbf{p} \cdot \mathbf{n}_s)^2/2 + p_s^2/2 + V(\mathbf{q}_s) \tag{20}$$

where $p_s$ is the projection of $\mathbf{p}$ along the curve $\mathbf{q}_s$ at $s$, we find

$$k_{TS} = kTQ^{-1} \int dp_s \, ds \, \exp\{-[p_s^2/2 + V(\mathbf{q}_s)]/kT\} \tag{21}$$

and what we have to calculate is a "reduced partition function" associated with the transition state curve $\mathbf{q}_s$.

A caution: Unless the force between any pair of atoms vanishes identically beyond some finite distance, it is conceivable that a reactive trajectory might extend arbitrarily far into the region of three nearly free atoms. Such trajectories contribute negligibly to the true thermal rate constant, because exchange reactions go at energies well below the energy of dissociation: A new bond is formed as the old is broken. Still, these trajectories must in principle be counted. To be certain that all reactive trajectories pass through the curve $\mathbf{q}_s$ we must extend this curve indefinitely far into the region of three free atoms. There the potential is bounded and the integrals (18) and (21) diverge.

It is useful, therefore, to define "microcanonical" rate constants $k(E)$ and "microcanonical" transition state theory $k_{TS}(E)$ by

$$Qk = \int dE \, k(E) \exp(-E/kT) \tag{22a}$$

$$k(E) = \int d\mathbf{p} \, ds (\mathbf{p} \cdot \mathbf{n}_s) \chi_r(\mathbf{p}, \mathbf{q}_s) \delta(E - H(\mathbf{p}, \mathbf{q}_s)) \tag{22b}$$

and

$$k_{TS}(E) = \int d\mathbf{p} \, ds (\mathbf{p} \cdot \mathbf{n}_s) \chi_+(\mathbf{p}, \mathbf{q}_s, \mathbf{n}_s) \delta(E - H(\mathbf{p}, \mathbf{q}_s)) \tag{23}$$

To calculate $k(E)$ we need only reactive trajectories with total energy $E$, that is, we need $\chi_r(\mathbf{p}, \mathbf{q}_s)$ only for momenta that satisfy $\mathbf{p}^2/2 + V(\mathbf{q}_s) = E$. To calculate $k_{TS}(E)$ we need only the portion of the curve $\mathbf{q}_s$ lying in the region $V(\mathbf{q}) \leq E$;

explicitly, on performing the momentum integral we find

$$k_{TS}(E) = 2 \int ds \{2[E - V(\mathbf{q}_s)]\}^{1/2} \qquad (24)$$

If $k_{TS}(E)$ is a good approximation to $k(E)$ up to an energy $E'$ many $kT$ above the classical threshold for reaction, the mutilated integral

$$k_{TS}^{E'} = Q^{-1} \int^{E'} dE \, k_{TS}(E) \exp(-E/kT) \qquad (25)$$

is a good approximation to the correct thermal rate constant $k$ [Eq. (22a)].

Notice that $k_{TS}(E) \geq k(E)$, for if $\mathbf{p} \cdot \mathbf{n}_s > 0$, we have $\chi_+(\mathbf{p}, \mathbf{q}_s, \mathbf{n}_s) = 1$ by definition but $\chi_r(\mathbf{p}, \mathbf{q}_s)$ may be zero, whereas if $\mathbf{p} \cdot \mathbf{n}_s < 0$, we have $\chi_+(\mathbf{p}, \mathbf{q}_s, \mathbf{n}_s) = 0$ by definition but $\chi_r(\mathbf{p}, \mathbf{q})$ may be unity. Transition state theory, then, may be looked on as a variational theory of reaction rates[11]: One attempts to minimize the transition state rate contant by varying the transition state, with the assurance that one can never underestimate the thermal rate constant.

The best choice of transition state in the one-dimensional model of the previous section was the point of highest potential energy between reactants and products. On the typical two-dimensional potential surface, with a single saddle point in the interaction region, the minimum of $V(\mathbf{q}_s)$ along any curve $\mathbf{q}_s$ separating reactants from products must be less than or equal to the saddle point energy, so for the sake of energy it would seem that the best transition state should pass through the saddle point and then climb the potential wall on either side as quickly as possible, that is, along the "path of steepest ascent" that follows the gradient of the potential. But it is not necessarily so. In two dimensions we want a transition state that is both high and narrow. The reader should ponder the potential maps of Figs. 5 and 6 to understand why in some cases entropic considerations—the length of the transition state—may shift the best transition state well off the saddle point.

As always, entropy is more important the higher the temperature, and for that matter classical mechanics is more accurate the higher the temperature, so

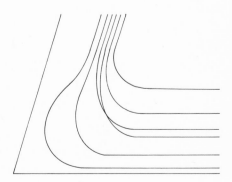

Fig. 5. Where is the transition state?

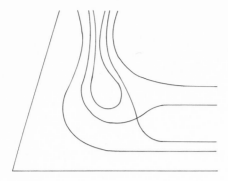

Fig. 6. Where is the transition state?

there is still interesting work to do on classical variational rate theory in high-temperature chemistry.

In the two-dimensional model of this section we can go fairly far with the problem of finding the "best" transition state. Return to Eqs. (22) and (23) defining microcanonical (single-energy) rate constants and their transition state approximants. Equation (22b) gives $k(E)$ independent of the curve $\mathbf{q}_s$, and no matter how we choose $\mathbf{q}_s$, $k_{TS}(E)$ can be no smaller than $k(E)$. We may in fact pick the best transition state for each energy—a different curve $\mathbf{q}_s$ for every $E$. Now to get reaction the total energy of the system must exceed the potential energy at the saddle point, and at any energy above the saddle point energy we shall usually have only two equipotentials, running along on either side of the saddle point. Draw a curve $\mathbf{q}_s$ from one equipotential to the other (Fig. 7); any reactive trajectory of energy $E$ must cross this curve. If we have happened by chance to pick the curve $\mathbf{q}_s$ for which the value of $k_{TS}(E)$ is a minimum, we must have [from Eq. (24)]

$$\delta \int_1^2 ds[E - V(\mathbf{q}_s)]^{1/2} = 0 \tag{26}$$

for all variations in curve leading from 1 to 2. We recognize Eq. (26) as one of the variational formulations of classical mechanics[12]; the curve $\mathbf{q}_s$ must be the

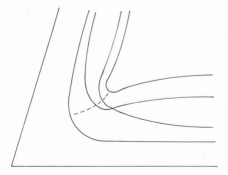

Fig. 7. Anharmonic vibration across the saddle point region (– –).

orbit of a classical trajectory, of total energy $E$, between points 1 and 2. The best transition state at given energy lies along a "vibration" of the system between the two equipotentials of that energy. Notice that at the end points, where $E = V$, the particle comes to a halt and therefore starts off at points 1 and 2 in the direction of the force: $\mathbf{q}_s$ meets the two equipotentials at right angles.

Do such vibrations exist? Certainly. In the immediate vicinity of the saddle point the potential may be expanded to quadratic terms in displacement from the saddle point, and at energies just above the saddle point energy the appropriate vibration is harmonic and in a straight line directly through the saddle point. At a higher energy we start a particle off with velocity zero on one of the equipotentials, somewhere in the neighborhood of the saddle point. It heads out toward the other equipotential but perhaps does not make it, falling away from the saddle point region in, say, the product direction. We move over and start the particle off on the equipotential again; now we find that it falls away to the reactant side. Somewhere between these two trajectories is a "dividing" trajectory that falls away neither to reactant nor to product; this is the required "vibration," across the saddle point region but not necessarily through the saddle point, and the curve executed on the plane by the vibration is the best transition state at that energy. (At least, it is the best in the vicinity of the saddle point. The potential surface may allow other such vibrations. Find one in Fig. 6 and convince yourself that at high energy it is the transition state of choice.)

As in one dimension, the best transition state is a state of dynamical instability. The vibration linking two equipotentials separates motion of qualitatively different character: A little push off to one side and we get reactant, a push to the other side and we get product.

The best transition state may of course be rather bad: The accuracy of the transition state approximation must depend on the details of the potential surface. Numerical calculations[13] on the Porter–Karplus $H + H_2$ surface[14] have shown that the transition state rate constant $k_{TS}(E)$ is an excellent approximation to the exact $k(E)$ up to energies 10–20 kcal/mole above the saddle point. It is interesting to ask why transition state theory is so good for this surface, and, more generally, how we can recognize surfaces for which transition state theory will be accurate.

We saw above that the necessary and sufficient condition that transition state theory be exact at energy $E$ is that $\chi_+ = \chi_r$ for all phase points of energy $E$ on the transition state. This is a dynamic condition and may be put in more picturesque terms as follows: $\chi_+ = \chi_r$ if and only if every trajectory that leaves the transition state never returns. The essential point here is that if a trajectory crosses the transition state and returns to cross it again the sign of $\mathbf{p} \cdot \mathbf{n}$—and therefore the value of $\chi_+$—differs at the two crossings, whereas $\chi_r$ is constant along the entire trajectory and therefore must differ from $\chi_+$ at one of the crossings.

It is instructive to examine the characterization given above of the best transition state at given energy in terms of the dynamic criterion that a trajectory leaving the transition state should not return. If the transition state is *not* the orbit of a classical trajectory, we can be certain that there exist trajectories that leave the transition state only to return and cross it again. For example, suppose that a phase point starts off at time zero along the transition state but falls away to, say, the product side by time $t$; classical position and momentum at time $t$ are continuous functions of classical position and momentum at time zero, so there is a phase space neighborhood of the original phase point all of whose trajectories reach the product side by time $t$, and this neighborhood necessarily contains points on the transition state with momentum initially directed to the reactant side. Transition state theory is automatically incorrect, then, unless trajectories that start along the transition state remain on it for all time.

One can check whether or not transition state theory is exact at energy $E$ by following all trajectories that leave the transition state with that energy and observing whether they return, but this is essentially a complete dynamical calculation. There is a simple geometric condition on potential surfaces sufficient (but not necessary!) to ensure that transition state theory is exact at all energies below a given energy $E$.[15] Suppose that we can draw a *straight* line through the saddle point and perpendicular to all the equipotentials up to energy $E$; this will be the transition state, the same for all energies $\leq E$. Consider the band of the plane lying between the two equipotentials at energy $E$, and suppose that we can cover the entire band with a family of straight lines in such a manner that the force $-\nabla V$, at any point on any of these straight lines, never lies to the transition state side of the line. Then no trajectory with energy less than or equal to $E$, and coming from the transition state, can ever return. The essence of the proof is that such a trajectory would have to "turn back" at a point of tangency to one of these straight lines, and it cannot do so if the forces will not help.

In practice this is an easy criterion to test. We have only to move a ruler over the map of the equipotentials, adjusting continuously as we go from

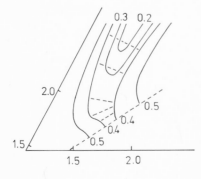

Fig. 8. Equipotentials on the Porter–Karplus $H_3$ surface. Energies in electron volts, distances in atomic units.

transition state to product or from transition state to reactant and checking whether the equipotentials can be made to "open out" as they cross the straightedge or, at worst, run perpendicular to the straightedge. Figure 8 illustrates the test on the Porter–Karplus $H_3$ surface, where symmetry ensures a straight transition state; we show typical members of a family of straight lines linking the 0.4 eV equipotentials, and the forces clearly lie as desired, so transition state theory is exact on the Porter–Karplus surface at least up to 0.4 eV (no great achievement, because the saddle point is at 0.396 eV).

Transition state theory is exact, then, if the potential surface "expands" in a well-defined sense as we move away from the transition state, in either product or reactant direction. This is just what one expect physically: Transition state theory is a theory for systems with a genuine "bottleneck" to reaction.

## 2.4. Multidimensional Potential Surfaces: Saddle Points, Counting Problems, and the Conservation of Nuclear Symmetry along Reaction Paths

We return to the general bimolecular reaction $A + B \rightleftharpoons C + D$. To calculate the exact equilibrium rate constant we need the characteristic function of reactive trajectories—trajectories that are separated $A + B$ in the indefinite past and separated $C + D$ in the indefinite future—and a surface, any surface, separating the asymptotic reactant and product regions in the $N$-atom phase space of the colliding pair. The rate constant is then proportional to the flux of the density $\chi_r \exp(-H/kT)$ across this surface, per unit volume allowed to the center of mass of the system. Transition state theory instructs us to approximate the reactive flux across the surface by the total thermal flux from reactant to product side and then to vary the surface until the best—that is, least—transition state rate is obtained.

This is easier said than done: There are many surfaces in phase space. Variational theory of reaction rates is hard enough in two dimensions and much more so in the real world even for a three-atom exchange reaction. Accordingly, we shall not dwell on the variational aspect of transition state theory in its most general setting.* The lesson of Section 2.3 is that when the potential "expands" in either direction away from a saddle point the best choice of transition state lies in the neighborhood of the saddle point and transition state theory can be quite accurate, so in this section we shall discuss saddle points on multidimensional potential surfaces and related matters of importance to rate theory.

---

*The interested reader should consult Keck's excellent 1967 review[11] for a full discussion of variational theory and for methods of calculating flux through a general surface in phase space. Also of interest is a recent article by Jaffe et al. on the $F + H_2$ reaction[16] in which the authors strike a compromise by choosing not the best surface but one easy to work with and supplement variational theory with exact trajectory calculations (see the remarks at the end of this section).

Let $V(\mathbf{q}_1, \mathbf{q}_2, \ldots, \mathbf{q}_N)$ be the potential energy of the $N$-atom system as a function of the positions $\mathbf{q}_1, \ldots, \mathbf{q}_N$ of the various atoms; we write $V(\mathbf{q}_1, \ldots, \mathbf{q}_N) \equiv V(\mathbf{q})$ for brevity. A stationary point of the potential is a point at which all first derivatives vanish: $\nabla V(\mathbf{q}) = 0$, where $\nabla$ is the gradient operator in $3N$-dimensional configuration space.

We classify stationary points according to the eigenvalues of the symmetric $3N \times 3N$ matrix of mixed second derivatives of $V$ at $\mathbf{q}$. Some of the eigenvalues will be zero, at least five if the stationary point is a linear configuration (i.e., $\mathbf{q}_1, \ldots, \mathbf{q}_N$ all lie on a line) and at least six otherwise, for the potential is unchanged by translation or rotation of the system as a whole, and if $(\mathbf{q}_1, \ldots, \mathbf{q}_N)$ is a stationary point, so is $(\mathbf{q}_1 + \mathbf{a}, \ldots, \mathbf{q}_N + \mathbf{a})$ and $(R\mathbf{q}_1, \ldots, R\mathbf{q}_N)$, where $\mathbf{a}$ is any vector and $R$ any rotation. The matrix may have negative eigenvalues. If there are none, the stationary point $\mathbf{q}$ is called a minimum, and in the immediate vicinity of $\mathbf{q}$ there is no direction in $3N$-dimensional configuration space along which the potential decreases. If the matrix has one negative eigenvalue, the stationary point $\mathbf{q}$ is called a simple saddle point, and there is one direction along which the potential is parabolic downward. If there are two negative eigenvalues, $\mathbf{q}$ is a double saddle point, and so on up to the case of no positive eigenvalues: a potential maximum.

Stable molecular configurations are minima, transition states are simple saddle points, and rate theory avoids all higher-order saddle points for reasons that will be clear in a moment.

A potential surface is a function of all possible configurations of the $N$-atom system, and any conceivable rearrangement process involving the $N$ atoms may occur on the surface; to which reaction does a given simple saddle point "belong"? We find out by following the "reaction path," defined loosely as the path of steepest descent from the saddle point and more formally as the path that follows the gradient of the potential away from the saddle point:

$$d\mathbf{q}/ds = -\nabla V(\mathbf{q})/|\nabla V(\mathbf{q})| \qquad (27)$$

where $s$ is arc length along the path. The reader should not confuse a solution of Eq. (27) with the transition state curves $\mathbf{q}_s$ of the preceding section or with solutions of Newton's equation. There is no dynamics here. The reaction path as defined by Eq. (27) is just a useful mathematical device for going from high ground, at the saddle point, to low ground, where the stable molecules are.

The reaction path through a simple saddle point in two dimensions is shown in Fig. 4.

One aspect of the mathematical theory of Eq. (27) is quite important. Notice that the right-hand side of the equation is indeterminate at a stationary point, where $\nabla V = 0$, so there is the possibility of many solutions to Eq. (27) emanating from a given stationary point. From a simple saddle point, however, there are only two solutions along which the potential decreases as we go out from the saddle point; the solutions start off in opposite directions along the "principal axis" with negative eigenvalue defined by the expansion of $V$ to second order about the saddle point. The reader who is unconvinced of this should sketch the force field around a simple saddle point in two dimensions or examine the behavior of solutions to Eq. (27) in the immediate vicinity of a simple saddle point, where the quadratic approximation is sufficient.

The two sections of reaction path will proceed out from the saddle point according to Eq. (27) until each reaches another stationary point at which $\nabla V = 0$ and the equation is again indeterminate. Usually each piece of the reaction path will by then have gotten outside the region in which all $N$ atoms are close together, and the configuration described by one piece will be, say, separated $A + B$ and by the other piece will be separated $C + D$. We have found a saddle point appropriate to the $A + B \rightleftharpoons C + D$ reaction.

What happens at a double saddle point? Then—quite apart from the fact that the approximate formulas below have no meaning—there is in general no unique reaction to which the saddle point belongs: Equation (27) has an infinite number of solutions passing through the saddle point and along which the potential decreases. A double saddle point is a "hill" with respect to the two directions in which the potential decreases, and we have to grope around the base of the hill until we locate some simple saddle points.

The contribution of a simple saddle point to the rate of the reaction to which it belongs is easy to calculate if the harmonic oscillator–rigid rotor approximation applies in the "thermal neighborhood" of the saddle point. We require that the potential be well approximated by an expansion to quadratic terms, along the directions in which the potential is parabolic upward, to an energy a few $kT$ in excess of the saddle point energy and also that the atomic displacements necessary to raise the energy a few $kT$ along these directions be small compared to the interatomic distances at the saddle point, so that we have essentially a rigid configuration of atoms. Then to describe the motion of the system in the thermal neighborhood of the saddle point it suffices to expand the potential to quadratic terms about the saddle point and make a standard transformation to normal coordinates (see any textbook on classical mechanics for the details). The Hamiltonian in normal coordinates is a sum of $3N$ independent harmonic oscillator Hamiltonians. The vibrational frequency associated with one normal coordinate, the so-called "reaction coordinate," will be imaginary: The potential is parabolic downward in this direction, which lies more or less (depending on the atomic masses) along the reaction path. The vibrational frequency associated with five (if the saddle point configuration is linear) or six (nonlinear) normal coordinates will be zero: an infinitesimal displacement along these coordinate directions is an infinitesimal translation–rotation of the saddle point configuration, with no change in potential energy. The vibrational frequencies associated with the other normal coordinates— $3N - 6$ if the saddle point is linear and $3N - 7$ otherwise—will all be positive if the simple saddle point occurs in the region of interaction where all $N$ atoms are close together.

Now we must choose the best transition state in the saddle point region. The lesson of Sections 2.2 and 2.3 is that the best transition state is a set of classical trajectories. This fixes it right off: We set the reaction coordinate and momentum to zero, and then we have rigid rotation–translation and harmonic vibrations in the remaining $3N-1$ degrees of freedom, an unstable "molecule" that hangs around forever if nothing disturbs it.

In the neighborhood of a simple saddle point, variational reaction rate theory leads us right back to the elementary derivation of Section 2.1. In the immediate vicinity of the transition state we have a Hamiltonian as in Eq. (5), with $p$ and $m^*$ the momentum and mass along the reaction normal coordinate, and the partition function of the transition state is calculated just as for an ordinary molecule in the harmonic oscillator–rigid rotor approximation, with one vibration missing. This is a standard calculation, and we shall simply quote the results. We shall have a product of translational, rotational, and vibrational partition functions. If the saddle point configuration is linear,

$$Q^{\ddagger} = \left(\frac{2\pi MkT}{h^2}\right)^{3/2} \frac{8\pi^2 IkT}{h^2} e^{-E^*/kT} \prod_{i=1}^{3N-6} \frac{kT}{h\nu_i} \tag{28}$$

where $M$ is the total mass of the $N$-atom system, $I$ the moment of inertia about the center of mass, $E^*$ the potential energy at the saddle point, and $\nu_i$ the various frequencies of vibration; if the saddle point configuration is nonlinear,

$$Q^{\ddagger} = \left(\frac{2\pi MkT}{h^2}\right)^{3/2} \pi^{1/2} \prod_{j=1}^{3} \left(\frac{8\pi^2 I_j kT}{h^2}\right)^{1/2} e^{-E^*/kT} \prod_{i=1}^{3N-7} \frac{kT}{h\nu_i} \tag{29}$$

where the $I_j$ are the principal moments of inertia of the saddle point configuration.

The reader familiar with the usual high-temperature approximation of molecular partition functions, which is in essence what a classical partition function is, will notice that Eqs. (28) and (29) lack nuclear spin degeneracies and symmetry numbers. Nuclear spin degeneracies can be ignored: The nuclei are the same in the transition state as in reactants or products, and the degeneracies cancel on top and bottom in the expression for the rate constant.

Symmetry numbers are another matter. It is possible to construct seeming paradoxes in rate theory by postulating transition states of particular symmetry for families of similar reactions where one knows from common sense the relative rates, and there has been discussion and some confusion in the literature as to whether one should use symmetry numbers for transition states in the same manner as for stable molecules.† Briefly, the answer is yes, absolutely. One cannot tamper with the laws of statistical thermodynamics, and the partition function of the transition state has to be evaluated in standard manner or the concentration of transition state intermediates at equilibrium

---

†See the text by Laidler[3] for an extensive and excellent discussion of the problem.

will not come out right; and if one has the concentration of transition state intermediates correct, the calculation in Section 2.1 of the rate at which they disappear to become product is flawless, within the dynamic approximation that a system that leaves the transition state can never return. But one has to be certain that a proposed reaction intermediate can in fact be a simple saddle point on the potential surface so that the theory will make sense (we shall see in a moment that there are severe restrictions on the symmetry of transition states), and one also has to be sure to find *all* transition states linking reactant and product, for all contribute to the rate.

Symmetry numbers are a concern when the $N$ atoms of colliding pair A + B are not all different: We may have many hydrogens, a couple of oxygens, a mess of carbons, and so on. The potential is unchanged by any permutation of the positions of identical atoms, and therefore if $\mathbf{q}$ is a stationary point of the potential, so is any point obtained from $\mathbf{q}$ by a permutation of identical atoms: If the system contains $n_i$ atoms of type $i$, we have automatically $\Pi_i n_i!$ stationary points, differing only in the numbering of identical atoms. Atomic arrangements differing only in the numbering of identical atoms are to be counted as the same molecule. However, we cannot automatically add all $\Pi_i n_i!$ contributions of the form (28) or (29) to the partition function of the arrangement defined by $\mathbf{q}$, for some of these stationary points may be obtained from $\mathbf{q}$ by rotation and therefore may already have been "counted" in the partition function assigned to $\mathbf{q}$. Specifically, let $\sigma - 1$ be the number of permutations of $\mathbf{q}$ that can be reached from $\mathbf{q}$ by a smooth curve, in $3N$-dimensional configuration space, along which the potential does not change; $\sigma$ is the symmetry number of the atomic arrangement defined by $\mathbf{q}$. If $\mathbf{q}$ is an arrangement in which the $N$ atoms are all close, as in a transition state, the only permutations of $\mathbf{q}$ that we can reach without changing the potential energy along the way are those obtained by rigid rotation of the $N$-atom arrangement. If $\mathbf{q}$ is a stationary point representing separated A and B, we can rotate A and B independently, and it is easy to see that the symmetry number of $\mathbf{q}$ is the product $\sigma_A \sigma_B$ of the symmetry numbers of isolated A and B.

We shall obtain a contribution of the form (28) or (29) to the partition function of the atomic arrangement defined by $\mathbf{q}$ for each *distinct* stationary point in the collection of $\mathbf{q}$ and its permutations; the number of such contributions is $\Pi_i n_i!/\sigma$. The ratio $Q^{\ddagger}/Q_A Q_B$ in the transition state rate constant is therefore equal to $(Q_1^{\ddagger}/Q_{1A} Q_{1B})(\sigma_A \sigma_B / \sigma^{\ddagger})$, where the $Q_1$ are partition functions associated with a *single* stationary point.

Methane is a useful molecule on which to exercise one's understanding of symmetry numbers. There are four equivalent hydrogens at the potential minimum. The number of minima related by permutation is $4! = 24$; the symmetry number is 12 (four threefold axes and three twofold axes); and so there are two distinct potential minima, one related to the other by reflection in the plane of two C—H bonds.

Question: Suppose that the Lord suddenly arranges, for pedagogic purposes, that at the potential minimum all four C—H bonds differ slightly in length by a few millionths of an Angstrom and that the symmetry number of methane becomes 1; does industry's chance of finding natural gas in an equilibrium mixture of hydrocarbons suddenly jump 12-fold? The answer is no: Distinct stationary points contribute independently to the partition function of an atomic arrangement only if the thermal neighborhoods of each do not overlap, and in this case we can go from 1 potential minimum to 11 others by a combination of rotation and change in bond length without changing the potential energy by anything approaching $kT$. Therefore caution is necessary: The distinct stationary points contributing to the partition function of a given atomic arrangement, whether stable molecule or transition state, must be truly distinct.

In a rate calculation we have to sum over all transition states connecting reactants and products. The saddle point of lowest energy (together with all its permutations) will usually dominate the rate, but there will be two such points if the saddle point configuration is optically active (that is, if the atomic configuration produced by reflecting the saddle point configuration in a plane cannot be made to coincide with the original configuration by any rotation). If the reactants and products are optically inactive, we shall have two contributions to the rate, from the saddle point enantiomers, equal in magnitude because the potential energy of an atomic configuration is not altered by reflection in a plane. We leave it to the reader to reflect on the various possibilities when reactants and/or products are optically active, bearing in mind that a given transition state configuration must lead to a unique configuration of reactants and products; for counting purposes, it is simplest to treat enantiomers as geometrical isomers—i.e., different molecules—just as one distinguishes and bottles and sells the various forms of hexane although all lie on the same potential surface.

As every organic chemist knows, an optically *inactive* transition state cannot connect optically active reactants and products. This is the most trivial consequence of a general result on conservation of symmetry along reaction paths that we shall now discuss.*

We label the nuclei 1 through $N$ (indistinguishable nuclei are nevertheless given different labels), and in the potential function $V(\mathbf{q}_1, \mathbf{q}_2, \cdots . \mathbf{q}_N)$ the first entry $\mathbf{q}_1$ refers always to the position of nucleus 1, the second entry $\mathbf{q}_2$ to the

---

*References for this discussion of nuclear symmetry are Pearson[17] and Metiu et al.[18] Both observe that $\nabla V$ has the symmetries of the configuration at which it is evaluated. Pearson concludes that nuclear symmetry is preserved along a reaction path until a stationary point is reached. Metiu et al. disagree, essentially because they do not accept Eq. (38) as the definition of the reaction path. They raise the possibility of interesting complications that we shall not consider here: First, in our language, the possibility that the solution of Eq. (38) is unstable—a little push off to one side and the reaction path finds an entirely new set of reactants or products; second, the possibility that a given set of reactants and products may not be linked by any solution of Eq. (38).

position of nucleus 2, and so on. We define permutation and rotation–reflection operators in the $3N$-dimensional configuration space as follows. If $P$ is a permutation of identical nuclei—for instance,

$$P = \begin{pmatrix} 1 & 2 & 3 & \cdots & 7 & \cdots \\ 3 & 2 & 7 & \cdots & 1 & \cdots \end{pmatrix} \tag{30}$$

in standard notation—then

$$P(\mathbf{q}_1, \mathbf{q}_2, \mathbf{q}_3, \ldots, \mathbf{q}_7, \ldots) = (\mathbf{q}_3, \mathbf{q}_2, \mathbf{q}_7, \ldots, \mathbf{q}_1, \ldots) \tag{31}$$

If $R$ is a rotation or reflection in three-dimensional space, then

$$R(\mathbf{q}_1, \mathbf{q}_2, \ldots) = (R\mathbf{q}_1, R\mathbf{q}_2, \ldots) \tag{32}$$

In the same manner we define $P$ and $R$ acting on an arbitrary $3N$-component vector $\mathbf{v} = (\mathbf{v}_1, \mathbf{v}_2, \ldots)$:

$$P\mathbf{v} = (\mathbf{v}_3, \mathbf{v}_2, \mathbf{v}_7, \ldots, \mathbf{v}_1, \ldots), \qquad R\mathbf{v} = (R\mathbf{v}_1, R\mathbf{v}_2, \ldots) \tag{33}$$

$P$ and $R$ are linear operators, commute with each other, and preserve dot products:

$$P\mathbf{u} \cdot P\mathbf{v} = \mathbf{u} \cdot \mathbf{v}, \qquad R\mathbf{u} \cdot R\mathbf{v} = \mathbf{u} \cdot \mathbf{v} \tag{34}$$

The potential is invariant under any $P$ or $R$:

$$V(P\mathbf{q}) = V(\mathbf{q}), \qquad V(R\mathbf{q}) = V(\mathbf{q}) \tag{35}$$

It follows that

$$P\boldsymbol{\nabla} V(\mathbf{q}) = \boldsymbol{\nabla} V(P\mathbf{q}), \qquad R\boldsymbol{\nabla} V(\mathbf{q}) = \boldsymbol{\nabla} V(R\mathbf{q}) \tag{36}$$

for if $d\mathbf{q}$ is any infinitesimal displacement, we have

$$\boldsymbol{\nabla} V(P\mathbf{q}) \cdot P\, d\mathbf{q} = V[P(\mathbf{q} + d\mathbf{q})] - V(P\mathbf{q}) = V(\mathbf{q} + d\mathbf{q}) - V(\mathbf{q})$$

$$= \boldsymbol{\nabla} V(\mathbf{q}) \cdot d\mathbf{q} = P\boldsymbol{\nabla} V(\mathbf{q}) \cdot P\, d\mathbf{q} \tag{37}$$

$$\boldsymbol{\nabla} V(R\mathbf{q}) \cdot R\, d\mathbf{q} = \cdots = R\boldsymbol{\nabla} V(\mathbf{q}) \cdot R\, d\mathbf{q}$$

and this can be true for arbitrary $d\mathbf{q}$ only if Eq. (36) holds.

Now consider the equation by which we define reaction paths,

$$d\mathbf{q}/ds = -\boldsymbol{\nabla} V(\mathbf{q})/|\boldsymbol{\nabla} V(\mathbf{q})| \tag{38}$$

$\mathbf{q}(s)$ follows along the gradient of the potential, and $V$ decreases as the arc length $s$ increases. There is a solution of Eq. (38) passing through any point $\mathbf{q}(s') \equiv \mathbf{q}'$ at which $\boldsymbol{\nabla} V(\mathbf{q}') \neq 0$, unique as long as $\boldsymbol{\nabla} V[\mathbf{q}(s)] \neq 0$. Further, if $\mathbf{q}(s)$ is a solution of Eq. (38), so are $P\mathbf{q}(s)$ and $R\mathbf{q}(s)$; for instance,

$$dP\mathbf{q}(s)/ds = -P\boldsymbol{\nabla} V(\mathbf{q})/|\boldsymbol{\nabla} V(\mathbf{q})| = -\boldsymbol{\nabla} V(P\mathbf{q})/|\boldsymbol{\nabla} V(P\mathbf{q})| \tag{39}$$

$R$ is a symmetry of the point $\mathbf{q}$ if there is a permutation $P$ such that $PR\mathbf{q} = \mathbf{q}$—in other words, if $\mathbf{q}$ and $R\mathbf{q}$ differ only in the labeling of indistinguishable particles. Consider a solution of Eq. (38) on which $\boldsymbol{\nabla} V[\mathbf{q}(s)] \neq 0$ and suppose that $R$ is a symmetry of one point lying on $\mathbf{q}(s)$—say, $\mathbf{q}(s') \equiv \mathbf{q}'$. Then $R$ is a symmetry of all points $\mathbf{q}(s)$, for $\mathbf{q}(s)$ and $PR\mathbf{q}(s)$ are both solutions of Eq. (38) and both pass through $\mathbf{q}'$; by uniqueness of the solution through $\mathbf{q}'$, they must in fact be identical. Symmetry is preserved along solutions of Eq. (38) as long as we do not meet a point at which $\nabla V = 0$.

What happens if $\nabla V(\mathbf{q}_0) = 0$ and $\mathbf{q}(s) \to \mathbf{q}_0$ as $s \to s_0$? If $R$ is a symmetry of the path $\mathbf{q}(s)$, $R$ must be a symmetry of the point $\mathbf{q}_0$, by continuity, for

$$\mathbf{q}_0 = \lim_{s \to s_0} \mathbf{q}(s) = \lim_{s \to s_0} PR\mathbf{q}(s) = PR\mathbf{q}_0 \qquad (40)$$

The point $\mathbf{q}_0$ may have other symmetries *not* shared by the path $\mathbf{q}(s)$ of which it is a limit point: If $R'$ is a symmetry of $\mathbf{q}_0$, $P'R'\mathbf{q}_0 = \mathbf{q}_0$, the path $P'R'\mathbf{q}(s)$ is a solution of Eq. (38) with limit $\mathbf{q}_0$, but this solution does not necessarily coincide with $\mathbf{q}(s)$, for there are in general *many* solutions of Eq. (38) passing through $\mathbf{q}_0$. The limit $\mathbf{q}_0$ is a point of indeterminacy for Eq. (38): A solution does not "know" in what direction to start.

Consider, however, the case where $\mathbf{q}_0$ is a simple saddle point. Then there are only two solutions of Eq. (38) starting at $\mathbf{q}_0$, one leading to reactants and one to products. If $R$ is a symmetry of the saddle point, $PR\mathbf{q}_0 = \mathbf{q}_0$, and $\mathbf{q}(s)$ leads down to reactants, the path $PR\mathbf{q}(s)$ also starts at $\mathbf{q}_0$, satisfies Eq. (38), and must therefore coincide with $\mathbf{q}(s)$ or with the second path, leading to products. The latter is a possibility only if the reaction is a symmetric exchange, $A + B \rightleftharpoons B + A$ (that is, no reaction at all); in this case the transition state may have more symmetry than points on the reaction path, as in the collinear $H + H_2$ reaction where the saddle point has a center. Otherwise, a symmetry of the transition state is a symmetry of the entire reaction path.

Transition state theory requires a simple saddle point as intermediate between reactants and products; the symmetry of the saddle point must be maintained at all points along the reaction path leading from reactants through the saddle point to products; therefore—and most important—the reactant and product configurations reached along the reaction path can have no less symmetry than the transition state.

This is in fact a severe restriction on mechanisms of reactions that proceed through a simple saddle point. The most immediate consequences are that linear transition states must give linear reactants and products, planar transition states must give planar reactants and products, and optically inactive transition states must give optically inactive reactants and products. Much more is true, because the reactant and product configurations are separated molecules, and there are not many symmetries available to a pair of spatially separated groups of atoms. We leave aside the case of a symmetric exchange, discussed above, and isomerization reactions of the form $A + A \rightleftharpoons B + B$. Then a transition state can have only one axis of rotation, for the only symmetry axis available to separated molecules that are not identical is a line through the centers of mass of the molecules. Similarly, the only reflections possible are through planes containing the centers of mass of the separated molecules. Finally, and most important, a symmetry of separated reactants and products can interchange indistinguishable particles only within the same molecule. In any proposed mechanism for the reaction $A+B \rightarrow C+D$—that is, any proposed reaction path from reactants to products—we can label each of the nuclei with one of the four pairs (A, C), (A, D), (B, C), (B, D) to indicate the reactant and the product molecule to which the nucleus belongs. Only those rotations or reflections that interchange nuclei with the same labels are allowed.

This ends our discussion of classical transition state theory per se. However, the notion of a "transition state"—a surface in the region of interaction through which every reactive trajectory must pass—has other uses, especially in the design of efficient classical dynamical calculations of exact thermal rate constants, final state distributions produced in reaction, and so on.[16] We need the characteristic function of reactive trajectories, $\chi_r(\mathbf{p}, \mathbf{q})$. If we attempt a Monte Carlo calculation of $\chi_r$ by sampling trajectories that start in the asymptotic reactant region, we often find that remarkably few trajectories manage to react. By contrast, if we start trajectories on a cleverly chosen transition state and integrate both forward and backward in time to determine their entire history, we can reasonably expect a far higher percentage of reactive trajectories and therefore a more rapid calculation of $\chi_r$.

We shall close this section by emphasizing an elementary point about the distributions generated by $\chi_r$. The probability distribution of reactive phase points on the transition state is $\chi_r \exp(-H/kT)$, *not* an equilibrium distribution: Each phase point carries the Maxwell–Boltzmann weight $\exp(-H/kT)$, but only phase points lying on reactive trajectories are "populated." If transition state theory is exact and $\chi_r$ is equal to the characteristic function of phase points about to leave the transition state to the product side, we have half an equilibrium distribution at the transition state. But even in this well-behaved case there is no assurance that the products produced in the reaction will be anywhere near equilibrium. Final state distributions are governed by the function $\chi_r \exp(-H/kT)$ in the asymptotic product region; again, each phase point carries its Maxwell–Boltzmann weight, but again only those phase points lying on reactive trajectories are populated. The energy released as products separate may go largely into product vibration or largely into product rotation or largely into product translation, depending on the shape of the potential surface from the transition state to the asymptotic product region.

We may also ask for the distribution of reactants *leading to reaction*. This is governed by the function $\chi_r \exp(-H/kT)$ in the asymptotic reactant region, and reaction may require mainly vibrational or mainly rotational or mainly translational energy, depending on the shape of the potential surface between the transition state and the asymptotic reactant region.

Transition state theory of direct reactions says nothing about initial and final state distributions. The distribution may be exactly equilibrium at the transition state, on phase points about to become product, but reaction may nevertheless proceed from a highly nonequilibrium distribution of reactants to a highly nonequilibrium distribution of products.


## 3. Quantum Transition State Theory of Direct Reactions

### 3.1. The Separable Approximation

To make a quantum transition state theory we reason that quantum effects should be more important the lower the temperature, and the lower the temperature the more the rate should be dominated by saddle points on the potential energy surface. Clearly we want to replace classical partition functions with quantum partition functions. In Eqs. (28) and (29) we can leave the rotational partition functions alone unless we are at very low temperature or have reactants and/or transition states with small moments of inertia. Vibrations are another matter; vibrational partition functions should certainly be evaluated quantum mechanically,

$$\prod_i \frac{kT}{h\nu_i} \to \prod_i \frac{\exp(-h\nu_i/2kT)}{1-\exp(-h\nu_i/kT)} \equiv \prod_i \frac{kT}{h\nu_i}\left[\frac{h\nu_i/2kT}{\sinh(h\nu_i/2kT)}\right] \tag{41}$$

The term $(h\nu_i/2kT)/\sinh(h\nu_i/2kT)$ can be regarded as "quantum correction" to the vibrational partition function of the $i$th normal mode. It is an enormous "correction": At room temperature thermal energy is about $200\,\text{cm}^{-1}$, whereas vibrational energies range up to $3000\,\text{cm}^{-1}$ or more.

This is not the only quantum correction necessary. The rate at which systems cross the transition state, from reactant to product side along the reaction coordinate, yields the classical factor $kT/h$ in the rate expression (8); quantum mechanically there is the possibility of tunneling through the saddle point barrier at energies below the classical threshold, as well as the possibility of reflection at energies above the classical threshold. To correct for these effects we look at the classical calculation:

$$\frac{kT}{h} = h^{-1}\int_0^\infty dp\,\frac{p}{m}\exp\left(-\frac{p^2}{2mkT}\right) = h^{-1}\int_0^\infty dE\,\exp\left(-\frac{E}{kT}\right)$$

$$= h^{-1}\int dE\,P(E)\exp\left(-\frac{E}{kT}\right) \tag{42}$$

where $P(E)$ is the classical probability of crossing the saddle point barrier at an energy $E$ measured with respect to the barrier maximum, $P(E) = 1$ if $E > 0$ and zero otherwise. This suggests that we replace $P(E)$ by the quantum probability of crossing the barrier at energy $E$, integrating over all energies for which barrier passage is possible. Because we are looking in the immediate vicinity of the saddle point, the barrier to reaction is parabolic downward along the reaction normal coordinate $q$, and the potential, as a function of reaction coordinate, is characterized by an "imaginary frequency" $i\nu^*$:

$$V(q) = m^*(2\pi i\nu^*)^2 q^2/2 \tag{43}$$

where $m^*$ is the mass associated with motion along the reaction coordinate in the vicinity of the saddle point. We reason, in the saddle point spirit, that only small $q$ count and use the barrier passage probability $P(E)$ for the infinite parabolic barrier [Eq. (43) for all $q$]. $P(E)$ is nonzero for all $E$ [because $V(q)$ goes down to $-\infty$ as $q \to \pm\infty$] and has the value[19]

$$P(E) = [1 + \exp(-2\pi E/h\nu^*)]^{-1} \tag{44}$$

The integral

$$h^{-1} \int_{-\infty}^{+\infty} dE \frac{\exp(-E/kT)}{1 + \exp(-2\pi E/h\nu^*)} \tag{45}$$

is not difficult (integrate in the complex energy plane around the rectangle with corners at $E = \pm\infty$, $\pm\infty + ih\nu^*$, and do not forget the pole at $E = ih\nu^*/2$). Instead of the classical factor $kT/h$, we obtain

$$\frac{kT}{h} \frac{h\nu^*/2kT}{\sin(h\nu^*/2kT)} \tag{46}$$

There is a marvelous symmetry here. The quantum correction for the reaction coordinate is the same as for the vibrational coordinates; it is just that the associated "vibrational frequency" is imaginary. However, there is an important difference between the functions $x/\sinh x$ and $x/\sin x$: The latter blows up every so often. In fact, the integral (45) diverges for any temperature below $h\nu^*/2\pi k$: The Maxwell–Boltzmann factor increases with decreasing energy more than fast enough to compensate the falloff in tunneling probability, and we obtain an infinite amount of tunneling through the barrier, coming from around $E = -\infty$.

That is clearly not right: We had better restrict quantum transition state theory, in the harmonic approximation, to temperatures well above $h\nu^*/2\pi k$. If the "vibrational energy" $h\nu^*$ is much above 500 cm$^{-1}$—as it usually is when the motion along the reaction coordinate involves transfer of a light atom such as hydrogen[20,21]—quantum transition state theory in the harmonic approximation cannot be used at room temperature.

This is distressing, because we wanted a low-temperature theory. Quantum transition state theory in the harmonic approximation will not work at low temperatures, for the tunneling is wrong, and will not work at high temperatures, for the harmonic approximation is wrong.

We try to do better by using tunneling corrections calculated from a one-dimensional barrier of more reasonable shape. First we have to choose the barrier.

The harmonic approximation is a one-parameter fit to the actual potential; we should do better with a two-parameter fit, so we "stick" an Eckart potential[22] on the saddle point, because the solution of the Schrödinger equation is known for this potential, and play with the parameters until the fit looks good.

But that seems a bit arbitrary, so next we extend the reaction normal coordinate out in a straight line from the saddle point and examine how the actual potential varies along it. We find that after falling for a while the potential turns around and shoots up, as two atoms squash together (see Fig. 4). That was a mistake.

Maybe we should look at the potential as a function of distance along the reaction path leading from saddle point to reactants and products. Now the barrier has a reasonable shape, but what is the effective mass in a one-dimensional barrier penetration calculation?

Therefore we use the normal coordinates defined by the saddle point and mass-weight them; now the motion of the system is that of a single particle of definite mass in $3N$ dimensions, but we have to reexpress the potential in these coordinates and calculate a new reaction path. The curve of steepest descent from the saddle point will be different in mass-weighted coordinates.

Then it occurs to us, for physical reasons, that the height of the barrier to be surmounted ought to be the activation energy for the reaction. This is not the distance from reactant minimum to saddle point, but evidently—from Eq. (41)—the distance from zero-point energy of the reactants to zero-point energy of the transition state, so at every point along the reaction path in mass-weighted coordinates we add in the zero-point energy of vibration normal to the reaction path. That seems logical, because what we must be thinking—when we calculate tunneling through a one-dimensional barrier defined by the reaction path—is that the dynamical problem is in some sense approximately separable, so the vibrations normal to the reaction path are independent of motion along the path, the system maintains the vibrational quantum numbers with which it started, and the energy left to motion along the reaction path is total energy less potential energy less the vibrational energy of the state.

Well, actually, maybe we should do this vibrationally adiabatic tunneling correction separately for each vibrational state.

But now there is another problem. Even if the energy were constant along the reaction path—no barrier—the reaction path would still, in general, curve in $3N$-dimensional space, and if we shoot a particle down a curved trough, it will not all come out the other end. Curvature of the reaction path must also affect tunneling. Possibly we should displace the reaction path until centrifugal force balances the potential force normal to the reaction path.[23,24]

We have chosen a potential, somehow; we now solve the one-dimensional Schrödinger equation at many energies and calculate the thermal average,

$$\int dE\, P(E) \exp(-E/kT) \tag{47}$$

To interpret the result we ask what energies contribute most to the integral. We calculate the position of the maximum of the integrand. If $h\nu^*/kT$ is substantially less than $2\pi$, where $\nu^*$ is the frequency associated with the top of our barrier, the maximum of the integrand is around $E = 0$ relative to the barrier top, and we have essentially the harmonic approximation. If $h\nu^*/kT$ is substantially greater than $2\pi$, however, we find the maximum around the energy at which $h\nu(|E|)/kT = 2\pi$, where $\nu(|E|)$ is the classical frequency of motion at energy $|E|$ in the well formed by turning the barrier upside down [to see this, use the semiclassical barrier penetration formula for $P(E)$[25]]. Because $\nu(|E|)$ should decrease as $E$ goes down ($|E|$ goes up), because the effective barrier to reaction should open out—relative to an inverted parabola—as we move away from the saddle point, the reaction rate is dominated by tunneling at lower and lower energy as temperature decreases. At low temperatures most of the reaction rate is tunneling through a very thick region extending far from the saddle point.

In the rate formula, then, should we still be using partition functions evaluated right at the saddle point, as if the saddle point were still dominant?

One begins to appreciate that quantum transition state theory is not so well founded as its classical counterpart.

For a clear discussion of the definition of various one-dimensional tunneling corrections, and of their respective merits as judged by comparison with exact quantum calculations for the collinear $H + H_2$ reaction, the reader should consult an important series of articles by Truhlar, Kuppermann, and Adams.[26–29] We shall not pursue the subject further; instead we shall outline, in Section 3.2, two recent attempts at a quantum transition state theory that go beyond the separable approximation and proceed by close analogy with the classical variational formulation.

### 3.2. Beyond the Separable Approximation

We need first a quantum formula for the equilibrium rate of the reaction $A + B \rightarrow C + D$ analogous to the classical flux integral expression. It is derived as in the classical theory: We insist that reactants be maintained at equilibrium, look for the steady state density operator arising from this equilibrium distribution, and calculate the steady state rate of product formation as the flux, across a surface separating reactant and product, associated with this density. The appropriate density operator is $P \exp(-H/kT)$, where $P$ projects onto states that in the indefinite past were pure reactant, that is, $P\psi = \psi$ if the wave function $\exp(-iHt/\hbar)\psi$ describes pure A+B as $t \to -\infty$ and $P\psi = 0$ if the wave function $\exp(-iHt/\hbar)\psi$ has no A+B component as $t \to -\infty$. The density function in configuration space defined by this operator is $\rho(\mathbf{q}) = \langle \mathbf{q}|P \exp(-H/kT)|\mathbf{q}\rangle$, and the flux vector at $\mathbf{q}$ associated with this density is obtained in standard fashion by casting Schrödinger's equation into the form of a continuity equation for $\rho(\mathbf{q})$.

We shall illustrate with one-dimensional model of Section 2.2 to fix the ideas. The rate of barrier passage is proportional to

$$h^{-1} \int_0^\infty dp \frac{p}{m} P(p) \exp\left(-\frac{p^2}{2mkT}\right) \tag{48}$$

where $P(p)$ is the quantum probability that a particle, incident from the left with momentum $p$, crosses the barrier [the classical limit of Eq. (48) is $(kT/h)\exp(-E^{\ddagger}/kT)$].† The relevant solution of the Schrödinger equation is denoted $|p^+\rangle$, with asymptotic behavior

$$\langle q|p^+\rangle = \begin{cases} e^{ipq/\hbar} + R(p)e^{-ipq/\hbar}, & q \to -\infty \\ T(p)e^{ip'q/\hbar}, & q \to +\infty \end{cases} \tag{49}$$

where $p'$ is the "product" momentum determined by conservation of energy, $R(p)$ and $T(p)$ are reflection and transmission coefficients, and the superscript $+$ on $|p^+\rangle$ is self-explanatory to anyone familiar with the standard confusing notation of scattering theory. The probability of crossing the barrier is§

$$P(p) = (p'/p)|T(p)|^2 \tag{50}$$

For the rate of barrier passage we can therefore write

$$h^{-1} \int_0^\infty dp \frac{p'}{m}|T(p)|^2 \exp\left(-\frac{p^2}{2mkT}\right) \tag{51}$$

Now consider the one-dimensional flux operator $j(q)$, defined by its matrix elements:

$$\langle \phi|j(q)|\psi\rangle = (-i\hbar/2m)(\phi^*\psi' - \phi^{*\prime}\psi)(q) \tag{52}$$

---

†There are a lot of $p$'s in this chapter, both capital and lowercase; we hope that the reader does not confuse $P(p)$ with the projection operator $P$ or with the permutations $P$ of Section 2.4.

§Not just $|T(p)|^2$. Think of sending a long wave packet $\psi$ of well-defined momentum $p$ against the barrier. The probability that it gets across is the integral of $|\psi|^2$ on the right after collision with the barrier, divided by the integral of $|\psi|^2$ on the left before collision with the barrier. If the incident packet takes, say, an hour to cross a fixed point to the left, the transmitted packet will take an hour to cross a fixed point to the right. But on the right the particle is traveling with momentum $p'$, on the left with momentum $p$, so the transmitted packet is $p'/p$ as long as the incident packet.

where $\psi' = d\psi/dq$, etc. If $q$ lies far to the right of the barrier, we have [from Eq. (49)]

$$\langle p^+|j(q)|p^+\rangle = (p'/m)|T(p)|^2 \tag{53}$$

But in fact $\langle p^+|j(q)|p^+\rangle$ is independent of $q$ (by direct computation, using the fact that $|p^+\rangle$ satisfies the Schrödinger equation). Therefore the rate is proportional to

$$h^{-1}\int_0^\infty dp\langle p^+|j(q)|p^+\rangle \exp\left(-\frac{p^2}{2mkT}\right) \tag{54}$$

evaluated at any $q$.

We need some elements of the formal theory of scattering, specialized to one dimension. First we define the scattering states $|p^+\rangle$ for negative $p$: $|p^+\rangle$ is the solution of Schrödinger's equation with asymptotic behavior

$$\langle q|p^+\rangle = \begin{cases} e^{ipq/\hbar} + R(p)e^{-ipq/\hbar}, & q \to +\infty \\ T(p)e^{ip'q/\hbar}, & q \to -\infty \end{cases} \tag{55}$$

where $p'$ is again determined by conservation of energy. The states $|p^+\rangle$ are eigenfunctions of the Hamiltonian,

$$H|p^+\rangle = (p^2/2m)|p^+\rangle \tag{56}$$

and if the potential has no bound states, they are complete, so that a state $|\psi\rangle_t$ evolving in time under the Hamiltonian $H$ can be expanded in the $|p^+\rangle$,

$$|\psi\rangle_t = h^{-1}\int dp\, a(p)e^{-ip^2t/2m\hbar}|p^+\rangle \tag{57}$$

Let $|p\rangle$ denote the plane waves,

$$\langle q|p\rangle = e^{ipq/\hbar}, \qquad \text{all } q \tag{58}$$

$|p^+\rangle$ "evolves" from $|p\rangle$ in the sense that, as $t \to -\infty$, $|\psi\rangle_t$ is indistinguishable from the state

$$h^{-1}\int dp\, a(p)e^{-ip^2t/2m\hbar}|p\rangle \tag{59}$$

The normalization of the scattering states is therefore identical to that of the plane waves,

$$\langle p'^+|p^+\rangle = \langle p'|p\rangle = h\delta(p'-p) \tag{60}$$

and we can calculate the expansion coefficients of any state $|\psi\rangle$:

$$|\psi\rangle = h^{-1}\int dp|p^+\rangle\langle p^+|\psi\rangle \tag{61}$$

We define a projector $P$ by

$$P|\psi\rangle = h^{-1}\int_0^\infty dp|p^+\rangle\langle p^+|\psi\rangle \tag{62}$$

From Eq. (59) we deduce that if $P|\psi\rangle_t = |\psi\rangle_t$, then $|\psi\rangle_t$ is pure reactant in the indefinite past; if $P|\psi\rangle_t = 0$, then $|\psi\rangle_t$ is pure product in the indefinite past. $P$ projects onto the subspace of states arising from pure reactant. We may write

$$P = h^{-1} \int_0^\infty dp |p^+\rangle\langle p^+| \tag{63}$$

and if $O$ is any operator, we have

$$\text{Tr } OP = h^{-1} \int_0^\infty dp \langle p^+|O|p^+\rangle \tag{64}$$

In particular, the rate (54) can be written as

$$\text{Tr } j(q)Pe^{-H/kT} \tag{65}$$

that is, as the flux through $q$ associated with the partial equilibrium density operator $P \exp(-H/kT)$ arising from reactants in equilibrium.

For the two-dimensional problem of Section 2.3—a particle of unit mass moving on the plane in a potential $V(\mathbf{q})$—the quantum rate is proportional to the flux integral

$$\int ds \, \mathbf{n}_s \cdot \text{Tr}[\mathbf{j}(\mathbf{q}_s)Pe^{-H/kT}] \tag{66}$$

where $\mathbf{q}_s$ is a curve separating reactants from products, $\mathbf{j}(\mathbf{q})$ is the two-dimensional flux operator with matrix elements

$$\langle\phi|\mathbf{j}(\mathbf{q})|\psi\rangle = (-i\hbar/2)[\phi^*(\nabla\psi) - (\nabla\phi)^*\psi](\mathbf{q}) \tag{67}$$

and $P$ projects onto the subspace of states arising from pure reactant. As in the one-dimensional case, we can construct $P$ from scattering states of the Schrödinger equation that "evolve" out of pure reactant states, that is, states that are products of a reactant vibrational state and a plane wave headed toward the region of interaction.

There is one difference between Eq. (66) and the classical formula (17), important in theory but not usually in practice: A state evolving from pure reactant may—if total energy permits—have amplitude for dissociation to free particles as well as amplitude for reaction. In classical theory this is no problem: We simply focus on the trajectories we want, the reactive trajectories, ignoring all trajectories that lead to dissociation or back to reactant. In quantum theory this is not possible: We cannot break $P$ into a sum of two projectors, one describing reaction and one describing nonreaction and each commuting with the Hamiltonian. We can require of a solution of Schrödinger's equation that it describe pure reactant in the indefinite past or pure product in the indefinite future; we cannot require both. The curve $\mathbf{q}_s$ must therefore be chosen so that Eq. (66) includes no contribution from dissociation, that is, $\mathbf{q}_s$ must "hug" the asymptotic product region in such a way that, at infinite separation of product, no part of the asymptotic free-atom region of the plane lies to the product side of $\mathbf{q}_s$.

So much for preamble. The transition state approximations suggested by this formulation of quantum rate theory are simply described.

McLafferty and Pechukas[30] looked for an upper bound to the quantum flux, because classical transition state theory provides an upper bound to the classical rate constant. We can reach classical transition state theory from the exact classical flux expression in two stages: First, we cast out the contribution from phase points on the transition state that are reactive but have momentum pointing the wrong way (i.e., back toward reactant); second, we add in the contribution from phase points that are nonreactive but have momentum pointing the right way (i.e., toward product). To see how this goes in quantum theory, look at the one-dimensional flux

$$\operatorname{Tr} j(q)Pe^{-H/kT} \tag{68}$$

$j(q)$ is a Hermitian operator; we break it into positive and negative parts,

$$j(q) = j_+(q) + j_-(q) \tag{69}$$

where $j_+$ is obtained from $j$ by setting all negative eigenvalues to zero and $j_-$ by setting all positive eigenvalues to zero. Because the average of a negative operator cannot be positive, we have

$$\operatorname{Tr} j(q)Pe^{-H/kT} \leqslant \operatorname{Tr} j_+(q)Pe^{-H/kT} \tag{70}$$

This is the quantum analogue of casting out reactive phase points with negative momentum. Now let $Q$ be the projector orthogonal to $P$, $P + Q = I$, $PQ = 0$. The average of the positive operator $j_+$ over the density $Q \exp(-H/kT)$ cannot be negative, so

$$\operatorname{Tr} j_+(q)Pe^{-H/kT} \leqslant \operatorname{Tr} j_+(q)e^{-H/kT} \tag{71}$$

and we have the quantum analogue of adding in nonreactive phase points with positive momentum. Quantum transition state theory, then should read

$$\operatorname{Tr} j(q)Pe^{-H/kT} \leqslant \operatorname{Tr} j_+(q)e^{-H/kT} \tag{72}$$

with interpretation as in classical theory: Transition state theory replaces the exact reactive flux through the transition state with the thermally averaged flux in the product direction.

The right-hand side of inequality (72) is indeed an upper bound to the quantum rate, but it is not one to write home about: It is always infinity. The problem is with the quantum current operator $j(q)$. This is a highly singular operator, as one would guess from the fact that its matrix element between any two wave functions depends only on the behavior of the wave functions in an infinitesimal neighborhood of $q$. To make progress it is necessary to "average" the current operator. If $f(q)$ is any positive function, normalized so that

$$\int dp\, f(q) = 1 \tag{73}$$

we can write the exact flux integral equally well as

$$\int dp\, f(q)\, \mathrm{Tr}\, j(q) P e^{-H/kT} \tag{74}$$

and finite upper bounds to Eq. (74) can be derived along the lines of inequalities (70) and (71). This quantum transition state theory, then, involves a transition state with a certain width, defined by the weighting function $f$, and we are instructed to vary $f$ so as to minimize the transition state rate. The theory can be easily generalized to more than one dimension without invoking separability.

The trouble is that the results are lousy: For the infinite parabolic barrier in one dimension McLafferty and Pechukas found that at all temperatures the best transition state rate overestimates the true rate of barrier passage by an even greater factor than the classical $kT/h$ underestimates the rate. Therefore one cannot look here for a solution to the low-temperature tunneling problem of transition state theory, and there are strong arguments of a physical nature that any transition state upper bound to the quantum rate must be similarly inaccurate.

Once one gives up the idea of bounding the quantum rate, the possibilities for a quantum transition state theory are endless: Any expression that reduces to the classical result in the limit $h \rightarrow 0$ is a possible transition state theory. A good theory is one that is easy to work with and that holds out the promise of accurate results. On both counts Miller's recent suggestion[31] seems inspired.

In essence Miller proposes that we use the flux integral rate expression of *classical* transition state theory and simply replace the classical Maxwell–Boltzmann factor $\exp(-H/kT)$ with the phase space function generated from the quantum operator $\exp(-H/kT)$ by Wigner's prescription[32]; in one dimension,

$$h^{-1}\exp[-H(p,q)/kT] \rightarrow h^{-1}\int dq'\, e^{ipq'/\hbar}\langle q-q'/2|e^{-H/kT}|q+q'/2\rangle \tag{75}$$

One comes to this theory from the quantum expression (65) by replacing the dynamic projector $P$ with the projector $P_+$ onto plane waves of positive momentum,

$$P_+ = h^{-1}\int_0^\infty dp\, |p\rangle\langle p| \tag{76}$$

the quantum analogue of replacing $\chi_r$ by $\chi_+$ (Section 2.2). However, the quantum rate is not unambiguously defined by this replacement; we have

$$\mathrm{Tr}\, j(q) P e^{-H/kT} = \mathrm{Tr}\, j(q) e^{-H/kT} P \tag{77}$$

because $P$ commutes with $H$, but

$$\text{Tr } j(q)P_+ e^{-H/kT} \neq \text{Tr } j(q)e^{-H/kT}P_+ \tag{78}$$

because $P_+$ does not commute with $H$ or with $j$. Miller defines the rate by interpreting $j(q)P_+$ as the operator associated by Weyl correspondence to the flux factor in the analogous classical expression; this definition is mathematically equivalent to using the classical flux factor in combination with the Wigner function associated to $\exp(-H/kT)$.

Miller's rate expression has many virtues. First, it passes the elementary test of getting right the rate for passage over an infinite parabolic barrier, and furthermore the classical phase space integral of the Wigner function associated to the quantum operator $\exp(-H/kT)$ gives the correct quantum partition function, so we are guaranteed that in the saddle point harmonic approximation we shall obtain the transition state rate of the previous section, correct in both the tunneling correction along the reaction coordinate and in the partition function for vibrations perpendicular to the reaction coordinate. Second, the Wigner function is well adapted to semiclassical approximation. This is important for the very low-temperature region where tunneling dominates the reaction rate and the semiclassical theory of multidimensional tunneling (also due to Miller[33]) should be accurate. Miller has in fact given a beautiful analysis of the semiclassical limit of his rate expression,[34] and we should soon have results on the $H + H_2$ collinear reaction to compare with accurate quantum calculations and with the various one-dimensional tunneling corrections mentioned in the preceding section. These calculations will indicate whether the Wigner function substitution should be regarded as the definitive version of quantum transition state theory or whether the problem is still unsettled.

The reader may be—in fact, should be—exasperated with all this uncertainty about what quantum transition state theory ought to be when tunneling dominates the rate. He will point out that what we need, after all, is just a simple dynamical theory of reactive cross sections near threshold. Look at the problem in the exothermic direction; we have only to follow a few low-energy reactant pairs as they tunnel through to become product in order to have a decent picture of threshold behavior. However, each wave function, when it reaches the product region, will divide into many branches, for the reaction is exothermic and many final states are accessible. In a simple theory we do not want to follow all the rivulets of the wave function, so we want to get reaction cross sections out of the wave while it is still "simple"—that is, before it gets to the product region. The only way to do it is by calculating the flux of the wave across some surface in the tunneling region, so we want a simple dynamical theory of the flux.

Go to it; when you find it, *that* will be quantum transition state theory.

## 4. Transition State Theory of Complex Reactions

### 4.1. Rates of Complex Formation and Decay

Again we shall consider a bimolecular reaction proceeding through an intermediate, $A+B \rightleftharpoons * \rightleftharpoons C+D$, but now we suppose that $*$ is a long-lived collision complex as produced, for example, by a deep hole in the potential surface between reactants and products. We imagine that a reactant pair $A+B$ or product pair $C+D$ enters the complex region and is subjected to strong forces and violent motion and quickly forgets its origins as $A+B$ or $C+D$. The complex may eventually break up to $A+B$—we shall call this possibility 1—or to $C+D$—possibility 2. It is the essence of the statistical theory of reactions proceeding through a complex that we assume that the probabilities $P_1$ and $P_2$ for the two modes of breakup are independent of how the complex is formed.[†]

The point of view here is exactly opposite that of transition state theory of direct reactions. There, we assume that at the transition state we can distinguish absolutely between trajectories coming from reactant $A+B$ and trajectories coming from product $C+D$ by looking at the direction of motion along the reaction coordinate; here, we assume that in the complex we absolutely cannot distinguish between trajectories from reactants and trajectories from products.

At equilibrium the rate at which complex is formed from $A+B$ is equal to the rate at which complex breaks up to $A+B$, and similarly the rate of complex formation from $C+D$ is balanced by the rate of breakup to $C+D$. If $k_1[A][B]$ is the rate of complex formation from reactants and $k_2[C][D]$ the rate from products, we must have

$$k_1[A][B] = P_1(k_1[A][B] + k_2[C][D])$$
$$k_2[C][D] = P_2(k_1[A][B] + k_2[C][D])$$

(79)

Suppose that to form a complex reactants must pass through transition state 1, whereas products must go through transition state 2, that is, the complex region is separated on either side from reactants and products by a transition state, and the theory of Section 2 can be used to calculate the rate constants $k_1$ and $k_2$. Then

$$k_1 = (kT/h)(Q_1^{\ddagger}/Q_A Q_B), \qquad k_2 = (kT/h)(Q_2^{\ddagger}/Q_C Q_D)$$

(80)

and because

$$[A][B]/Q_A Q_B = [C][D]/Q_C Q_D$$

(81)

we have

$$P_1 = Q_1^{\ddagger}/(Q_1^{\ddagger} + Q_2^{\ddagger}), \qquad P_2 = Q_2^{\ddagger}/(Q_1^{\ddagger} + Q_2^{\ddagger})$$

(82)

[†]The theory of this section is in essence just the theory of unimolecular reactions (see Chapter 3) applied to bimolecular reactions. Keck first gave the transition state formulation of the theory for bimolecular reactions[35]; see also the 1965 articles by Nikitin.[36,37]

The probability that the system leaves the complex region through a given exit is proportional to the width of the exit, as measured by its partition function.

$P_1$ and $P_2$ define "branching ratios" for breakup of the complex, that is, relative rates for the possible modes of decomposition. From transition state theory we can calculate as well the absolute rate constants for breakup of the complex. At equilibrium the concentration of complexes is proportional to the partition function per unit volume, $Q^*$, associated with the complex region of phase space:

$$[*] = Q^*[A][B]/Q_A Q_B = Q^*[C][D]/Q_C Q_D \qquad (83)$$

The rate at which complex breaks up to, say, C+D is equal to the rate of formation of complex from C+D,

$$k_2[C][D] = (kT/h)Q_2^{\ddagger}[C][D]/Q_C Q_D = (kT/h)Q_2^{\ddagger}[*]/Q^* \qquad (84)$$

so the rate constant for breakup of complex to C+D is

$$(kT/h)Q_2^{\ddagger}/Q^* \qquad (85)$$

But we have made an error. A particular complex will of course have a definite energy, and the probabilities $P_1$ and $P_2$ will in general depend on energy. If the energy is too low, for example, it may be impossible for a complex formed from A+B to decompose to give C+D. We should use microcanonical transition state theory (see Section 2.3) and calculate branching ratios for complex breakup as a function of energy.

The branching ratios will depend only on internal energy of the complex, not on the motion of the center of mass, and the same is true of the energy-dependent rate constants for complex formation from reactants and products. For example, at transition state 1 the Hamiltonian is

$$H = \mathbf{P}^2/2M + p_1^2/2m_1^* + h(\gamma_1) \qquad (86)$$

where $\mathbf{P}$ is center of mass momentum, $M$ is total mass, $p_1$ is the momentum along the reaction coordinate, $m_1^*$ is the associated mass, and $h(\gamma_1)$ is the internal energy of the transition state as a function of the remaining $3N-4$ coordinates and $3N-4$ momenta of the $N$-particle system, collectively denoted $\gamma_1$. For the rate of complex formation through transition state 1 we can write

$$\frac{kT}{h}Q_1^{\ddagger} = \frac{1}{h^{3N}} \int_0^{\infty} dp_1 \int d\mathbf{P} \int d\gamma_1 \frac{p_1}{m_1^*} \exp\left(-\frac{H}{kT}\right)$$

$$= \frac{1}{h^3} \int d\mathbf{P} \exp\left(-\frac{\mathbf{P}^2}{2MkT}\right) \int dE\, k_1(E) \exp\left(-\frac{E}{kT}\right) \qquad (87)$$

where

$$k_1(E) = \frac{1}{h^{3N-3}} \int_0^\infty dp_1 \int d\gamma_1 \frac{p_1}{m_1^*} \delta\left(E - \frac{p_1^2}{2m_1^*} - h(\gamma_1)\right)$$

$$= \frac{1}{h^{3N-3}} \int_0^\infty dE' \int d\gamma_1 \delta(E - E' - h(\gamma_1))$$

$$= \frac{1}{h^{3N-3}} \int_{h(\gamma_1) \le E} d\gamma_1$$

$$= \frac{N_1^\ddagger(E)}{h} \tag{88}$$

Here $N_1^\ddagger(E)$ is the "volume" $\int d\gamma_1$ of the transition state phase space up to energy $E$, divided by $h^{3N-4}$, that is, the semiclassical approximation to the number of internal quantum states at transition state 1 with energy $\le E$. Clearly, the only alteration necessary for a quantum theory of complex breakup is replacement of $N_1^\ddagger(E)$ by the correct quantum count of states.

At equilibrium, flux into the complex region of phase space through transition state 1 will be balanced, at every energy, by flux out, so we have

$$k_1(E) = P_1(E)[k_1(E) + k_2(E)] \tag{89}$$

and therefore

$$P_1(E) = N_1^\ddagger(E)/[N_1^\ddagger(E) + N_2^\ddagger(E)] \tag{90}$$

The lifetime of the complex will also depend on energy. Because

$$Q^* = (1/h^{3N}) \int d\Gamma \exp(-H/kT)$$
$$= (1/h^3) \int d\mathbf{P} \exp(-\mathbf{P}^2/2MkT) \int dE \, N(E) \exp(-E/kT) \tag{91}$$

where the integrals are over the region of phase space defining the complex and $N(E)$ is the density of internal states of the complex (number of states per unit energy), the rate constant for breakup through transition state 1 of complexes with energy in range $dE$ around $E$ is

$$k_1(E) \, dE/N(E) \, dE = (1/h)N_1^\ddagger(E)/N(E) \tag{92}$$

$N_1^\ddagger$ is dimensionless, the units of $N(E)$ are energy$^{-1}$, so the units of Eq. (92) are time$^{-1}$, as they should be.

Further refinement is possible: A given complex has definite angular momentum about the center of mass, and the branching ratios and lifetimes of the complex will in general depend on both energy and angular momentum. We shall put off this discussion until Section 4.3. In Section 4.2 we shall discuss the dynamical foundations of the theory.

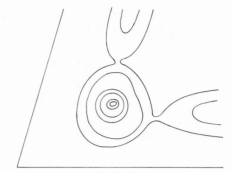

Fig. 9. Two-dimensional model for transition state theory of complex reactions.

## 4.2. Dynamical Foundations

The dynamical justification of statistical theory for complex reactions, crudely stated, is that a trajectory falling into a deep potential well will execute a motion so complicated, and so unstable with respect to its initial conditions, that "on average" the behavior of the trajectory will be characteristic of the entire region of phase space defining the complex, independent of the point of entry into this region.

It is an interesting but difficult problem to make this argument more precise.* We suspect that statistical behavior, and the branching ratios calculated in Section 4.1, must follow if the complex is sufficiently long-lived (that is, if the exits from the complex are sufficiently narrow), but we know of no demonstration that this is in fact the case. Statistical theory for complex reactions is less secure, in its dynamical foundations, than transition state theory of direct reactions.†

The problems one encounters in analyzing the dynamics of complex formation and decay can already be seen in the simplest model, motion of a unit mass in two dimensions [there is no (classical) one-dimensional model for complex formation] under the influence of a potential with a deep well separating region 1 from region 2 and "protected" on either side by transition states (Fig. 9). We shall consider trajectories of fixed energy $E$ and assume that outside the region of the well the potential is such that a particle leaving the complex region through either transition state will never return. No trajectory, then, can "visit" the complex region more than once in its history.

Let $\chi_1(\mathbf{p}, \mathbf{q})$ be the characteristic function of phase points lying on trajectories that in the indefinite past are in region 1 and that enter the complex region at some time in their history; $\chi_2(\mathbf{p}, \mathbf{q})$, similarly, is the characteristic

*Computer "experiments" are therefore most informative; see the trajectory calculations of Brumer and Karplus on alkali halide exchange reactions.[38]

†On the other hand, transition state theory of complex reactions may give accurate results, by accident, which transition state theory of direct reactions never can: When it fails, the latter always fails to the high side of the true reaction rate.

function of phase points lying on trajectories that enter the complex from region 2. So that a trajectory of energy $E$ will be reactive—that is, get from region 1 to region 2—the trajectory must pass through the complex region, and the rate of reaction $k_{12}(E)$ is the flux associated with $\chi_1$ across any curve separating regions 1 and 2 [nonreactive trajectories—trajectories that enter the complex from region 1 and exit back to region 1—contribute to $\chi_1$ but not to the flux associated with $\chi_1$ (see Sections 2.2 and 2.3)]. Therefore, in the notation of Section 2.2,

$$k_{12}(E) = \int d\mathbf{p} \, ds (\mathbf{p} \cdot \mathbf{n}_s) \chi_1(\mathbf{p}, \mathbf{q}_s) \delta(E - H(\mathbf{p}, \mathbf{q}_s)) \tag{93}$$

The total rate of formation of complex, from region 1, is the rate at which trajectories from region 1 cross transition state 1 *into the complex region*; that is, again in the notation of Section 2.2,

$$k_1(E) = \int d\mathbf{p} \, ds (\mathbf{p} \cdot \mathbf{n}_s) \chi_1(\mathbf{p}, \mathbf{q}_s) \chi_+(\mathbf{p}, \mathbf{q}_s, \mathbf{n}_s) \delta(E - H(\mathbf{p}, \mathbf{q}_s)) \tag{94}$$

where the integral is along transition state 1 and the unit normal $\mathbf{n}_s$ points to the complex region. By our assumption on the behavior of the potential outside the complex region, any trajectory that crosses transition state 1 into the complex must originate, in the indefinite past, from region 1, so $\chi_1 \chi_+ \equiv \chi_+$ at the transition state, and $k_1(E)$ is just the rate of complex formation calculated by ordinary transition state theory of direct reactions.

Now $k_{12}(E) = k_1(E) P_{12}(E)$, where $P_{12}(E)$ is by definition the probability that a complex formed from region 1 will decay to region 2. The question is whether this probability is given correctly by the formula of the previous section,

$$P_{12}(E) = k_2(E) / [k_1(E) + k_2(E)] \tag{95}$$

where $k_2(E)$ is the rate of complex formation from region 2, calculated as in Eq. (94) by ordinary transition state theory of direct reactions at transition state 2.

It is presumably the random and irregular motion inside the complex that leads to statistical behavior, so we want to examine the functions $\chi_1$ and $\chi_2$ inside the complex region. First, we ask whether the entire set of states of the complex is accessible, either to reactant or to product, that is, does $\chi_1 + \chi_2 = 1$ at every point in the complex region for which $H(\mathbf{p}, \mathbf{q}) = E$? The answer in general is no; a phase point $(\mathbf{p}, \mathbf{q})$ in the complex region is accessible from outside only if the trajectory through it manages eventually to escape the complex. There may be classical trajectories that never leave the complex, although energy permits, because they never get close to either of the transition states. In the theory of unimolecular reactions one speaks of "inactive modes" that are insulated from the rest of the molecule for times comparable to the lifetime of the molecule, with complete energy sharing only among the remaining modes, but classical-mechanical reality may be more complicated: Rather than dividing into "active" and "inactive" modes, the classical motion may be such that some of the trajectories on the energy surface are "irregular" and wander all over the surface while the rest are "regular" and confined forever to a corner of the surface (see the excellent review by Ford[39]). Intuition and computer calculations agree that irregular motion dominates if the potential well is deep, and in any case *quantum* bound states at energies above dissociation are rare—the "tail" of the wave function extends over all space, in particular out the exits from the complex, and sooner or later the wave function is pulled out of the complex by its tail. But this may take a while, and the rate of decomposition of the complex can be greater than that calculated by Eq. (92) if the density of states in effective communication with the outside is reduced by the occurrence of long-lived quasi-bound states.

So be it; for bimolecular reactions we need only the branching ratios, not the absolute rate of decomposition of the complex, and we can calculate probabilities with respect to the distribution that actually obtains on the energy surface, determined by $\chi_1 + \chi_2$, rather than with respect to the full micro-canonical distribution. Let us calculate, for example, the probability that a complex, selected without regard to origin, decays to region 2. The trajectory through a phase point $(\mathbf{p}, \mathbf{q})$ leads to region 2 if and only if the time-reversed trajectory, through $(-\mathbf{p}, \mathbf{q})$, comes from region 2, so

$$P_2(E) = \frac{\int d\mathbf{p} \, d\mathbf{q} \chi_2(-\mathbf{p}, \mathbf{q}) \delta(E - H)}{\int d\mathbf{p} \, d\mathbf{q} [\chi_1(\mathbf{p}, \mathbf{q}) + \chi_2(\mathbf{p}, \mathbf{q})] \delta(E - H)} \tag{96}$$

where the integrals are over the region of phase space defining the complex. Because $H$ is an even function of momentum, $P_2$ is also the probability that a complex, selected at random, comes from region 2.

Is $P_2(E)$ equal to $k_2(E)/[k_1(E) + k_2(E)]$, the probability that a complex will exit through transition state 2, irrespective of origin, as calculated by the flux integrals? Not necessarily; particles may enter the complex region at equal rates from sides 1 and 2, but if those from side 2 spend, on average, more time in the complex region than those from side 1, we shall find that $P_2(E)$ as calculated from Eq. (96)—the probability that a complex chosen at random comes from side 2—is greater than one half.

The moral is that probabilities calculated by phase space integration inside the complex region may not be the probabilities one needs for rate calculations. However, let us hope for the best; there is no reason to suspect a difference in lifetimes of complexes originating from the two sides, and that would in any case be out of keeping with the basic tenet of statistical theory, that complexes are essentially the same, no matter what their origin.

What we need, then, is $P_{12}(E)$, the probability that a complex originating from region 1 will decay to region 2,

$$P_{12}(E) = \frac{\int d\mathbf{p} \, d\mathbf{q} \chi_1(\mathbf{p}, \mathbf{q}) \chi_2(-\mathbf{p}, \mathbf{q}) \delta(E - H)}{\int d\mathbf{p} \, d\mathbf{q} \chi_1(\mathbf{p}, \mathbf{q}) \delta(E - H)} \tag{97}$$

and the question is whether Eq. (97) equals Eq. (96).

$P_{12}$ is $\chi_2(-\mathbf{p}, \mathbf{q})$ averaged over the distribution $\chi_1$, and $P_2$ is $\chi_2(-\mathbf{p}, \mathbf{q})$ averaged over the distribution $\chi_1 + \chi_2$. We see how the argument ought to go: If the motion inside the complex region is complicated and irregular, we expect that the distribution $\chi_1$ will spread itself uniformly over the accessible region of the energy surface and be essentially indistinguishable from $\chi_1 + \chi_2$—not uniformly in the literal sense, of course; $\chi_1$ is either one or zero, so it winds over the energy surface like the plague through Europe, missing some points while hitting their neighbors, and what we can expect is that the average of any *smooth* function over $\chi_1$ should be identical to the average of that function over the full distribution $\chi_1 + \chi_2$.

There are two problems with this argument: First, it is not really correct, and second, if it were, it nevertheless would not apply to Eq. (97).

Calculate, for instance, the average momentum using each of the three distributions $\chi_1$, $\chi_2$, and $\chi_1 + \chi_2$. The averages all differ, because on balance points from region 1 are moving to region 2 and vice versa, whereas the average complex is going nowhere.

These differences in the averages of smooth functions are in fact small if the potential well is deep and $\chi_1$ and $\chi_2$ both cover a considerable part of the energy surface, but one still cannot infer that Eqs. (97) and (96) are equal: The function averaged over $\chi_1$ in Eq. (97)—the function $\chi_2(-\mathbf{p}, \mathbf{q})$—is *not* a smooth function. If the motion of complexes arising from region 1 is complicated, so equally is the motion of complexes going to region 2, and Eq. (97) is in fact the overlap of two highly irregular functions. Simple analytical arguments about the magnitude of this overlap are suspect.

"Highly irregular functions" are perhaps not well correlated. What happens if we suppose that $\chi_1(\mathbf{p}, \mathbf{q})$ and $\chi_2(-\mathbf{p}, \mathbf{q})$ are uncorrelated with respect to the distribution $\chi_1 + \chi_2$? We obtain

$$P_{12} = \frac{\int d\mathbf{p}\, d\mathbf{q} \chi_1(\mathbf{p}, \mathbf{q}) \chi_2(-\mathbf{p}, \mathbf{q}) \delta(E - H)}{\int d\mathbf{p}\, d\mathbf{q} \chi_1 \delta(E - H)}$$

$$= \frac{\int d\mathbf{p}\, d\mathbf{q} \chi_1(\mathbf{p}, \mathbf{q}) \chi_2(-\mathbf{p}, \mathbf{q}) \delta(E - H)}{\int d\mathbf{p}\, d\mathbf{q} (\chi_1 + \chi_2) \delta(E - H)} \frac{\int d\mathbf{p}\, d\mathbf{q} (\chi_1 + \chi_2) \delta(E - H)}{\int d\mathbf{p}\, d\mathbf{q} \chi_1 \delta(E - H)}$$

$$= \frac{\int d\mathbf{p}\, d\mathbf{q} \chi_1 \delta(E - H)}{\int d\mathbf{p}\, d\mathbf{q} (\chi_1 + \chi_2) \delta(E - H)} \frac{\int d\mathbf{p}\, d\mathbf{q} \chi_2 \delta(E - H)}{\int d\mathbf{p}\, d\mathbf{q} (\chi_1 + \chi_2) \delta(E - H)}$$

$$\times \frac{\int d\mathbf{p}\, d\mathbf{q} (\chi_1 + \chi_2) \delta(E - H)}{\int d\mathbf{p}\, d\mathbf{q} \chi_1 \delta(E - H)}$$

$$= P_1 \cdot P_2 \cdot \frac{1}{P_1} = P_2 \tag{98}$$

Marvelous; we have restated, in obscure notation, the basic premise of statistical theory, that the future of a complex is uncorrelated with its past.

Statistical theory is easier to justify quantum mechanically, up to a point.[40] The properties of an isolated resonance—that is, a quasi-bound state with lifetime long compared to $h/\Delta E$, where $\Delta E$ is the distance in energy to the next resonance—are independent of the mode of formation: You cannot tell one uranium atom from another. If the resonance is *not* isolated, quantum theory encounters much the same difficulties as classical theory. Suppose, for instance, that $\phi_1$ and $\phi_2$ are neighboring quasi-bound states with energies $E_1$ and $E_2$ and that the system initially is in state $a_1\phi_1 + a_2\phi_2$; then

$$\psi(t) = a_1 e^{-iE_1 t/\hbar}\phi_1 + a_2 e^{-iE_2 t/\hbar}\phi_2$$
$$= e^{-iE_1 t/\hbar}(a_1\phi_1 + a_2 e^{-i\Delta E t/\hbar}\phi_2) \tag{99}$$

and the system will not "sample" all possible states, consistent with the absolute magnitudes $|a_1|$ and $|a_2|$, unless it lives for a time of order $h/\Delta E$.

In the statistical theory of complex formation and decay, we are dealing with isolated resonances when the lifetime of the complex is greater than $hN(E)$, where $N(E)$ is the density of states of the complex (that is, the inverse of the energy spacing between neighboring states). Because the lifetime of the complex is the inverse of the total rate constant for decay in all possible modes, from Eq. (92) we must have

$$\sum_i N_i^\ddagger(E) \simeq 1 \tag{100}$$

That is, statistical theory of complex decay is justified quantum mechanically when the exits from the complex are so narrow that no more than a few final states are possible. This makes sense: One state in, one state out, and therefore one state in the middle, and its properties must of course be independent of how it is formed. But the theory of isolated resonances rarely applies to molecular collisions, where the transition states leading to and from the complex region typically accommodate many states even at low energy.

In short, the dynamical foundations of statistical theory for complex reactions are not firmly settled. Nevertheless, physical intuition insists that the theory is correct. If a complex lives long enough, its dynamical state—either classical or quantum; it does not matter—will be an exceedingly sensitive function of initial state, and there is always in rate theory some averaging over initial state. It is not necessarily essential that statistical theory have a foundation in dynamics: The averaging over initial conditions may suffice. Consider, for example, a perfect spherical cavity, silvered on the inside, with a number of tiny holes to let light in and out; we shine a bulb for an instant behind one of the holes and photons start off in all directions around the cavity; the motion is perfect "regular"—there is no chance, in this case, that a photon trajectory "samples" all of phase space—yet the illumination leaking out any hole will certainly be proportional to the size of the hole, and any man who claims otherwise should be put away. Shouldn't he?

### 4.3. Loose Transition States and the Statistical Theory of Final State Distributions†

If the exit from a complex lies at large separation of the product molecules, where the interaction between them is weak or at any rate weakly dependent on their internal state, we have a "loose" transition state in which the product molecules vibrate and rotate freely. The counting of internal states at the exit [Eq. (88)] then has a significance beyond serving as a measure of the total rate of decay to product: These states are the possible final states of product, and the flux from complex to each of them is equal, so we can calculate the final state distributions arising from decay of the complex.

If the transition state entrance to the complex is also loose, we can calculate as well the rate of formation of complex from given internal states of the reactants. In keeping with the spirit of statistical theory we assume that decay of the complex is independent of its mode of formation, so we can determine state-to-state rates of reaction. If the transition states separating complex from reactants and products are loose, then, we should be able to calculate—within the statistical model—cross sections from given vibration–rotation states of reactants to vibration–rotation states of products.

How do loose transition states arise? A transition state is a surface of dynamical instability: Trajectories leaving to one side behave in a qualitatively different manner from those leaving to the other side. Suppose that we have two molecules with *no* potential barrier hindering approach but rather, at moderate to large distances, an attractive potential dependent to good approximation only on the distance $r$ between the molecular centers of mass. There is a surface of dynamical instability here, even though the potential decreases smoothly toward small $r$: At given relative energy $\varepsilon$ there is a distance $r^*$ at which the two molecules will orbit about each other, each moving in a circle about the common center of mass. At this distance the attractive force between the molecules is just balanced by the centrifugal "force" driving them apart as they orbit each other. A trajectory of energy $\varepsilon$, starting at $r^*$ and heading out, will escape to infinity; a trajectory of energy $\varepsilon$, starting at $r^*$ and heading in, spirals to smaller and smaller $r$ until the region of strong interaction is reached.

To determine $r^*$ we look at the effective potential for radial motion in a central field,

$$V(r) + l^2/2\mu r^2 \tag{101}$$

where $l$ is the orbital angular momentum and $\mu$ the reduced mass of the two molecules. If $V(r)$ is attractive and falls off faster than $1/r^2$ as $r \to \infty$, and if $l$ is sufficiently small, this function will have a maximum at large $r$; call the position of the maximum $r^*(l)$. As $l$ increases, $r^*(l)$ decreases until the maximum

---

†The statistical theory of final state distributions is due to Light and co-workers.[41-44] It is often called the *phase space* theory of chemical kinetics, because that is what Light named it.

disappears into a point of inflection where both the first and second derivatives of Eq. (101) vanish, or until the approximation of the intermolecular potential by a spherically symmetric function becomes untenable. The maximum is an "angular momentum barrier." A molecular pair with angular momentum $l$ and relative energy of translation

$$\varepsilon = V(r^*) + l^2/2\mu r^{*2} \tag{102}$$

will not move off the top of the barrier once placed there: all the kinetic energy is in the angular direction, forever.

The orbiting parameters $r$, $l$, and $\varepsilon$ are all interrelated: Any one determines the other two. In what follows the important variable is energy, so we define $r^*(\varepsilon)$ and $l^*(\varepsilon)$ as the orbiting distance and angular momentum at relative energy $\varepsilon$; $l^*(\varepsilon)$ is the maximum orbital angular momentum permitted to the complex at relative energy $\varepsilon$, for a colliding pair with greater angular momentum cannot reach the boundary of the complex at $r^*(\varepsilon)$.

This is clearly a model for reactions without activation energy. There is no potential barrier to formation of the complex, and in fact as the relative energy decreases, the size of the complex, as measured by its radius $r^*(\varepsilon)$, increases.

It is also possible to model reactions with activation energy by a potential permitting loose transition states. We simply put potential barriers in the entrance and exit channels, again dependent only on intermolecular separation.† To reach the complex region reactant molecules have to cross the barrier, so complex formation cannot take place at relative energy below the barrier maximum. At higher energies we get complex formation at any angular momentum low enough that the maximum in the effective potential (101) lies below the relative energy of translation. Again the transition state separation $r^*(\varepsilon)$ is determined by the maximum of Eq. (101), and the theory proceeds as indicated below.

As a model for reactions with activation energy this is very useful, because we obtain state-to-state cross sections, but not especially realistic. Reactions with appreciable activation energy usually have steric as well as energy requirements. If the reaction in fact proceeds through a complex, the entrance to the complex is generally well into the region where the potential depends strongly on relative orientation of the reactant molecules, and we have a "tight" transition state: The collision partners do not rotate and vibrate freely. Statistical theory can still be used to get state-to-state cross sections if the internal states at the entrance and exit to the complex correlate directly with vibration–rotation states of separated reactants and products, but this is an assumption about the dynamics outside the complex and not part of basic

---

†More generally, dependent on intermolecular separation and also on conserved dynamical properties of the separate molecules such as vibrational and rotational energy, to model collisions in which vibrational and rotational energy helps the colliding molecules surmount the barrier to reaction; see Reference 43.

statistical theory of complex formation and decay. We refer the reader to a recent stimulating article by Marcus for a theory of product energy distributions from tight transition states[45] and also to earlier articles by the same author for a discussion of final state distributions in the transition state theory of direct reactions.[46,47]

We shall now return to calculations on loose transitions states and look first at the rate of complex decay to, say, C+D. Each complex has definite energy and angular momentum, and the rate of complex decay to C+D will depend on the angular momentum of the complex as well as on energy, but for the moment we shall ignore that complication and calculate the full microcanonical rate $k(E)$.

The transition state is the set of all phase points with intermolecular separation $r^*(\varepsilon)$ and orbital angular momentum less than or equal to $l^*(\varepsilon)$, where $\varepsilon$ is the relative translational energy of C+D. This is a rather complicated surface in phase space, and to calculate the flux through it in a straightforward fashion is a challenge; Eq. (88) is not directly applicable. Instead we use the fact that a system passing through the transition state, from the complex side out, inevitably becomes separated molecules, so we can calculate the flux just as well through a surface of fixed and large intermolecular separation $r$ provided we restrict the integration to phase points that arise from complex decay.

At large $r$ the interaction between C and D is negligible, and the Hamiltonian is

$$H = \mathbf{P}^2/2M + \mathbf{p}^2/2\mu + h_C(\gamma_C) + h_D(\gamma_D) \qquad (103)$$

where $\mathbf{P}$ is center of mass momentum, $M$ is total mass, $\mathbf{p}$ is the relative momentum of the two molecules, $\mu$ is the reduced mass, and $h_C(\gamma_C)$ and $h_D(\gamma_D)$ are the Hamiltonians of isolated molecules C and D as a function of "internal" phase points $\gamma_C$ and $\gamma_D$ [that is, $\gamma_C$ is a point in the $(6N-6)$-dimensional phase space of coordinates and momentum relative to center of mass and total momentum of molecule C]. The microcanonical rate out of the complex at energy $E$ is [see Eqs. (87) and (88)]

$$k(E) = \frac{1}{h^{3N-3}} \int \int d\gamma_C \, d\gamma_D \int d\mathbf{p} \int d\sigma \frac{\mathbf{p} \cdot \hat{\mathbf{r}}}{\mu} \chi(\mathbf{p}, \mathbf{r}, \gamma_C, \gamma_D)$$
$$\times \delta\left(E - \frac{\mathbf{p}^2}{2} - h_C(\gamma_C) - h_D(\gamma_D)\right) \qquad (104)$$

where $d\sigma$ is an element of area on the sphere of radius $r$, $\hat{\mathbf{r}}$ is the unit vector in the outward radical direction, and $\chi$ is the characteristic function of phase points lying on trajectories that come from the complex. A phase point has come from the complex if the radial momentum $\mathbf{p} \cdot \hat{\mathbf{r}}$ is positive and the orbital angular momentum is less than $l^*(\varepsilon)$, where the relative translation energy $\varepsilon$ is now simply $\mathbf{p}^2/2\mu$.

Fix **r** and do the integral over momentum first, using spherical polar coordinates with axis along $\hat{\mathbf{r}}$. The volume element is $p^2\, dp \sin\theta\, d\theta\, d\phi$, $\mathbf{p}\cdot\hat{\mathbf{r}}$ is $p\cos\theta$, and the angular momentum, in magnitude, is $rp\sin\theta$. The momentum integral is therefore

$$\int_{rp\sin\theta<l^*} d\mathbf{p}\,\frac{\mathbf{p}\cdot\hat{\mathbf{r}}}{\mu} = 2\pi \int dp\,\frac{p^3}{\mu} \int_0^{l^*/rp} \sin\theta\, d(\sin\theta)$$

$$= \int \pi\frac{l^{*2}}{r^2}\frac{p}{\mu}\,dp \equiv \int \pi\frac{[l^*(\varepsilon)]^2}{r^2}\,d\varepsilon \qquad (105)$$

From the integral over the surface of the sphere we obtain a factor $4\pi r^2$, and therefore

$$k(E) = (1/h^{3N-3}) \iiint d\gamma_C\, d\gamma_D\, d\varepsilon \cdot 4\pi^2 l^*(\varepsilon)^2 \delta(E-\varepsilon-h_C(\gamma_C)-h_D(\gamma_D)) \qquad (106)$$

which we can write in more suggestive form as

$$k(E) = h^{-1} \int dn(\varepsilon')N(E-\varepsilon') \qquad (107)$$

using the functions

$$n(\varepsilon') = (1/h^{3N_C+3N_D-6}) \iint_{h_C(\gamma_C)+h_D(\gamma_D)\leq\varepsilon'} d\gamma_C\, d\gamma_D \qquad (108a)$$

$$N(E-\varepsilon') = 4\pi^2 l^*(E-\varepsilon')^2/h^2 \qquad (108b)$$

$n(\varepsilon')$ is the semiclassical approximant to the number of internal states of separated C+D with total energy less than or equal to $\varepsilon'$. Going immediately to quantum mechanics, the analogue to Eq. (107) is therefore

$$k(E) = h^{-1} \sum_{i,j} N(E-E_C(i)-E_D(j)) \qquad (109)$$

where $i$ denotes a vibration–rotation state of C with energy $E_C(i)$, $j$ denotes a vibration–rotation state of D with energy $E_D(j)$, and the sum is over all possible states of C and D with total energy less than $E$. The total rate $k(E)$ is the sum of rates to the various possible final states, and the rate to a given final state is

$$N(E-E_C(i)-E_D(j))/h \qquad (110)$$

The essential feature of Eq. (110) is that the rate of complex decay to a given final state of separated C+D depends only on the total energy of this state. Two quanta in a stretching vibration or 7 in a bending mode or 30 in rotation—it does not matter so long as the total energy is the same: Statistical theory predicts equal decay rate from the complex to each of these alternatives.

Relative decay rates to the possible final states of a given molecular pair are entirely determined, according to Eqs. (108b) and (110), by the behavior of the orbiting angular momentum $l^*(\varepsilon)$ as a function of energy. $l^*(\varepsilon)$ increases with relative energy $\varepsilon$ (raise $\varepsilon$, and the effective potential at the maximum must increase, which can be arranged only by increasing $l$). At given total energy, therefore, the more energy one has in vibration–rotation of C and D, the less energy one has available to relative translation and the lower the rate of complex decay to the state in question. Statistical theory never predicts population inversion as a result of complex decay. In fact, if the total energy is large compared to the spacing of internal states of C and D, the low-lying states of the molecules are produced in a Maxwell–Boltzmann distribution,

$$N(E - E_C(i) - E_D(j))/N(E) = \exp\{-[E_C(i) + E_D(j)]/kT\} \qquad (111)$$

where the energies $E_C(i)$ and $E_D(j)$ in Eq. (111) are measured relative to the lowest vibration–rotation state of each molecule and $T$ is determined "thermodynamically" by

$$T^{-1} = dS/dE \qquad (112)$$

where the "entropy" is

$$S = k \ln N(E) \qquad (113)$$

$E$ is the maximum energy available to relative translation, that is, essentially the exothermicity if we are looking at products from a low-energy collision of unexcited molecules. If the potential $V(r)$ goes as $1/r^6$ at large distances (neutral products), $l^*(\varepsilon)$ goes as $\varepsilon^{1/3}$, and therefore the temperature is determined by $kT = 3E/2$; if $V(r)$ goes as $1/r^4$ at large distances (ion–molecule products), $l^*(\varepsilon)$ goes as $\varepsilon^{1/4}$, and the relation is $kT = 2E$.

The distribution of relative translational energy is determined by Eq. (107); we see that $P(\varepsilon)\, d\varepsilon$, the probability that translational energy lies in the range $d\varepsilon$ around $\varepsilon$, is proportional to

$$d\varepsilon N(\varepsilon)\rho(E - \varepsilon) \qquad (114)$$

where $\rho$ is the density of internal product states. $N(\varepsilon)$ goes as $\varepsilon^{2/3}$ (neutral products) or $\varepsilon^{1/2}$ (ion–molecule products). If $E$ is large compared to the level spacings of C and D and $\varepsilon \ll E$, we have

$$\rho(E - \varepsilon) \propto (E - \varepsilon)^n \propto \exp(-n\varepsilon/E) \qquad (115)$$

where the exponent $n$ is one less than the power of $T$ in the product of the classical vibration–rotation partition functions for the two molecules, because this product goes as

$$\int dE \rho(E) \exp(-E/kT) \propto \int dE E^n \exp(-E/kT) \propto T^{n+1} \qquad (116)$$

We get a $T$ for each vibrational degree of freedom and $T^{1/2}$ for each rotational degree of freedom [see Eqs. (28) and (29)], and the reader can work out the various possibilities according to whether products are linear or nonlinear, atoms or molecules, and so on.

This translational energy distribution was derived by Safron et al. in 1972[48] under the assumptions that the angular momentum of the complex is large compared to rotational angular momentum of reactants or products and that the lifetime of the complex is essentially independent of total angular momentum. Provided the orbital angular momentum of product at given relative energy $\varepsilon$ is not limited to a value less than $l^*(\varepsilon)$ by the mode of formation of the complex, this is equivalent to using the full microcanonical rate $k(E)$ to calculate product distributions (see below).

We have been speaking of relative rates of complex decay to the possible final states of a particular arrangement channel (that is, fixed product molecules C+D). Relative rates to states of identical energy in different arrangement channels (A+B, C+D, etc.) are determined, in statistical theory, by the long-range forces and reduced masses in the various channels, for these are what determine $l^*(\varepsilon)$. Other things being equal, statistical theory favors products with high reduced mass and strong attractive interactions. As for the total decay rate to various channels, the more exothermic a given reaction, the more of that product we shall get (other things being equal): The greater the energy available for distribution among final states, the greater the number of final states available to accept that energy and the higher the total rate to the lot of them.

There is, however, a problem, at high energy, with the definition of the transition state as the maximum of the angular momentum barrier. If the relative energy of translation is too high, the position of the maximum, as determined by the long-range forces, moves into the region of the short-range forces.

We shall not pause to discuss the various possibilities for patching the theory in the very exothermic channels but instead shall consider now the role of angular momentum in statistical theory.

Ignoring the angular momentum of the complex can lead to serious errors. For instance, if the complex can be formed with at most 20 units of angular momentum from reactants but the complete energy surface of the complex "communicates" with states of product possessing up to 80 units of angular momentum, we shall probably be off by quite a bit in using the full microcanonical rate to product for calculating reaction cross sections. We need the probability of decomposition, to all possible final states, of a complex having definite energy $E$ and definite angular momentum quantum numbers $(J, M_J)$.

Look back at the full rate to C+D, Eq. (109). $N(\varepsilon)$ is the semiclassical approximant to the number of quantum states with orbital angular momentum quantum number $L \leqslant l^*(\varepsilon)/\hbar$, because

$$N(\varepsilon) = \frac{4\pi^2 l^*(\varepsilon)^2}{h^2} = \frac{l^*(\varepsilon)^2}{\hbar^2} \simeq \sum_0^{l^*(\varepsilon)/\hbar} 2L + 1 \tag{117}$$

We can therefore write

$$k(E) = h^{-1} \sum_{i,j} \sum_{L, M_L} 1 \tag{118}$$

where the sum over $L$ is restricted to values less than $l^*(E - E_C(i) - E_D(j))/\hbar$.

Equation (118) says that the complex decays at equal rate to any of the possible final states $(i, j, L, M_L)$ accessible from the complex. From these states we can construct linear combinations of definite $(J, M_J)$; the total number of states will not change, and clearly statistical theory for decay of a complex of given $(J, M_J)$ should require equal rate to each of the possible final states with these quantum numbers.

What we have, then, is just a counting problem in the addition of angular momentum, messy but straightforward, and the computer does it well. We are interested in the rate of production of product states lying in various subspaces labeled with quantum numbers $(v_C, v_D, L_C, L_D, L)$, where $v_C$ and $v_D$ are vibrational quantum numbers, $L_C$ and $L_D$ rotational quantum numbers, and $L$ the orbital angular momentum quantum number (if C or D is nonlinear, the third rotational quantum number is included in $v_C$ or $v_D$). There are $(2L_C + 1)$ $(2L_D + 1)(2L + 1)$ states in this subspace, and from then we can form a certain number of states with total angular momentum quantum numbers $(J, M_J)$; let this number be $N(f, L|J, M_J)$, where $f$ stands for the composite $(v_C, v_D, L_C, L_D)$ specifying a vibration–rotation energy level of the product pair. We define

$$N(f|E, J, M_J) = \sum_{\hbar L < l^*(E - E_F)} N(f, L|J, M_J) \tag{119}$$

The rate of decay of complex with energy $E$ and quantum numbers $(J, M_J)$ to the vibration–rotation level $f$ is proportional to $N(f|E, J, M_J)$; the probability that the complex decays to level $f$ is therefore

$$P(f|E, J, M_J) = N(f|E, J, M_J) / \sum_f N(f|E, J, M_J) \tag{120}$$

where the sum in the denominator is over all accessible vibration–rotation levels of all possible molecular pairs arising from complex decay.

The cross section at total energy $E$ from an initial state of reactant A + B in vibration–rotation level $i$ to vibration–rotation level $f$ of product C + D, averaged over $M_{L_A}$ and $M_{L_B}$, is

$$\sigma(f|i) = \sum_{J, M_J} P(f|E, J, M_J)\sigma(E, J, M_J|i) \tag{121}$$

where $\sigma(E, J, M_J|i)$ is the cross section for formation of complex with quantum numbers $(E, J, M_J)$ from a state of $i$ averaged over $M_{L_A}$ and $M_{L_B}$. Two molecules colliding with orbital angular momentum $L'$ and relative momentum $\hbar k$ will contribute [25] $\pi(2L'+1)/k^2$ to the cross section for complex formation if $\hbar L'$ is less than $l^*(E-E_i)$ and zero otherwise. The probability that the pair gives a complex with quantum numbers $(J, M_J)$ is

$$N(i, L'|J, M_J)/(2L'+1)(2L_A+1)(2L_B+1) \tag{122}$$

and so

$$\sigma(E, J, M_J|i) = \frac{\pi/k^2}{(2L_A+1)(2L_B+1)} \sum_{\hbar L' < l^*(E-E_i)} N(i, L'|J, M_J)$$

$$= \frac{\pi N(i|E, J, M_J)}{k^2(2L_A+1)(2L_B+1)} \tag{123}$$

The reaction cross section $\sigma(f|i)$, averaged over $M_{L_A}$ and $M_{L_B}$, is therefore

$$\sigma(f|i) = \frac{\pi}{k^2(2L_A+1)(2L_B+1)} \sum_{J, M_J} \frac{N(f|E, J, M_J)N(i|E, J, M_J)}{\sum_f N(f|E, J, M_J)} \tag{124}$$

This is Light's statistical theory of reaction cross sections. The reader should consult the original papers [41-44] for detailed presentation and discussion of reaction cross sections calculated from Eq. (124). We shall make two remarks.

First, the reaction cross section is still indifferent to the nature of final vibrational energy: Two product levels differing only in the distribution of energy among vibrational modes will be reached with the same cross section. Rotation is another matter. Shifting energy from vibration to rotation affects the angular momentum coupling, in a complicated manner.

Second, when the rotational angular momentum of reactant and product molecules is small compared to the values of $J$ that are important in Eq. (124)—that is, when the angular momentum in both entrance and exit channels is mainly orbital angular momentum of relative motion—the final state distribution can be calculated rather simply provided the sum in the denominator of Eq. (124) can be regarded as independent of $(J, M_J)$. This sum is proportional to the total rate of decay of a complex with angular momentum quantum numbers $(J, M_J)$ [the density of states of the complex entering the rate expression (92) will be independent of $(J, M_J)$ for values of $J$ that are important in Eq. (124) if the complex region is in fact a deep potential hole in the energy surface]. We assume, therefore, that the lifetime of the complex is independent of $(J, M_J)$, and then the final state distribution is governed by

$$\sum_{J,M_J} \frac{N(f|E, J, M_J)N(i|E, J, M_J)}{(2L_A+1)(2L_B+1)} \tag{125}$$

Because $J$ is essentially the orbital angular momentum in the entrance and exit channels, the sum over $J$ is restricted to $\hbar J < l^*(E-E_i)$ and $\hbar J < l^*(E-E_f)$;

the upper limit in the sum is therefore the lesser of these two values. Call it $J^*$. For $J < J^*$ we can calculate the numbers $N(f|E, J, M_J)$ and $N(i|E, J, M_J)$, defined by Eq. (119), without restriction on the initial and final orbital angular momenta, for these are essentially $J$. One finds easily that

$$\sum_L N(f, L \,|\, J, M_J) = (2L_C + 1)(2L_D + 1)$$

$$\sum_L N(i, L \,|\, J, M_J) = (2L_A + 1)(2L_B + 1)$$

(126)

and therefore Eq. (125) becomes

$$\sum_{J, M_J} (2L_C + 1)(2L_D + 1) \simeq J^{*2}(2L_C + 1)(2L_D + 1) \tag{127}$$

The product $(2L_C + 1)(2L_D + 1)$ is of course just the degeneracy of the final vibration–rotation level. If the upper limit $J^*$ is set by the orbital angular momentum restriction in the exit channel, then we have just the final state distribution given by the full microcanonical rate calculation [Eq. (110)]: Angular momentum of the complex places no restriction on the range of states available in the exit channel. If the upper limit $J^*$ is set by the orbital angular momentum restriction in the entrance channel, however, the cross section to a given final state is independent of the energy of that state.

In particular we can calculate the distribution of low-lying product vibration–rotation states using Eq. (127), for these states, of course, have little rotational angular momentum. If $J^*$ is set by the exit channel, we have the Maxwell–Boltzmann distribution derived above [Eq. (111)]; if $J^*$ is set by the entrance channel we have a uniform distribution—that is, a Maxwell–Boltzmann distribution with infinite temperature.

ACKNOWLEDGMENTS

It is a pleasure to acknowledge the hospitality of the Theoretical Chemistry Department at Oxford, where this chapter was written. I would like to thank Professors R. A. Marcus and W. H. Miller for sending me preprints of their recent work prior to publication.

## References

1. D. L. Bunker, *Theory of Elementary Gas Reaction Rates*, Pergamon Press, Inc., Elmsford, N.Y. (1966).
2. H. S. Johnston, *Gas Phase Reaction Rate Theory*, The Ronald Press Company, New York (1966).
3. K. J. Laidler, *Theories of Chemical Reaction Rates*, McGraw-Hill Book Company, New York (1969).

4. R. D. Levine and R. B. Bernstein, *Molecular Reaction Dynamics*, Oxford University Press, Inc., New York (1974).
5. E. E. Nikitin, *Theory of Elementary Atomic and Molecular Processes in Gases*, trans. by M. J. Kearsley, Clarendon Press, Oxford (1974).
6. R. D. Levine, *Quantum Mechanics of Molecular Rate Processes*, Clarendon Press, Oxford (1969).
7. R. B. Bernstein, A. Dalgarno, H. Massey, and I. C. Percival, Thermal scattering of atoms by homonuclear diatomic molecules, *Proc. R. Soc. London Ser. A* **274**, 427–442 (1963).
8. R. D. Levine and B. R. Johnson, Rotational excitation in molecular collisions: Statistical features of the $S$-matrix, *Chem. Phys. Lett.* **4**, 365–368 (1969).
9. J. B. Anderson, Statistical theories of chemical reactions. Distributions in the transition region, *J. Chem. Phys.* **58**, 4684–4692 (1973).
10. B. Widom, Reaction kinetics in stochastic models. II, *J. Chem. Phys.* **61**, 672–680 (1974).
11. J. C. Keck, Variational theory of reaction rates, *Adv. Chem. Phys.* **13**, 85–121 (1967).
12. H. Goldstein, *Classical Mechanics*, Addison-Wesley Publishing Company, Inc. Reading, Mass. (1951).
13. S. Chapman, S. M. Hornstein, and W. H. Miller, Accuracy of transition state theory for the threshold of chemical reactions with activation energy: Collinear and three-dimensional $H + H_2$, *J. Am. Chem. Soc.* **97**, 892–894 (1975).
14. R. N. Porter and M. Karplus, Potential surface for $H_3$, *J. Chem. Phys.* **40**, 1105–1115 (1964).
15. P. Pechukas and F. J. McLafferty, On transition-state theory and the classical mechanics of collinear collisions, *J. Chem. Phys.* **58**, 1622–1625 (1973).
16. R. L. Jaffe, J. M. Henry, and J. B. Anderson, Variational theory of reaction rates: Application to $F + H_2 = HF + H$, *J. Chem. Phys.* **59**, 1128–1141 (1973).
17. R. G. Pearson, Symmetry rules for chemical reactions, *Acc. Chem. Res.* **4**, 152–160 (1971).
18. H. Metiu, J. Ross, R. Silbey, and T. F. George, On symmetry properties of reaction coordinates, *J. Chem. Phys.* **61**, 3200–3209 (1974).
19. D. L. Hill and J. A. Wheeler, Nuclear constitution and the interpretation of fission phenomena, *Phys. Rev.* **89**, 1102–1145 (1953).
20. H. S. Johnston, Large tunnelling corrections in chemical reaction rates, *Adv. Chem. Phys.* **3**, 131–170 (1961).
21. H. S. Johnston and D. Rapp, Large tunnelling corrections in chemical reaction rates. II, *J. Am. Chem. Soc.* **83**, 1–9 (1961).
22. C. Eckart, The penetration of a potential barrier by electrons, *Phys. Rev.* **35**, 1303–1309 (1930).
23. R. A. Marcus, Generalization of the activated complex theory of reaction rates. I. Quantum mechanical treatment, *J. Chem. Phys.* **41**, 2614–2623 (1964).
24. R. A. Marcus, On the analytical mechanics of chemical reactions. Quantum mechanics of linear collisions, *J. Chem. Phys.* **45**, 4493–4499 (1966).
25. L. D. Landau and E. M. Lifshitz, *Quantum Mechanics*, Addison-Wesley Publishing Company, Inc., Reading, Mass. (1958).
26. D. G. Truhlar and A. Kuppermann, Exact tunneling calculations, *J. Am. Chem. Soc.* **93**, 1840–1851 (1971).
27. D. G. Truhlar and A. Kuppermann, A test of transition state theory against exact quantum mechanical calculations, *Chem. Phys. Lett.* **9**, 269–272 (1971).
28. D. G. Truhlar and A. Kuppermann, Exact and approximate quantum mechanical reaction probabilities and rate constants for the collinear $H + H_2$ reaction, *J. Chem. Phys.* **56**, 2232–2252 (1972).
29. D. G. Truhlar, A. Kuppermann, and J. T. Adams, Exact quantum mechanical reaction probabilities and rate constants for the isotopic collinear $H + H_2$ reactions, *J. Chem. Phys.* **59**, 395–402 (1973).
30. F. J. McLafferty and P. Pechukas, Quantum transition state theory, *Chem. Phys. Lett.* **27**, 511–514 (1974).
31. W. H. Miller, Quantum mechanical transition state theory and a new semiclassical model for reaction rate constants, *J. Chem. Phys.* **61**, 1823–1834 (1974).

32. E. Wigner, On the quantum correction for thermodynamic equilibrium, *Phys. Rev.* **40**, 749–759 (1932).
33. W. H. Miller, Classical limit quantum mechanics and the theory of molecular collisions, *Adv. Chem. Phys.* **25**, 69–177 (1974).
34. W. H. Miller, Semiclassical limit of quantum mechanical transition state theory for non-separable systems, *J. Chem. Phys.* **62**, 1899–1906 (1975).
35. J. C. Keck, Statistical theory of chemical reaction rates, *J. Chem. Phys.* **29**, 410–415 (1958).
36. E. E. Nikitin, Statistical theory of endothermic reactions. Part I. Bimolecular reactions, *Theor. Exp. Chem. USSR* **1**, 83–89 (1965).
37. E. E. Nikitin, Statistical theory of exothermic ion–molecule reactions, *Theor. Exp. Chem. USSR* **1**, 275–280 (1965).
38. P. Brumer and M. Karplus, Collision complex dynamics in alkali halide exchange reactions, *Faraday Discuss. Chem. Soc.* **55**, 80–92 (1973).
39. J. Ford, The transition from analytic dynamics to statistical mechanics, *Adv. Chem. Phys.* **24**, 155–185 (1973).
40. W. H. Miller, Study of the statistical model for molecular collisions, *J. Chem. Phys.* **52**, 543–551 (1970).
41. J. C. Light, Phase-space theory of chemical kinetics, *J. Chem. Phys.* **40**, 3221–3229 (1964).
42. P. Pechukas and J. C. Light, On detailed balancing and statistical theories of chemical kinetics, *J. Chem. Phys.* **42**, 3281–3291 (1965).
43. J. Lin and J. Light, Phase-space theory of chemical kinetics. III. Reactions with activation energy, *J. Chem. Phys.* **45**, 2545–2559 (1966).
44. J. C. Light, Statistical theory of bimolecular exchange reactions, *Discuss. Faraday Soc.* **44**, 14–29 (1967).
45. R. A. Marcus, On the theory of energy distributions of products of molecular beam reactions involving transient complexes, *J. Chem. Phys.* **62**, 1372–1384 (1975).
46. R. A. Marcus, Chemical-reaction cross sections, quasiequilibrium, and generalized activated complexes, *J. Chem. Phys.* **45**, 2138–2144 (1966).
47. R. A. Marcus, On the theory of chemical-reaction cross sections. I. A statistical-dynamical model, *J. Chem. Phys.* **45**, 2630–2638 (1966).
48. S. A. Safron, N. D. Weinstein, D. R. Herschbach, and J. C. Tully, Transition state theory for collision complexes: Product translational energy distributions, *Chem. Phys. Lett.* **12**, 564–568 (1972).

## Note Added in Proof

Two papers discussing nuclear symmetry restrictions on transition states (Section 2.4) have recently appeared: R. E. Stanton and J. W. McIver, Jr., Group theoretical selection rules for the transition states of chemical reactions, *J. Am. Chem. Soc.* **97**, 3632–3646 (1975); and P. Pechukas, On simple saddle points of a potential surface, the conservation of nuclear symmetry along paths of steepest descent, and the symmetry of transition states, *J. Chem. Phys.* **64**, 1516–1521 (1976).

# Thermodynamic Approach to Collision Processes

*R. D. Levine*

*and*

*R. B. Bernstein*

*Entia non sunt multiplicanda praeter necessitatem.* Occam, William of. *Dialogus* (1343)

## 1. Introduction

The thermodynamic approach to collision processes, which makes use of information theory, is an attempt to bridge the gap between the equilibrium and disequilibrium (i.e., thermodynamic and kinetic) points of view. It deals with the phenomena of kinetics, retaining the concern with time-evolution, yet adopts from thermodynamics the concept of a state function (and thus allows for the construction of thermodynamic cycles), which is valid irrespective of approximation or detailed models of the dynamics. The realization that the overriding concern of traditional thermodynamics with equilibrium, static, situations has been out of choice rather than out of necessity makes our approach possible. In what follows we shall often draw parallels between equilibrium and disequilibrium systems.

The need for a thermodynamic approach stems from the recent experimental and theoretical developments in the field of molecular collision

*R. D. Levine* • Department of Physical Chemistry, The Hebrew University, Jerusalem, Israel, and *R. B. Bernstein* • Chemistry and Physics Departments, The University of Texas, Austin, Texas

dynamics.[1] The new techniques are providing a wealth of detailed dynamical information[2,3] (e.g., the collision products' polarization, internal state, angular, and recoil velocity distributions as a function of the reactants' internal state and collision energy). There is now emerging a picture of elementary reactions that can be characterized by highly *specific* product state distributions and very *selective* reactant energy utilization.[4] When alternative sets of reaction products are possible, the branching ratios are often found to differ from prior expectation, based on statistical considerations.

The first task is to cope with this body of detail, to find optimal means of characterizing the product distributions (or branching ratios), and to compact the data. Quantitative measures for the specificity and selectivity shown by elementary chemical reactions have been provided.[5]

The second stage consists of developing consistency relations among the different measures. This can obviate the need of measuring related manifestations of the collision dynamics. Ultimately, one hopes to arrive at the purely predictive stage where these measures can be computed directly from the first principles.

The aim of this chapter is to show the thermodynamic-like approach to these goals. In Section 2 we shall provide the conceptual background, which consists of three key ingredients: the thermodynamic weight, the approach of maximal entropy (minimal information content, i.e., the so-called Occam's razor), and the technique of a constrained maximum. As an illustration we shall consider both physical equilibrium and disequilibrium in macroscopic systems. Using rather simple arguments we shall analyze vibrational relaxation and show accord with more elaborate descriptions.

Nowhere in the general theory are we restricted to macroscopic systems. Thus in Section 3 we shall consider applications to physical disequilibrium processes at the microscopic (molecular) level. We shall apply thermodynamic-like arguments to systems with few degrees of freedom. Much of the analysis of experimental (or theoretical) data has been reviewed before[4,5] and will not be repeated. Rather, we shall stress the spirit and implications of the approach.

By a logical extension of the methods developed in Section 3 it becomes possible to treat problems of *chemical* equilibrium, as well as disequilibrium, on the macroscopic and microscopic level. Branching ratios for different product sets are briefly considered in Section 4.

## 2. Information Content and Entropy

### 2.1. Introduction

The general approach to be outlined in this section provides a method of inference that is not limited to molecular collision phenomena. It is essentially a quantitative version of Occam's razor, a method of reasoning that attempts to use all the available evidence but otherwise be least committal. It is a search for the simplest and most reasonable basis to explain the known observations and predict future results. However, such an approach cannot provide details that are not somehow contained in the available (or assumed; see Section 2.3) evidence.

One can derive the method from an axiomatic point of view.[6–10] We shall follow an alternative route, assuming, from the very beginning, that we are concerned with an experiment that consists of very many repetitions of some event.* Our concern with molecular collision phenomena ensures the validity of this approach. Any scattered flux detected in the lab usually represents the outcome of a very large number of independent binary molecular collisions.

## 2.2. The Thermodynamic Weight

We shall consider an event with $n$ distinct possible outcomes and an experiment that consists of many (say $N$) repetitions of this event. For any given experiment we can record the sequence of outcomes, one outcome per event. The number of different possible sequences is $n^N$. The number of possible sequences increases exponentially with $N$, $n^N = \exp(N \ln n)$.

Usually we are not interested in the order of outcomes within the sequence. Instead we merely record the number of events, $N_i$, that resulted in the $i$th possible outcome, $i = 1, 2, \ldots, n$. Of course the $N_i$ need not equal one another. Any set of integers $\{N_i\}$ that satisfy

$$\sum_{i=1}^{n} N_i = N \tag{1}$$

is a possible result of the experiment. We shall refer to any particular set $\{N_i\}$ as the distribution of outcomes in the experiment.

The thermodynamic weight[6,9–13] $W$,

$$W = N! \Big/ \prod_{i=1}^{n} N_i! \tag{2}$$

is the number of different sequences of outcomes that correspond to a given distribution (i.e., to a given set $\{N_i\}$). $W$ is the number of original sequences subject to the constraint that the sequence contains the first outcome $N_1$ times, the second outcome $N_2$ times, and so on. It can be seen that even though $W$ corresponds to a subset of the original sequences, as $N$ becomes larger, $W$ also increases exponentially with $N$.

*In the language of statistical mechanics, we are dealing with an ensemble.

The proof follows. The fraction of trials that a particular distribution of outcomes is observed is $P(\{N_i\})$,

$$P(\{N_i\}) = W/n^N \tag{3}$$

[Using the multinomial theorem one readily verifies that summing Eq. (3) over all sets $\{N_i\}$ that satisfy Eq. (1) yields unity.] One thus needs to show that, for large $N$, $P(\{N_i\}) \rightarrow \exp(-N\,\Delta S)$, where $\Delta S \leq \ln n$ so that $W \rightarrow \exp[N(\ln n - \Delta S)]$.

Using Stirling's approximation, in the limit $N \gg n$ we obtain

$$\ln P(\{N_i\}) = N \ln N - \sum_i N_i \ln N_i - N \ln n + \mathcal{O}(\ln N)$$

$$= -N(\ln n + \sum_i x_i \ln x_i) + \mathcal{O}(\ln N) \tag{4}$$

Here $x_i = N_i/N$ is the fraction of trials leading to the $i$th outcome in a sequence of $N$ trials. [With the convention $x \ln x \rightarrow 0$ as $x \rightarrow 0$ or 1, $\ln(P\{N_i\})$ is well defined even if some outcomes are never observed.]

It is easy to see [cf. Eq. (A.4) of the Appendix] that as $N \rightarrow \infty$, $\Delta S$,

$$\Delta S = -N^{-1} \ln P\{N_i\} = \ln n + \sum_i x_i \ln x_i \tag{5}$$

is nonnegative (it is the negative of a logarithm of a proper fraction) and attains its smallest value (zero) when $x_i = 1/n$ (so that all outcomes are equally probable). It is also convenient to define $S$,

$$S = \ln n - \Delta S = -\sum_i x_i \ln x_i \tag{6}$$

as a nonnegative quantity that attains its maximal value ($\ln n$) for the uniform distribution of outcomes. From Eqs. (3)–(6)

$$S = N^{-1} \ln W \tag{7}$$

or $W \rightarrow \exp(NS)$ as $N \rightarrow \infty$.

$S$ is the quantity known as the entropy. The more probable the distribution, the higher is its entropy. In practical work or numerical computations we shall always measure $S$ in entropy units (e.u., or gibbs, i.e., calories degree$^{-1}$ mole$^{-1}$) obtained by multiplying the dimensionless expressions (6) or (7) by the gas constant $R$. Thus, if in Eq. (7) $N$ is Avogadro's number $N_A$, then $S$ (e.u.) = $(R/N_A) \ln W = k \ln W$, where $k$ is Boltzmann's constant. $\Delta S$ is the entropy deficiency. It is the nonnegative difference between the maximal ($\ln n$) and actual value of $S$ [cf. Eq. (6)].

The asymptotic properties that we have just established have a very clear physical interpretation. $P(\{N_i\})$ is, by construction, the fraction of all possible sequences of outcomes that give rise to a particular, $\{N_i\}$, distribution of outcomes. In the limit $N \to \infty$ (or, more precisely, $N \gg n$), one particular subset of sequences becomes overwhelmingly more probable. This is the sequence corresponding to the distribution with the minimal value of $\Delta S$ where [cf. Eq. (4)]

$$P(\{N_i\}) \propto \exp(-N\,\Delta S), \qquad N \to \infty$$

The thermodynamic weight $W$,

$$W \propto \exp[N(\ln n - \Delta S)] = \exp(NS)$$

of this distribution (of smallest $\Delta S$) is larger than that of any other distribution with a higher value of $\Delta S$ (i.e., lower value of $S$). No matter how near two distributions are in their value of $S$, after a sufficient number of repetitions (i.e., one can always find a sufficiently large $N$ for which) the distribution with the higher value of $S$ will correspond to overwhelmingly more sequences than any other.

When an event with $n$ possible outcomes is repeated $N$ (independent*) times and $N \gg n$, the distribution of outcomes tends to stabilize. Only one distribution is observed despite the fact that many other distributions are possible in principle. (This does, of course, conform to our intuition that, say, in $10^3$ tosses of a coin the fraction of heads will serve as a reliable guide for that fraction expected in future tosses of this coin.)

The argument is not yet complete, however. If the distribution of maximal $S$ is the favored one, then we might expect that the only distributions that should be observed when $N \gg n$ are the uniform ones, where $x_i = (N_i/N) = 1/n$ and $S$ has its highest possible value ($\ln n$). Yet any visitor to Las Vegas has found out that this is not the case. The observed distribution does indeed stabilize but not necessarily to the uniform ($x_i = 1/n$) one. There is obviously a "slight" bias imposed by the house. How can we explicitly take into account any such constraints?

## 2.3. Constraints

The entropy is a function of the $x_i$ (where $x_i = N_i/N$ is the fractional number of trials that resulted in the $i$th outcome). Hence any set of $x_i$ must satisfy the normalization condition [cf. Eq. (1)]

$$\sum_{i=1}^{n} x_i = 1 \tag{8}$$

When we seek a maximum value of $S$ we must limit the variation of the $x_i$ such that the constraint (8) always obtains. What we require is not the set of numbers

---

*The assumption of independent repetitions is merely a convenience. Our discussion applies even if there are correlations between the different trials, provided only that the correlations are of a finite range,[14] that is, that the $(m+r)$th repetition is independent of the $m$th repetition and $N \gg r$.

$\{x_i\}$ that leads to a maximal value of $S$ but rather the particular set of (nonnegative) numbers $\{x_i\}$, among all such sets that satisfy (8), that gives a maximal $S$. [Our search is then limited to those sets of numbers that satisfy Eq. (8).]

The mathematical procedure for seeking a maximum among a limited set of functions ("a maximum subject to a constraint") has been introduced by Lagrange.[6,8-10] It consists of the introduction of an auxiliary function (the Lagrangian $\mathscr{L}$); rather than seeking a constrained maximum of $S$, we seek an unconstrained maximum of $\mathscr{L}$. For the problem at hand the Lagrangian is obtained by adding to $S$ a multiple of the constraint (8). Writing (for convenience) this *Lagrange multiplier* as $\lambda_0 - 1$, we have

$$\mathscr{L} = S - (\lambda_0 - 1) \sum_i x_i \qquad (9)$$

When we determine the maximum of $\mathscr{L}$ the answer will depend on $\lambda_0$. In the final stage we choose the value of $\lambda_0$ to ensure that Eq. (8) is satisfied. Substituting Eq. (6) for the entropy and differentiating Eq. (9), we obtain

$$\delta\mathscr{L} = \sum_i (-\ln x_i - 1 - \lambda_0 + 1) \, \delta x_i \qquad (10)$$

Because we seek an unconstrained extremum, the $x_i$ can be varied independently. The maximum then obtains when the coefficient of each $\delta x_i$ separately vanishes, i.e., when

$$x_i = \exp(-\lambda_0) \qquad (11)$$

Now, substituting into Eq. (8), $\lambda_0$ is determined by the normalization condition

$$\exp(\lambda_0) = \sum_{i=1}^{n} 1 = n \qquad (12)$$

or $\lambda_0 = \ln n$ and $x_i = 1/n$.

Of course, we already knew from Section 2.2 that when the only constraint is normalization the most probable distribution is the uniform one. Suppose, however, that we know more than just the normalization condition. For example, assume that the first moment $\langle i \rangle$ of the distribution, defined as

$$\langle i \rangle = \sum_i ix_i \qquad (13)$$

is available. Then the search is for that set of $x_i$ chosen from among all sets that satisfy Eqs. (8) *and* (13) that gives rise to the maximal value of $S$. This will be the most probable distribution that is (1) normalized *and* (2) reproduces the known magnitude of the first moment.

The Lagrangian to be used when both Eqs. (8) and (13) are imposed is

$$\mathscr{L} = S - (\lambda_0 - 1)\langle 1 \rangle - \lambda_1 \langle i \rangle \qquad (14)$$

Here $\langle 1 \rangle$ is used to denote the zeroth moment, $\sum_i x_i$, and $\lambda_1$ is an additional Lagrangian multiplier. At the final stage the value of $\lambda_1$ will be assigned so as to yield the correct magnitude of the first moment $\langle i \rangle$. Upon variation of the $x_i$,

$$\delta \mathscr{L} = \sum_i (-\ln x_i - \lambda_0 - \lambda_1 i) \, \delta x_i \tag{15}$$

Now the condition for the maximum is $\ln x_i = -\lambda_0 - \lambda_1 i$ or

$$x_i = \exp(-\lambda_0 - \lambda_1 i) \tag{16}$$

It is convenient to determine $\lambda_1$ first by the solution of the implicit equation

$$\langle i \rangle = \sum_i i \exp(-\lambda_1 i) / \sum_i \exp(-\lambda_1 i) \tag{17a}$$

Once the magnitude of $\lambda_1$ is assigned, $\lambda_0$ is determined from Eq. (16), using Eq. (8), i.e.,

$$\exp(\lambda_0) = \sum_i \exp(-\lambda_1 i) \tag{18a}$$

The geometric series in Eq. (18a) can be summed to give $\exp(\lambda_0) = [1 - \exp(-\lambda_1)]^{-1}$. Substituting this into Eq. (17a) and summing the numerator, we finally obtain

$$\langle i \rangle = [\exp(\lambda_1) - 1]^{-1} \tag{17b}$$

Having established the dependence of $\lambda_1$ on $\langle i \rangle$ (and vice versa), we can express $\lambda_0$ as a function of $\langle i \rangle$ as well, leading to

$$\exp(\lambda_0) = \langle i \rangle + 1 \tag{18b}$$

One can already anticipate that $\exp(\lambda_0)$ will play the role of the partition function (i.e., the Zustandsumme).

In the general case, there can be any finite number (up to $n$) of constraints that we require the $x_i$ to satisfy. We can represent this case by considering that the average values of a set of different observables is known,

$$\sum_i g_m(i) x_i = \langle g_m \rangle \tag{19}$$

$m = 0, 1, 2, \ldots$. Here $g_m(i)$ is the magnitude of the $m$th property for outcome $i$. We take $m = 0$ to be the normalization condition so that $g_0(i) = 1$. $m = 1$ might be the first moment, in which case $g_1(i) = i$, etc. In principle, the $g_m(i)$, $m = 1, 2, \ldots$, are any possible properties of the outcomes. Then

$$\mathscr{L} = S - (\lambda_0 - 1)\langle g_0 \rangle - \sum_{m=1} \lambda_m \langle g_m \rangle \tag{20}$$

and, using the same route as before, the set of $x_i$ that maximizes $\mathscr{L}$ is of the form

$$x_i = \exp\left[-\lambda_0 - \sum_m \lambda_m g_m(i)\right] \tag{21}$$

It is easy to verify (see the Appendix) that among all the distributions $\{x_i\}$ that satisfy the conditions (19), the distribution (21) has the maximal value of $S$.

Using Eq. (21), the average value of $g_m(i)$ is

$$\langle g_m \rangle = \sum_i g_m(i) \exp\left[-\sum_m \lambda_m g_m(i)\right] \Big/ \exp(\lambda_0) \tag{22a}$$

where, from Eq. (19) for $m = 0$,

$$\exp(\lambda_0) = \sum_i \exp\left[-\sum_m \lambda_m g_m(i)\right] \tag{23}$$

Substituting this value of $\exp(\lambda_0)$ into Eq. (22a), one obtains a set of equations relating the average values $\langle g_m \rangle$, $m = 1, 2, \ldots$, and the Lagrange multipliers $\lambda_m$, $m = 1, 2, \ldots$,

$$\langle g_m \rangle = \sum_i g_m(i) \exp\left[-\sum_m \lambda_m g_m(i)\right] \Big/ \sum_i \exp\left[-\sum_m \lambda_m g_m(i)\right] \tag{22b}$$

These provide a set of implicit equations that need to be solved for the $\lambda_m$. Once the magnitudes of the $\lambda_m$, $m = 1, 2, \ldots$, have been determined, $\lambda_0$ is available from Eq. (23). We can summarize Eqs. (22)–(23) by

$$\langle g_m \rangle = -\partial \lambda_0 / \partial \lambda_m, \qquad m = 1, 2, \ldots \tag{24}$$

The value of the entropy for the distribution (21) is readily evaluated [using Eqs. (6) and (19)] to be

$$S = \lambda_0 + \sum_m \lambda_m \langle g_m \rangle \tag{25}$$

In the language of thermodynamics, $S$ and $\lambda_0$ are Legendre transforms of one another.

It is important to clarify one point, which we shall illustrate by making a particular choice for the averages $\langle g_m \rangle$. We take them to be the different moments [i.e., $\langle g_m \rangle = \langle i^m \rangle$ or $g_m(i) = i^m$] of the distribution. Now any well-defined distribution $\{x_i\}$, $i = 1, 2, \ldots, n$, has a complete set of moments $\langle i^m \rangle$, $m = 0, 1, \ldots, n$. However, we do not need all the $n$ moments in order to define the distribution $\{x_i\}$. In practice, one knows that a limited number ($m < n$) of moments suffices to characterize a given, experimentally measured distribution. The equivalent theoretical statement is that all we require are the *independent* moments.[12] How do we know which moments are independent? The answer is that we do not really need to know. The formalism automatically handles this aspect of the problem.

To see this we note the following formal definition: An independent moment $\langle i^m \rangle$ is any moment with a nonvanishing Lagrange multiplier $\lambda_m$ in Eq. (21). We take as an example the exponential distribution (16). All its moments are well defined, but only the first moment is independent.

Given the magnitude of $\langle i \rangle$, all other moments can be computed in terms of $\langle i \rangle$. In fact using Eqs. (17b) and (18b) one can rewrite Eq. (16) as

$$x_i = [\langle i \rangle / (\langle i \rangle + 1)]^i / (\langle i \rangle + 1)$$

showing that the exponential distribution can be explicitly characterized by the magnitude of the single independent moment, $\langle i \rangle$. Very well, but how do we know that we are dealing with an exponential distribution? Again, we do not have to know that. If we are given, say, both $\langle i \rangle$ and $\langle i^2 \rangle$, and the (unknown) distribution is, in fact, exponential, then, employing Eqs. (22), $\lambda_2$, the Lagrange multiplier of $\langle i^2 \rangle$, would automatically have the value zero.*

To conclude, one does not have to know a priori which moments are independent. One simply uses in Eqs. (22) *all* the available moments (or other expectation values). When Eqs. (22) are solved, only the multipliers of the independent moments will have a nonvanishing value. This is the *synthetic route*.[12] One inputs the available data (in the form of expectation values) and obtains the distribution, expressed in terms of the Lagrange parameters of the independent moments.

Finally, we note that the $\langle g_m \rangle$ need not be the moments. Any set of expectation values can be used.†

A very important corollary should also be noted.[12] If we know that the distribution is of the form (21), then we know the set of independent functions $g_m$, $m = 1, 2, \ldots$, that characterizes the distribution. This is the analytic approach. Given the distribution, one identifies the independent moments—the constraints—that govern the distribution.

Applications of both methods, the analytic and the synthetic, to molecular collision problems will be presented in Section 3.

We can summarize the method of maximal entropy, as discussed in the last two subsections, as follows.[12] The most probable distribution of the outcomes of an experiment (and the only distribution that will be realized in practice if $N$ is large enough) is the one with maximal entropy, where the maximum is to be determined not among all possible distributions (in which case the uniform distribution always wins) but among all the distributions that are consistent with the available information [i.e., those satisfying the constraints (19)]. A

---

*One can prove this in complete generality. A quicker route is to note that the statement "$\lambda_2 = 0$" and the statement "the distribution is exponential" are identical and that hence either one implies the other.

†An even stronger statement is possible. Any set of inequalities for the $\langle g_m \rangle$ [i.e., relations of the type $\sum_i g_m(i)x_i \leq \langle g_m \rangle$] can also be used.[15]

knowledge of these constraints is then sufficient to determine the most probable distribution. This is the *synthetic route*: from the known (or conjectured) constraints to the distribution. The search for the distribution with the largest thermodynamic weight is the search for the most random (i.e., most uniform) distribution. By limiting the search to those distributions that are consistent with the available constraints, we are determining the distribution that reproduces the data at hand and is otherwise least biased. It is the most *uniform* distribution that is *consistent* with what we know. This is our quantitative version of Occam's razor.

In the information theoretic approach,[5] $\Delta S$ is the so-called information content, and $S$ is the missing information. Further justification for these names is available in Section 3 and elsewhere.[5,10]

We can also turn the argument around. Having observed a distribution of outcomes in an experiment for which $N \gg n$, we are safe in concluding that it is the most probable distribution and hence is of the form (21). By examining the observed distribution, we can ascertain the constraints, i.e., the $g_m(i)$. This is the so-called *analytical route*: from the observed distribution to the underlying constraints.

### 2.4. Equilibrium Statistical Thermodynamics

Before turning to collision phenomena we shall illustrate the method by a more familiar case, that of a bulk system at equilibrium.[9,10] As a specific example we choose a dilute gas of diatomic molecules where, for simplicity, the vibrational motion is taken to be harmonic, with $v$ the vibrational quantum number. A molecule in the vibrational state $v$ will have vibrational energy $\varepsilon_v = v\hbar\omega$ (with a choice of the energy scale such that $\varepsilon_0 = 0$). At equilibrium the mean vibrational energy is unchanging with time and hence has a constant magnitude. If $P(v)$ is the fraction of molecules in the vibrational state $v$,

$$\sum_v \varepsilon_v P(v) = \langle \varepsilon_v \rangle = \text{const.} \tag{26}$$

Besides normalization there are no other obvious constraints on $P(v)$, and hence Eq. (26) provides that [cf. Eqs. (13)–(18)]

$$P(v) = \exp(-\lambda_0 - \beta\varepsilon_v) \tag{27}$$

Here $\beta$ is a Lagrange parameter chosen to ensure a definite, constant first moment, i.e. [cf. Eqs. (17)],

$$\langle \varepsilon \rangle = \sum_v \varepsilon_v \exp(-\beta\varepsilon_v) / \sum_v \exp(-\beta\varepsilon_v) \tag{28}$$

For harmonic energy levels, the summations are readily performed to yield

$$\langle \varepsilon \rangle = \hbar\omega / [\exp(\beta\hbar\omega) - 1] \qquad (29)$$

Similarly, $\lambda_0$, as determined by the normalization constraint [cf. Eqs. (18)] is given by

$$\exp(\lambda_0) = [1 - \exp(-\beta\hbar\omega)]^{-1} \qquad (30)$$

Using the fact that at equilibrium $\langle \varepsilon \rangle$ has a constant value, we could derive the functional form of the distribution of vibrational states, Eq. (27), in terms of a single Lagrange parameter $\beta$. But how can we find the magnitude of $\beta$? This would require knowledge of $\langle \varepsilon \rangle$ [cf. Eq. (29)]. It is at this point that equilibrium thermodynamics has a practical advantage because it can appeal to the so-called *zeroth law*. In the present context this is the statement[11] that two systems, which can only exchange energy, have a common value of $\beta$ in equilibrium. The proof is as follows. Let $\langle \varepsilon_a \rangle$ be the average energy of system $a$ and $\langle \varepsilon_b \rangle$ the average energy of system $b$. Because energy can be exchanged, only the overall mean energy $\langle E \rangle$,

$$\langle E \rangle = \langle \varepsilon_a \rangle + \langle \varepsilon_b \rangle \qquad (31)$$

is conserved. The Lagrangian for the combined system is*

$$\mathcal{L} = S_a + S_b - \beta(\langle \varepsilon_a \rangle + \langle \varepsilon_b \rangle) - (\lambda_{0a} - 1)\langle 1_a \rangle - (\lambda_{0b} - 1)\langle 1_b \rangle \qquad (32)$$

By maximizing $\mathcal{L}$ it can be seen that both systems will have an identical value of $\beta$.

The zeroth law used in conjunction with an ideal gas thermometer will then provide that $\beta = 1/kT$, where $k$ is Boltzmann's constant and $T$ is defined such that the average energy of an ideal gas (i.e., structureless) molecule at equilibrium is $\langle \varepsilon \rangle = 3kT/2$. The results (29) and (30) will then be recognized as the well-known expressions for the mean vibrational energy and the partition function $Q(T)$ of a harmonic oscillator of frequency $\omega$ at a temperature $T$, where $T = 1/k\beta$. In what follows, we shall reserve the notation $Q$ for the "sum over states" (the partition function) of a thermal equilibrium distribution.

For the distribution (27), i.e., $P(v) = \exp(-\beta\varepsilon_v)/\exp(\lambda_0)$, we obtain, using Eq. (6),

$$S = -\sum_v P(v) \ln P(v) = \beta\langle \varepsilon \rangle + \lambda_0 \qquad (33)$$

which, up to a multiplicative (Boltzmann) constant, is just the vibrational contribution to the entropy of the gas. In terms of $S$ [Eq. (33)] we can rewrite

---

*Only energy can be exchanged. Hence the probabilities of the states *within* each system are separately normalized; each system will have its own value of $\lambda_0$. This situation is to be contrasted with that of Section 4, where both energy *and* molecules can be exchanged. There $\lambda_0$ is the same for both systems.

Eq. (27) as

$$P(v) = \exp[-S - \beta(\varepsilon_v - \langle \varepsilon \rangle)] \tag{34}$$

One can indeed derive the entire machinery of thermodynamics and of equilibrium statistical mechanics continuing in this fashion.[6,9,10,13] We should note, however, that the state of equilibrium has no unique place in the general approach. Equilibrium is simply a particular physical situation for which we can readily determine the constraints. Equilibrium is the situation when we know the average values of the constants of the motion.[6,10] This definition ensures that the equilibrium distribution is unchanging in time, because if the constraints are time independent, so are the Lagrange multipliers. For the equilibrium situation the Lagrange multiplier $\beta$ can readily be determined via the zeroth law. As yet there is no such convenience available for the disequilibrium case.

## 2.5. Bulk Disequilibrium

The equilibrium distribution and hence the entire machinery of equilibrium thermodynamics can be so readily derived because of the unique definition of the constraints. This is no longer the case once the populations can change with time, i.e., in a disequilibrium situation. Here our procedure remains the same: Maximize the entropy subject to all the available information (past and present) on the system.* We shall consider only the simplest of all situations, namely, that of a system without "memory." By this we mean that the magnitude of the constraints at any given time suffices to determine the distribution. The past history provides no additional information. This is a special situation, yet it is sufficiently common to warrant a separate discussion.

Let us return to the dilute diatomic gas and displace it from vibrational equilibrium, say by irradiating it (pumping) with infrared light at the vibrational frequency.[16–18] It is observed that after the irradiation (the perturbation) is

---

*These seemingly innocuous instructions suffice to introduce a time asymmetry into our description. Because we have information only for the past (and present) time, the resulting theory will not be *reversible*. Irreversibility is built into the theory as due to the asymmetric operational definition of past and future. (The past is that region of time over which we can gather prior information.) An objection to the extremization procedure described in this section has been raised by I. Prigogine (private communication, 1975). He disagrees with the claim that, in general, a time-independent distribution function can be obtained by maximization of entropy under suitable constraints. He called attention to two possible counterexamples. One of them is a dilute gas, obeying the Boltzmann equation, subjected to a temperature gradient. The distribution function in the steady state is given by the Chapman–Enskog expansion. It is yet to be shown that it is equivalent to the distribution that would be obtained through extremization of the entropy. The second counterexample refers to steady states in the frame of the linear theory of irreversible processes, which are normally described in terms of the minimum entropy production theorem. Similarly, K. E. Shuler (private communication, 1975) suggested that the procedure of this section might be limited to so-called *canonically invariant*[62] problems.

turned off, the gas relaxes back to equilibrium. As $t \to \infty$, the mean vibrational energy $\langle \varepsilon(t) \rangle$ returns to a constant, $\langle \varepsilon(\infty) \rangle$, value. The simplest assumption (and one that is known to hold often, in practice) is that $\langle \varepsilon(t) \rangle$ satisfies a first-order relaxation equation, i.e.,

$$\frac{d\langle \varepsilon(t) \rangle}{dt} = -\frac{\langle \varepsilon(t) \rangle - \langle \varepsilon(\infty) \rangle}{\tau} \tag{35}$$

with a relaxation time $\tau$. The boundary value, $\langle \varepsilon(\infty) \rangle$, in Eq. (35) is the equilibrium value of the mean vibrational energy and can be computed (Section 2.4) from the temperature of the buffer gas. Hence, given the magnitude of $\langle \varepsilon(t) \rangle$ at some time $t$, say $t = 0$, one knows the magnitude of $\langle \varepsilon(t) \rangle$ at any subsequent value of the (reduced) time $t/\tau$.

Now consider the implications of the maximal entropy procedure of Section 2.3. Given that at any time $t$ the populations $P(v, t)$ are normalized and that the specified value of $\langle \varepsilon(t) \rangle$ obtains, it provides[19] that [cf. Eq. (27)]

$$P(v, t) = \exp[-\lambda_0(t) - \beta(t)E_v] \tag{36}$$

The difference between Eq. (36) and Eq. (27) is that now the Lagrange multipliers are time dependent. For harmonic energy levels, $\beta(t)$ is determined from [cf. Eq. (29)]

$$\langle \varepsilon(t) \rangle = \hbar\omega / \{\exp[\beta(t)\hbar\omega] - 1\} \tag{37}$$

Once $\beta(t)$ is known [cf. Eq. (30)],

$$\exp[\lambda_0(t)] = \{1 - \exp[-\beta(t)\hbar\omega]\}^{-1} \tag{38}$$

Such an approach[19] has been successful in accounting for the decay of the vibrational population disequilibrium (to equilibrium), as shown in Fig. 1.

This is an example of a synthesis. Subject to the constraint that $\langle \varepsilon(t) \rangle$ is known, the populations (and their time evolution) were determined and then

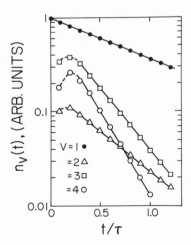

Fig. 1. Temporal evolution of the populations of the $v = 1$–4 levels of HF in vibrational disequilibrium.[19] Points [18]: exact numerical solution of five coupled rate equations (Figure 1 of Reference 18). The population ratios are not to scale. Solid lines: the maximal entropy prediction, Eq. (36). By using a "reduced" time $t/\tau$, one avoids the use of any kinetic input in the predictions. To account for the initial, rise region (broken curve), one needs to introduce an additional constraint.[19]

verified by comparison to the actual behavior. Not only has one obtained a quantitative comparison but also a qualitative understanding. For example, under the experimental conditions appropriate[18] for Fig. 1, $\langle \varepsilon(t) \rangle \ll \hbar\omega$, so that $\beta(t)\hbar\omega \gg 1$ and hence [from Eq. (38)] $\lambda_0(t) \simeq 0$. Then, Eq. (36) (recalling that $E_v = v\hbar\omega$) implies that $P(v, t) \simeq [P(v = 1, t)]^v$. The higher vibrational levels relax faster $[-d \ln P(v, t)/dt = -vd \ln P(v = 1, t)/dt]$ than the $v = 1$ level, as is clear from the figure.

### 2.6. Epilogue and Prologue

The approach taken in this section is by no means entirely new. The application to equilibrium situations was well understood by the end of the nineteenth century[6] and even served to usher in quantum mechanics.[20] The relation between entropy and thermodynamic weight is engraved on Boltzmann's tombstone, and it is said[21] that Gibbs suggested the use of such methods to predict the outcome of lotteries. Even our next topic, the application of thermodynamic methods to systems with very few degrees of freedom, was considered.[6] But then with the advent of quantum mechanics, as is often the case with scientific revolutions,[22] attention was centered on questions of a different kind.

The present approach has, therefore, historical precedents. Moreover, the theory is fully consistent with quantum mechanics.

## 3. Thermodynamic Approach to Collision Processes

### 3.1. Microdisequilibrium

In this section we shall deal first with the analysis and synthesis of disequilibrium on the molecular level. (Additional details and working equations can be found elsewhere.[5]) We shall proceed to show that this method not only provides the optimal means of characterization and compaction of the data but also (inherently and hence "automatically") provides measures of the specificity and selectivity shown by chemical reactions.

The first question we need to consider (Section 3.2) is the nature of the *reference* or *prior expectation* distribution. When the energy disposal of a chemical reaction is described as *specific* we mean specific when compared with some standard, i.e., we must have an idea of what a nonspecific distribution looks like. We already know from Section 2 that a useful "standard" is provided by the distribution of largest thermodynamic weight (the maximal entropy distribution). What does that imply for binary molecular collisions?

## 3.2. The Prior Distribution

The type of phenomena we wish to consider is illustrated by the following binary reactive collision[1]:

$$F + D_2 \rightarrow FD(v') + D$$

Immediately after the collision, before the products have time to relax (by collision or radiatively), what is the fraction of DF molecules in any final vibrational $(v')$ state? The question is not an elementary one, as we must recognize that a complete (quantum mechanical) specification of the state of the DF molecule requires knowledge of not just $v'$ but also $J'$ and $m_{J'}$ (the rotational quantum number of DF and its projection on some axis; the prime identifies the product variables). Even then, the final state of the products is not completely specified. For a DF molecule in a given internal state one still needs to determine the final relative momentum $\mathbf{p}'$ of the DF and D products.

With $k(v')$ the rate coefficient for the formation of DF in a given final vibrational state $(v')$,

$$P(v') = k(v') \Big/ \sum_{v'} k(v') \tag{39}$$

where

$$k(v') = \sum_{J'} \sum_{m_{J'}} k(v', J', m_{J'}) \tag{40}$$

Here $k(v', J', m_{J'}) \equiv k(n)$ is the rate of formation of DF in a given final internal state $n$.

To evaluate the entropy (or the thermodynamic weight) one now needs to specify the relative motion. We have often discussed this route before.[4,5,22,23] Here we shall provide an alternative derivation. (However, the final results are identical.)

We assume that in the absence of any constraints the rate constant $k(n')$ (the flux of products in a specified internal state $n'$) is simply proportional to the relative velocity (speed) $v'_n$ with which the products separate,

$$k^0(n') = Cv'_n \tag{41}$$

$C$ is a (dimension-bearing) constant that is the same for all product internal states. The superscript 0 on $k(n')$ designates the "no-constraint" limit (i.e., of maximal entropy). To determine $P(n')$ we need to normalize $k(n')$. To do so it is convenient to consider first the case of collisions at a given total energy $E$. The products' relative recoil energy $E'_T = \mu v'^2_n/2$ is determined by conservation of total energy, $E' = E'_T + E'_n$, where $E'_n$ is the products' internal energy. It is also convenient to introduce the translational density of states $\rho_T(E'_T)$, defined by[5,24,25]

$$d\mathbf{p}'/h^3 = \rho_T(E'_T)\, dE'_T(d\Omega'/4\pi) \tag{42a}$$

or

$$\rho_T(E'_T) = A'_T E'^{1/2}_T \tag{42b}$$

where $\mathbf{p}' = \mu \mathbf{v}'$, $E'_T = p'^2/2\mu'$, and $A'_T = (\mu'^{3/2}/2^{1/2}\pi^2\hbar^3)$. To derive Eqs. (42) one needs to put $d\mathbf{p}' = p'^2\, dp'\, d\Omega'$, where $\Omega'$ is the (solid) angle in the direction of $\mathbf{p}'$.

Because $\rho_T \propto E'^{1/2}_T \propto v'$, one can now rewrite Eq. (41) as*

$$k^0(n') = R\rho_T(E' - E'_n) \tag{43}$$

where $\rho_T(E - E_n) = 0$ for $E \le E_n$. Hence

$$k^0 = \sum_{n'} k^0(n') = R \sum_{n'} \rho_T(E' - E'_n) = R\rho(E') \tag{44}$$

where $\rho(E')$, defined by the last equality, is the total density of states of the products. Now, from Eqs. (43) and (44),

$$P^0(n') = k^0(n')/k^0 = \rho_T(E' - E'_n)/\rho(E') \tag{45}$$

and

$$\sum_{n'} P^0(n') = 1 \tag{46}$$

The actual product state distribution, $P(n')$, will not, in general, be identical to the prior expectation distribution $P^0(n')$. In Section 3.3 we shall make use of a convenient measure of the deviation of $P(n')$ from $P^0(n')$, known as the surprisal,[4,5,23]

$$I(n') = -\ln[P(n')/P^0(n')] \tag{47}$$

Alternatively, using the omega function notation of Reference 25,

$$P(n') = \omega(n')P^0(n') \tag{48}$$

where $\omega(n') = \exp[-I(n')]$.

When the distribution is such that any final state is equiprobable, $P(n') = P^0(n')$, i.e., $\omega(n') = 1$ and $I(n') = 0$.

Next we need to examine two more points: first, that experiments are seldom able to fully resolve the complete set of internal quantum numbers for the products. What, then, is our prior expectation for, say, the vibrational distribution $P^0(v')$? Having determined $k^0(n')$ we merely need go up the ladder of Eq. (39) to Eq. (40). Thus

$$P^0(v', J') = (2J' + 1)\rho_T(E' - E'_{vJ})/\rho(E') \tag{49}$$

*This is the form used previously.[22-24] $R$ is a unit-bearing constant related to $C$; $\rho_T(E' - E'_n)$ is the number of translational states (per unit volume) of the products when $E'_T$ is in a narrow interval about $E' - E'_n$. In words, it states that all product states (i.e., of specified translational and internal states) are formed with the same rate $R$.

and

$$P^0(v') = \sum_{J=0}^{J'^*(v')} P^0(v', J') \qquad (50)$$

where $J'^*(v')$ is the highest allowed value of $J'$ when the maximal energy in the rotation is $E' - E'_v$.

In the RRHO approximation,[23,25] the energy levels of the diatom are taken to be those of a classical rigid rotor–harmonic oscillator. In this case Eq. (50) can be explicitly evaluated to give[5,23]

$$P^0(E'_v) = \tfrac{5}{2}(E' - E'_v)^{3/2}/E'^{5/2} \qquad (51)$$

There are no modifications of these results even when the initial internal state of the reactants is specified. To see this, recall the collision theory canon[24]: Sum over final states, and average over initial states. Hence, if $k(n \to n'; E)$ is the reaction rate constant from reactants in the state $n$ to products at the state $n'$ at the total energy $E$, then the prior rate constant is of the form

$$k^0(n \to n'; E) = R\rho_T(E'_T) \qquad (52)$$

In terms of the prior state-to-state cross section,[5,26] $k^0(n \to n'; E) = v_n \sigma^0(n \to n')$, where

$$\sigma^0(n \to n') = (\pi/k_n^2)q\rho_T(E_T)\rho_T(E'_T) \qquad (53)$$

Here $k_n$ is the initial wave vector $E - E_n = E_T = \hbar^2 k_n^2/2\mu$, and $q$ is a constant. The total reaction rate, in the absence of constraints, is

$$k^0(n, E) = R\rho(E') \qquad (54)$$

where $\rho(E')$ is the products' total density of states and $E' = E - \Delta E_0$, where $\Delta E_0$ is the zero-point to zero-point endoergicity of the reaction. The total reaction cross section out of state $n$ is then

$$\sigma_R^0(n, E) = (\pi/k_n^2)q\rho_T(E_T)\rho(E') \qquad (55)$$

Second, we need to consider a distribution of reactants over both internal states *and* total energy.[27] For simplicity we shall consider thermal reactants.[28,29] First, at a given total energy $E$, the fraction of reactants in the state $n$ is merely $P^0(n)$ [Eq. (45)]. At a given temperature $T$ we thus have[28,29]

$$P^0(n, E) = P^0(n)\rho(E') \exp(-\beta E)/Q(n)$$

Then, the prior rate constant at the temperature $T$ is the Boltzmann average

$$k^0(n \to n'; T) = \frac{R}{Q(n)} \int_0^\infty dE \, \exp(-\beta E)\rho_T(E - E_n)\rho_T(E' - E'_n) \qquad (56)$$

Here $Q(n)$ is the normalization constant

$$Q(n) = \int_0^\infty dE \exp(-\beta E)\rho_T(E - E_n) \tag{57}$$

(i.e., the required average over initial states). $Q(n)$ will be recognized as the (thermal) partition function for reactants in the internal state $n$ at the temperature $T$.

Prior rate constants for less detailed distributions are obtained from Eq. (52) by an application of the canon.[25,26] In general if the internal states of the reactants are selected from a group $\Gamma$, and the internal states of the products are collected into groups $\Gamma'$, then the prior, thermal rate for $\Gamma \to \Gamma'$ is of the form[29]

$$k^0(\Gamma \to \Gamma'; T) = \sum_{n \in \Gamma} \sum_{n' \in \Gamma'} \frac{1}{Q(\Gamma)} \int_0^\infty dE \exp(-\beta E)\rho_T(E - E_n)v_n\sigma^0(n \to n') \tag{58}$$

where

$$Q(\Gamma) = \sum_{n \in \Gamma} \int_0^\infty dE \exp(-\beta E)\rho_T(E - E_n) \tag{59}$$

The summations in Eqs. (58) and (59) are over the internal states that belong to the specified groups. Comparison with Eq. (56) shows that, as expected,

$$k^0(\Gamma \to \Gamma'; T) = \left[ \sum_{n \in \Gamma} \sum_{n' \in \Gamma'} \exp(-\beta E_n)k^0(n \to n'; T) \right] \bigg/ \sum_{n \in \Gamma} \exp(-\beta E_n) \tag{60}$$

However, for an actual evaluation of the prior rate, the form (58) is more convenient.

As a particular example, the prior, $v \to v'$, thermal rate is of the form[28,29]

$$k^0(v \to v'; T) = \frac{R}{Q(v)} \int_0^\infty dE \exp(-\beta E)\rho(v, E)\rho(v', E') \tag{61}$$

where

$$Q(v) = \int_0^\infty dE \exp(-\beta E)\rho(v, E) \tag{62}$$

Here

$$\rho(v, E) = \sum_{J=0}^{J^*(v)} (2J+1)\rho_T(E - E_{vJ}) \tag{63}$$

The summations over $n$ and $n'$ in Eq. (58) were carried out before the evaluation of the integral. It is now possible to evaluate the integral explicitly and to show that,[28,29] in the RRHO approximation, Eq. (61) can be expressed as

$$k^0(v \to v'; T) = A(T)\Delta^2 \exp(\Delta)K_2(\Delta) \tag{64}$$

where

$$\Delta = \beta(E_v - E'_v - \Delta E_0)/2$$

and $-\Delta E_0$ is the reaction exoergicity. $K_2(\Delta)$ is the modified Bessel function of the second order. $A(T)$ is, in the RRHO model used to derive Eq. (64), a temperature-dependent factor (but independent of $v$ and $v'$).

The result (64) was first derived[30] for purely inelastic ($\Delta E_0 = 0$) collisions. It is important to note that due to a property of the Bessel functions, i.e.,

$$\Delta^n K_n(\Delta) = (-\Delta)^n K_n(-\Delta) \tag{65}$$

Eq. (64) is independent of the sign of $\Delta$. Also, for highly exoergic reactions, such that $\beta|\Delta E_0| > 1$, one can invoke the asymptotic limit of the Bessel functions, i.e.,

$$\Delta^n \exp(\Delta) K_n(\Delta) \to (\pi/2)^{1/2} \Delta^{n-1/2}, \qquad \text{as } \Delta \to \infty$$

to conclude that the product prior vibrational state distribution $P^0(v')$ is (in the RRHO approximation) of the form $P^0(v') \propto \Delta^{3/2}$ or

$$P^0(v') \propto (E' - E'_v)^{3/2} \tag{66}$$

This is the same form that obtains [cf. Eq. (51)] at a given total energy $E'$. Thus (as previously argued[23]), for highly exoergic reactions the thermal spread in the reactants' translational energy can be neglected in obtaining the prior product distribution.

## 3.3. Entropy and Surprisal

In this section we shall discuss measures[23] for the deviations of the observed from the prior products' internal state distribution and shall relate such measures to the thermodynamic formulation of Section 2. One must recognize that even in the absence of any constraints (apart from normalization) the internal states of the products are not equiprobable [cf. Eq. (45)]. We need to develop explicit expressions for the entropy when the prior (reference) distribution is not uniform. Our discussion follows Reference 31.

Each internal state of the products corresponds to an entire set of possible translational states for the relative motion (e.g., different directions of recoil). It can then be shown[5,31] that the entropy of such a distribution will be of the form

$$S[n] = -\sum_n P(n) \ln P(n) + \sum_n P(n) S(n) \tag{67}$$

Here $S(n)$ is the entropy of the distribution of the different translational states that correspond to a given internal state $n$. $S[n]$ is the overall entropy of the internal state distribution. The square brackets indicate that the entropy is computed by a summation over all internal states.*

For the case discussed in Section 3.2, where every event $n$ is an elementary one (i.e., does not correspond to a group of events), $S(n) = \ln 1 = 0$.

In the absence of constraints the distribution (i.e., the "prior") is determined by maximizing the Lagrangian

$$\mathcal{L} = S[n] - (\lambda_0 - 1) \sum_n P(n) \tag{68}$$

This leads to [cf. Eq. (11)]

$$P^0(n) = \exp[S(n) - \lambda_0] \tag{69}$$

where, as usual, $\lambda_0$ ensures the normalization

$$\exp(\lambda_0) = \sum_n \exp[S(n)] \tag{70}$$

The entropy $S[n]$ can now be expressed in terms of the observed, $P(n)$, and the prior, $P^0(n)$, distributions as

$$S[n] = \lambda_0 - \sum_n P(n) \ln[P(n)/P^0(n)] \tag{71}$$

To derive Eq. (71) we have written Eq. (69) as $S(n) = \lambda_0 + \ln P^0(n)$ and substituted into Eq. (67).

In the limit of no constraints $[P(n) = P^0(n)]$, $S[n]$ is indeed maximal (cf. the Appendix) and equals $\lambda_0$.

The entropy deficiency of the distribution is the (nonnegative) difference between $S^0[n]$ (the maximal value of $S[n]$, i.e., $S^0[n] = \lambda_0$) and its actual value

$$\Delta S[n] = \sum_n P(n) \ln[P(n)/P^0(n)] \tag{72}$$

The entropy deficiency is a measure of the deviation between $P$ and $P^0$, averaged over all outcomes. It is the overall measure of the specificity of the energy disposal (or the selectivity of the energy consumption) for the reaction.

The surprisal is a local measure of the deviation:

$$I(n) = -\ln[P(n)/P^0(n)] \tag{73}$$

It is a measure of the deviation of the actual probability (here, of an internal product state) from the prior distribution as defined in Section 3.2. The entropy deficiency [Eq. (72)] is seen to be the negative of the average value of the surprisal† [cf. Eq. (73)].

---

*The need for such a notation will become evident in Section 3.4, where we shall consider lower resolution distributions.

†We note again that in actual computations entropy is measured in entropy units [i.e., Eqs. (71) and (72) are multiplied by the gas constant $R$], whereas the surprisal is invariably dimensionless.

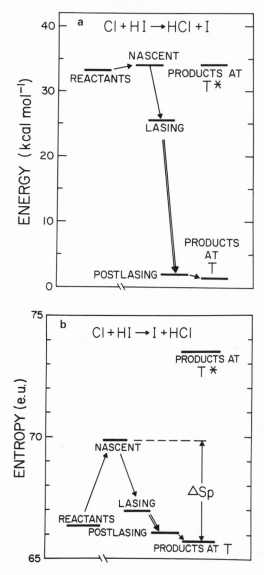

Fig. 2. Thermodynamic cycles[38] for a chemical laser pumped by the Cl + HI reaction at 300°K. (a) Energy cycle. $T^*$ is the temperature of the products that would obtain if all the reaction exoergicity were used to heat the products. (b) Entropy cycle. Note that whereas the energy content of the nascent products equals that of the products at $T^*$, their entropy is much lower.

The entropy function, in the context of disequilibrium processes, can be employed in the traditional thermodynamic role as a state function.[12,32-34]

One area where this idea can be applied is chemical lasers. Consider, for example, the so-called *pumping efficiency* $\eta_p$,

$$\eta_p = W/E_p \tag{74}$$

of a chemical reaction.[32] $E_p$ is the energy provided by the exothermic reaction (i.e., $-\Delta H_T^0$). $W$ is the maximum work that can be extracted (via any isothermal process) from the *nascent* reaction products (i.e., from the product molecules before they have suffered any collisional relaxation). If $\Delta S_p$ is the difference in entropy between the nascent and the thermal (at a temperature $T$) products, $W = E_p - T\Delta S_p$. Thus from its definition, Eq. (74),

$$\eta_p = 1 - T\Delta S_p/E_p \tag{75}$$

In an actual chemical laser, work is extracted from a translationally (and, often, rotationally) relaxed product distribution,[35-37] known as the *lasing distribution*. Also, not all of the energy can be extracted because lasing requires a finite gain.[35] The thermodynamic energy and entropy cycles[38] for a chemical laser pumped by the Cl + HI reaction are shown in Fig. 2.

## 3.4. Surprisal Analysis

The simplest application of the maximal entropy procedure is to analysis.[4,5,17,23,27-30,37,39-44] How can we compact a given distribution and characterize it, preferably in terms of its deviance from prior expectation? Taking the F + D$_2$ reaction as an example,[37,45] we shall consider the vibrational surprisal $I(v')$,

$$I(v') = -\ln[P(v')/P^0(v')] \tag{76}$$

(see Fig. 3). We observe that for this reaction, and for many other reactions,[4,5] the vibrational surprisal can be well represented by a linear relationship

$$I(v') = \lambda_0 + \lambda_v f_{v'} \tag{77}$$

where $f_{v'}$ is the fraction of the total energy $E$ in the products' vibration,

$$f_{v'} = E_{v'}/E$$

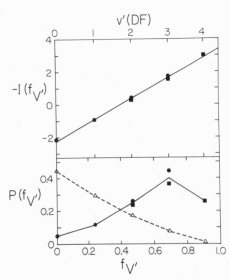

Fig. 3. Vibrational energy disposal surprisal analysis of the F + D$_2$ → D + DF($v'$) reaction. Lower panel: experimental results $P(v')$ (solid circles, data of Reference 37; solid squares, data from Reference 45) and the prior expectation $P^0(v')$ [Eq. (51), open triangles] vs. $v'$. Upper panel: vibrational surprisal $I(v')$ vs. $f_{v'}$.

The implications are clear.[46] Because the observed distribution is the one that maximizes the entropy subject to some constraint, and because [cf. Eq. (21)] the constraints are just the expectation values of the functions multiplying the Lagrange parameters in the surprisal, it follows that $P(v')$ is generated subject to the constraints

$$\sum_{v'} P(v') = 1 \tag{78a}$$

and

$$\sum_{v'} E_{v'} P(v') = \langle E_{v'} \rangle \tag{78b}$$

These are the same two constraints previously encountered in the treatment of bulk vibrational equilibrium (Section 2.4) and disequilibrium (Section 2.5). Indeed, as in the bulk case, one can seek the reasons $\langle E_{v'} \rangle$ should be the constraint.[46–48] There is, however, a very essential practical difference between the microscopic and macroscopic cases. For the bulk vibrational distribution we not only identified the constraint (Section 2.4) but also had comparatively simple means to determine its numerical magnitude without the need to observe the distribution. For the molecular disequilibrium problem this is still not the case. We do not know how to predict the magnitude of $\langle E_{v'} \rangle$ except via a (theoretical or experimental) determination of the distribution $P(v')$.

Detailed experiments can determine not just the energy disposal into one mode (e.g., vibration) but also joint distributions. Similarly, joint recoil energy and angular distributions or internal energy and angular distributions* can be determined, as in the case of the collision[49]

$$Li^+ + H_2 \rightarrow Li^+ + H_2(v', J')$$

studied by the molecular beam method.

Given the vibrational product state distribution $P(v', J')$, one can compute the entropy $S[v', J']$. Moreover, one can always generate the vibrational distribution $P(v')$ from a knowledge of the more detailed vibrotational distribution $P(v', J')$. The converse is, of course, not true. One has lost information in going from $P(v', J')$ to $P(v')$. Thus $S[v'] \geq S[v', J']$, or, in quantitative terms,[5,23,50]

$$\Delta S[v', J'] = \Delta S[v'] + \Delta S[J'|v'] \tag{79}$$

where $\Delta S[J'|v']$ is the (strictly nonnegative) amount of information lost when

---

*The prior angular distribution is isotropic in space,[5] $d\sigma^0(\theta)d\omega = (4\pi)^{-1}$ for three-dimensional (3D) collisions and $d\sigma^0(\theta)/d\theta = (2\pi)^{-1}$ for two-dimensional (2D) (coplanar) collisions. Thus the surprisal of the angular distribution is proportional to the differential cross section in both cases. In the approximation that the surprisal is dimensionally invariant,[26] $(d\sigma/d\omega)_{3D} \propto (d\sigma/d\theta)_{2D}$.

$P(v', J')$ is collapsed to form $P(v')$. Using the appropriate definitions and introducing the conditional probability $P(J'|v')$,

$$P(J'|v') = P(J', v')/P(v') \tag{80}$$

with

$$\sum_{J'} P(J'|v') = 1$$

we find[50] that

$$\Delta S[J'|v'] = \sum_{v'} P(v') \, \Delta S[J'|v']$$

$$= \sum_{v'} P(v') \sum_{J'} P(J'|v') \ln[P(J'|v')/P^0(J'|v')] \tag{81}$$

The quantity $\Delta S[J'|v']$ is defined by the second equality of Eq. (81).

When considering the vibrotational (joint) product state distributions one needs to consider two surprisals. One is $I(v')$, as discussed previously; the other is the conditional surprisal $I(J'|v')$,

$$I(J'|v') = -\ln[P(J'|v')/P^0(J'|v')] \tag{82}$$

When angular momentum limitations are absent it is usually the case[5,39] that $\Delta S[J'|v']$ is only very weakly $v'$ dependent.

Energy consumption can be analyzed in a similar fashion.[29] As an example,[29,51] consider the following reactions[52] of $H_2^+(v)$ with He:

$$H_2^+(v) + He \rightarrow \begin{cases} HeH^+ + H \\ He + H^+ + H \end{cases}$$

where the collision-induced dissociation path is accessible only at sufficiently high collision energy. The surprisal of the consumption is given by

$$I(v) = -\ln[k(v; E)/k^0(v; E)] \tag{83}$$

in the notation of Eqs. (52)–(55). Again (Fig. 4) the surprisal plot is linear, leading to the representation

$$k(v; E) = k^0(v; E) \exp(-\lambda_0 - \lambda_v f_v) \tag{84}$$

where $\lambda_v$ is negative because $k(v; E)$ is a strongly *increasing* function of $v$ [in contrast to $k^0(v; E)$]. Note also that $\lambda_v$ is essentially the same for both reaction paths.

The correlations between energy consumption and energy disposal (leading to the concept of relevance[5,27,50]) are discussed in Section 3.6. Such experimental data are still scarce. Most of the available data are either for inelastic processes[53,54] or are results of trajectory computations.[3]

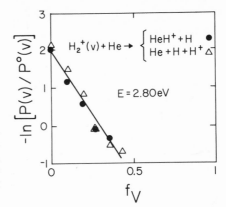

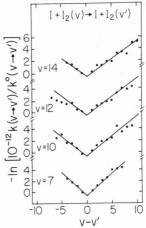

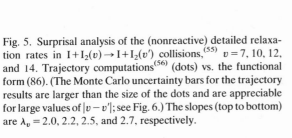

Fig. 4. Vibrational energy consumption surprisal analysis[29] of the data of Chupka et al.[52] for the $H_2^+(v) + He$ reaction. For both paths, Eq. (54) was used to define $k^0(n, E)$, which was then converted to $k^0(v, E)$. It then follows that for consumption at a given energy $P^0(v) \propto (E - E_v)^{-1/2}$.

The surprisal analysis of the products' state distribution for state-selected reactants can be carried out using the state-to-state prior distributions discussed in Section 3.2. As an example, consider

$$I + I_2(v) \rightarrow I + I_2(v')$$

where a $T \leftrightarrow V$ exchange (translational–vibrational energy transfer) can take place. For thermal reactants the use of Eq. (64) as the prior is appropriate.[30,55] Figure 5 shows the surprisal of the energy disposal for different initial vibrational states of $I_2(v)$ at $T = 1100°K$ using trajectory-generated[56] rate constants.

For the case when $\hbar\omega/kT < 1$ it is found that the surprisal can be well represented by a linear function of the (dimensionless) energy transfer $\Delta E_{\text{vib}}/kT$,

$$I(v \rightarrow v') = -\ln[k(v \rightarrow v'; T)/k^0(v \rightarrow v'; T)]$$

$$= \lambda_0 + \lambda_v |v - v'|(\hbar\omega/kT) \qquad (85)$$

Fig. 5. Surprisal analysis of the (nonreactive) detailed relaxation rates in $I + I_2(v) \rightarrow I + I_2(v')$ collisions,[55] $v = 7, 10, 12,$ and 14. Trajectory computations[56] (dots) vs. the functional form (86). (The Monte Carlo uncertainty bars for the trajectory results are larger than the size of the dots and are appreciable for large values of $|v - v'|$; see Fig. 6.) The slopes (top to bottom) are $\lambda_v = 2.0, 2.2, 2.5,$ and 2.7, respectively.

or

$$k(v \rightarrow v'; T) = k^0(v \rightarrow v'; T) \exp[-\lambda_0 - \lambda_v |v - v'|(\hbar\omega/kT)] \qquad (86)$$

Here $k^0(v \rightarrow v'; T)$ is defined by Eq. (61).

As discussed in Section 3.5, the constraint leading to the functional form (85) can be identified[55,57] [cf. Eq. (88) below] and related to bulk equilibrium relaxation.

One can summarize[28] the available[28,30,55,57,58] surprisal analysis of energy transfer collisions by the following *exponential gap* law.[1,16,59–62] For any internal–translational energy transfer in a buffer gas at a temperature $T$, the surprisal is essentially linear in $|\Delta E|/kT$:

$$I(\Delta E) = \lambda_0 + \lambda |\Delta E|/kT \qquad (87)$$

Here $\Delta E$ is the *energy mismatch*, i.e., the amount of energy transferred into (or out of) the translational motion.

### 3.5. Surprisal Synthesis

Next we shall consider two examples of the procedure for going from a constraint to an actual distribution. Both show the predictive power of the maximal entropy approach. First we shall consider the case in which we know both the functional *form* and the *magnitude* of the constraint. This will enable us to predict the magnitude of the individual probabilities. Then we shall examine the situation where only the functional *form* of the constraint is known. This will suffice to determine the functional form of the distribution but not the magnitude of the Lagrange parameter. (There is no inherent difference between the two examples. It just so happens that the needed magnitude of the constraint is a physically more familiar quantity in the first case.)

Consider bulk vibrational disequilibrium. It is known[16,17] that the final (i.e., slowest) step in the relaxation to equilibrium is a vibrational to translational ($V$–$T$) energy transfer. It is also known that during that final stage the (bulk) mean vibrational energy $\langle \varepsilon(t) \rangle$ often decays exponentially to its equilibrium value $\langle \varepsilon(\infty) \rangle$ [Section 2.4, Eq. (28) in particular]. One can not show that a necessary condition for the exponential decay of $\langle \varepsilon(t) \rangle$ is[55,62] that the vibration–translation detailed rate constants (Section 3.4) satisfy the constraint[55,57]

$$\sum_{v'} (E_{v'} - E_v) k(v \rightarrow v'; T) = \alpha[\langle \varepsilon(\infty) \rangle - E_v] \qquad (88)$$

Here $\alpha$ is the bulk relaxation rate constant, i.e., $\tau = 1/\rho\alpha$, where $\rho$ is the density of the gas and $\tau$ is the relaxation time, defined in Eq. (35). It thus follows that, because Eq. (88) is a necessary condition, an observation of an exponential relaxation of $\langle \varepsilon(t) \rangle$, i.e., Eq. (35), necessarily implies the validity of Eq. (88).

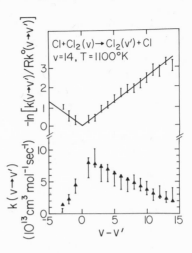

Fig. 6. Synthetic vs. (trajectory) computed rate constants[57] for $Cl + Cl_2(v) \rightarrow Cl_2(v') + Cl$ collisions. Bottom panel, detailed rates: synthetic (triangles) vs. computed[56] (bars). Here $\lambda_v = 0.25$. (The bars span the stated[56] statistical uncertainty limits of the trajectory computation.) Upper panel: surprisal representation. The straight line is not a fit but the prediction of Eq. (86) with the synthetic value of $\lambda_v$. $\alpha = 1.79 \times 10^{14}$ cm$^3$ sec$^{-1}$ mole$^{-1}$ was determined by fitting Eq. (88) to the computed[56] rates. The same value of $\alpha$ accounts for all the computed[56] detailed $(v \rightarrow v')$ rates for this system (to within the stated uncertainty limits).

Examination of experimental and trajectory–generated detailed, $k(v \rightarrow v'; T)$, rate constants for vibration–translation transfer has indeed verified[55] the form [Eq. (88)] of this constraint.

It is therefore possible to use bulk relaxation measurements of the average vibrational (or, for that matter, of the rotational[28]) energy to synthesize the entire manifold of detailed rate constants.

As an example[57] we shall consider the (trajectory-generated[56]) detailed rates for the

$$Cl + Cl_2(v) \rightarrow Cl_2(v') + Cl$$

reaction.* Using these detailed rates in Eq. (88) [and the thermal equilibrium value of $\langle \varepsilon(\infty) \rangle$], $\alpha$ could be determined. The substitution of Eq. (86) into Eq. (88) led to a procedure for the determination of $\lambda_v$, given the value of $\alpha$. Thereby, all the detailed rate constants could be recovered (Fig. 6).

The second synthesis is of a translational (recoil) energy distribution of the products of an exoergic reaction. One could assume[23,42,50] that in conformity with the vibrational energy distribution from thermal reactants (Section 3.4) the translational energy distribution would also be characterized by a simple first-moment constraint, i.e., by a given value of $\langle E'_T \rangle$. This would then lead to a distribution [cf. Eq. (16)]

$$P(E'_T) = P^0(E'_T) \exp(-\lambda_0 - \lambda_T f'_T) \tag{89}$$

Here $f'_T = E'_T/E'$, and in the RRHO approximation[5,23]

$$P^0(E'_T) = \tfrac{15}{4} E_T'^{1/2} (E' - E'_T)/E'^{5/2} \tag{90}$$

*The fact that the process under consideration is a chemical reaction rather than a purely inelastic collision is immaterial from the point of view of the synthesis. If we know the constraint (and, in this case, we do[57]), we can perform the synthesis. Note, by the way, that a linear surprisal plot is not limited to exoergic reactions (Fig. 6).

The molecular beam method and trajectory computations have provided $P(E'_T)$ distributions for a wide class of reactions and for a significant range of initial conditions. One knows by now that Eq. (89) is only a limiting form. What additional information can we input to the problem? To examine the situation in detail we choose the well-studied[63-65] reaction

$$Cl + HI \rightarrow I + HCl$$

Here one can derive a functional form for the constraint from physical arguments (as we shall do below) or from an approximate solution of the collision dynamics.[65]

During the $Cl + HI$ collision the heavy iodine atom is nearly unperturbed; it plays the role of a "drowsy spectator," i.e., the momentum of the iodine is essentially unchanged. (It takes a force to change the momentum.) This is the so-called spectator stripping model.[1,66] If $\mathbf{p}$ is the initial relative momentum of the reactants, then the initial momentum of the iodine atom (in the center of mass system) is $\gamma_i \mathbf{p}$, where $\gamma_i = m_I/(m_H + m_I)$. If $\mathbf{p}'$ is the final relative momentum of the products, it is also the final momentum of iodine. Hence, in the strict spectator limit $\mathbf{p}' = \gamma_i \mathbf{p}$. But this is not meant to be taken literally. What one means is that the average (over all the product energy states) of the magnitude of $\mathbf{q} = \mathbf{p}' - \gamma_i \mathbf{p}$ is small, i.e.,

$$\langle(\mathbf{p}' - \gamma_i\mathbf{p})^2\rangle \simeq 0 \tag{91}$$

Unfortunately, we cannot assign a precise numerical value* to $\langle(\mathbf{p}' - \gamma_i\mathbf{p})^2\rangle$ and hence shall not be able to determine the magnitude of the Lagrange parameter. We can proceed, however, to determine the functional form of $P(E'_T)$ by setting

$$\langle(\mathbf{p}' - \gamma_i\mathbf{p})^2/2\mu_{HI}\rangle = \text{const} \tag{92}$$

without specifying the magnitude of the constant.

Using the maximal entropy procedure one concludes that[65]

$$P(E'_T) = P^0(E'_T)\exp[-\lambda_0 - \lambda'_T(p' - \gamma_i p)^2/2\mu_{HI}] \tag{93}$$

Putting $E'_T = p'^2/2\mu'$, this can be expressed in the form[65]

$$P(E'_T) = P^0(E'_T)\exp\{-\lambda_0 - \lambda_T[(E'_T)^{1/2} - \varepsilon^{1/2}]^2\} \tag{94}$$

In terms of the ubiquitous skewing angle[1] $\beta$, $\lambda'_T = \lambda_T\sin^2\beta$ and $\varepsilon = E_T\cos^2\beta$. We see for low $E_T$ (and for reactions where $\cos^2\beta$ is small) that $\varepsilon \rightarrow 0$ and that Eq. (94) reduces to Eq. (89). Otherwise, Eq. (94) will have a bell-shaped form peaked at about $E'_T \simeq \varepsilon$ (Fig. 7).

The functional form (94) can now be compared to experiments or trajectory-generated results. (Even collinear computations can be

---

*In particular, note that the I atom will also have a momentum due to the internal energy $E_I$ of HI. Hence, a more reasonable form of the constraint is[24,66] $\langle q^2/2\mu_{HI}\rangle = E_I$. Here $\mu_{HI}$ is the reduced mass of HI.

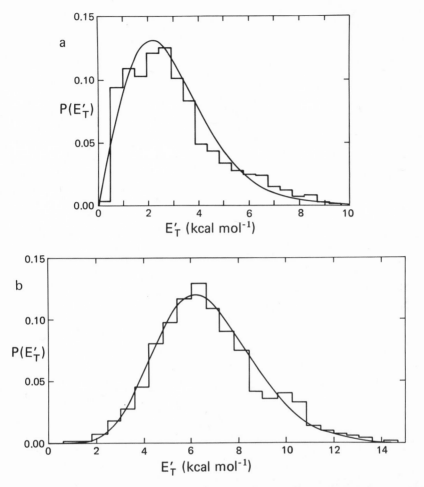

Fig. 7. The fit of the synthetic recoil energy distribution [Eq. (94)] to 3D trajectory-generated histograms of $P(E'_T)$ for the Cl+HI reaction.[65] (a) $E_T = 0.93_5$ kcal mole$^{-1}$; (b) $E_T = 5.72$ kcal mole$^{-1}$. Note the shift in the location of the peak: $\varepsilon = 2.7$ and $6.0$ kcal mole$^{-1}$ and $\lambda_T = 1.8$ and $2.9$ (kcal mole$^{-1}$)$^{-1}$, respectively. (For both computations HI is initially in the ground vibrational and fourth rotational states.) $\lambda_T = 1.8$ (kcal mole$^{-1}$)$^{-1}$ for collinear collisions at $E_T = 0.93_5$ kcal mole$^{-1}$. This is an example of the validity of the following approximation[67]: The $E'_T$ dependence of the surprisal $I(E'_T)$ is invariant in going from collinear to 3D collisions. This enables one to approximate a full 3D $P(E'_T)$ distribution from collinear computations.

examined.[26,67]) It can account not only for the often-observed shape of the distribution but also for the shift in the location of the most probable value of $E'_T$ with $E_T$. Of course, because angular momentum conservation has not yet been imposed as a constraint, it will fail to reproduce the "banded" structure of $P(E'_T)$ that one expects[1,68] for such cases.

The discussion above was motivated by the spectator stripping model, with preferential forward scattering [as implied by the constraint (92)] of the products. However, there is also a class of back scattering "rebound" reactions.[1,66] As would be suggested by, say, the DIPR model,[2,69] one should add the repulsive release of the exoergicity to the constraint, replacing Eq. (92) by[65]

$$\langle (\mathbf{p}' - \gamma_i \mathbf{p} - \mathbf{I})^2 / 2\mu_i \rangle = \text{const.} \tag{95}$$

Here $\mathbf{I}$ is the impulse imparted to the departing atom and $\mu_i$ the reduced mass of the reactant diatom. This would merely require a redefinition of $\varepsilon$ in Eq. (94). We thus expect that for a class of similar reactions, e.g., alkalis plus organic halides,

$$M + RI \rightarrow MI + R$$

the product recoil momentum distributions would (at low $p$) be very similar, as is found to be the case experimentally.[2]

### 3.6. Concept of Relevance

When detailed experiments (or computations) are available the concept of entropy can be extended to explore correlations.[50] Such correlations can be of two types. The first, leading to the concept of relevance, is the interdependence between the selectivity and specificity of the reaction.[50]

Consider, for example, the reaction[70]

$$F + HCl(v) \rightarrow Cl + HF(v')$$

Both experiments and trajectory computations have shown[70] that the product vibrational state distribution and the overall rate are distinctly dependent on the initial vibrational level $v$. Similar evidence is available for other reactions, e.g.,[70]

$$H + Cl_2(v) \rightarrow HCl(v') + Cl$$

In general, therefore, the detailed rate constants $k(v \rightarrow v')$ will not be of the simple, factorizable form $k(v)P(v')$. The entropic measure of this correlation between the initial and final distributions is termed the *relevance*.

Consider an experiment in which there is a distribution $P(n)$ of initial, reactant internal states. The products' internal state distribution $P(n')$ can then be expressed as

$$P(n') = \sum_n P(n'|n)P(n) \tag{96}$$

Here $P(n'|n)$ is the so-called *conditional* probability of the final state $n'$ given that the initial state is $n$. Note that $P(n'|n)$ is, by definition, normalized:

$$\sum_{n'} P(n'|n) = 1 \tag{97}$$

The entropy deficiency $\Delta S[n']$ of the distribution $P(n')$ can now be computed as usual [cf. Eq. (72)].

Assume now that the product state distribution could be measured (or computed) for each reactant internal state $n$. One would then know the distribution $P(n'|n)$. For each $n$ one could then compute the entropy deficiency $\Delta S[n'|n]$ from $P(n'|n)$:

$$\Delta S[n'|n] = \sum_{n'} P(n'|n) \ln[P(n'|n)/P^0(n')] \qquad (98)$$

$\Delta S[n'|n]$ is the entropy deficiency for the product state distribution from a specified initial reactant state. Its average value is $\Delta S[n'|n]$,

$$\Delta S[n'|n] = \sum_n P(n) \Delta S[n'|n) \qquad (99)$$

It can now be readily seen that $\Delta S[n'|n] \geq \Delta S[n']$. The entropy deficiency of the averaged distribution, $P(n')$, is lower than the averaged entropy deficiency. This result is a particular example of a general theorem[23] that any averaging increases the entropy (or decreases the entropy deficiency). One always loses information by an averaging procedure.

The relevance $R[n, n']$ is defined by[50]

$$R[n, n'] = S[n'|n] - \Delta S[n'] \qquad (100)$$

It is the measure of the additional information on the product state distribution provided by a knowledge of the reactant state distribution (and vice versa). It is nonnegative and symmetric in $n$ and $n'$, taking the value zero in the uncoupled limit, i.e., when $P(n'|n) = P(n')$, irrespective of $n$. In this case, $P(n', n)$ $= [P(n'|n)P(n)]$ assumes the uncoupled form, $P(n', n) = P(n')P(n)$. Note that the prior distribution has zero relevance [cf. Eq. (52) or (53)], and hence $P^0(n'|n) = P^0(n')$.

The relevance is thus a measure of the deviation of the detailed rate constant $k(n \to n')$ from the uncoupled (factorizable) form $k(n)P(n')$.

Although the prior distribution has, by definition, zero relevance, the converse does not follow. Low relevance does not imply that, say, $P(v') \simeq P^0(v')$.

Computations show[27] that exoergic reactions often have low relevance, because the dynamics are usually dominated by the early release of the exoergicity, and hence the precise distribution of the reactants' energy does not matter much. Yet such reactions often lead to extensive population inversion. Table 1 is a summary of results obtained[27] (using classical trajectory computations) for three types of (model) potential energy surfaces. For the exoergic reaction, there is considerable population inversion (and thus entropy deficiency), yet the relevance is low. For the thermoneutral reaction (on a "flat" surface) the relevance is quite high: The product state distribution is quite

*Table 1. Entropy Deficiency and Relevance (in e.u., i.e., cal deg$^{-1}$ mole$^{-1}$)*
*for a Model Problem$^{(27)}$*

| System | $\Delta S[E'_I]^a$ | $R[E_I, E'_I]$ |
|---|---|---|
| Exoergic | 2.62 | 0.02 |
| Thermoneutral, no well, | | |
| no barrier | 0.03 | 2.67 |
| Thermoneutral, deep well | 0.03 | 0.07 |

$^a$The entropy deficiency was computed over the distribution of internal energy $E'_I$ of the products.

sensitive to the initial conditions. Finally, for a surface with a deep attractive "well," resulting in many "snarled trajectories" (i.e., "collision complexes"), both the relevance and the entropy deficiency are low.

The second type of correlation, the so-called[50] *mutual information*, is between product distributions in two different modes, say vibration and rotation or recoil energy and angular distributions. We have briefly touched on this Section 3.3, and the reader is referred to Reference 50 for further details.

## 4. Chemical Disequilibrium

### 4.1. The Well-Reasoned Choice

In Sections 2 and 3 we have adopted a *high-resolution* point of view. For an experiment where there is a distribution over a set of possible outcomes, we have asked "What is the rule by which the most probable *entire* distribution is determined?" This has led to the formulation of the maximal entropy procedure. The most probable distribution is the one with the maximal thermodynamic weight (subject to whatever constraints operate).

In this section we shall consider a lower-resolution question, motivated by the need to treat species disequilibrium (and equilibrium). As an example, consider the reaction(s)

$$H + ICl \rightarrow \begin{cases} HI + Cl \\ HCl + I \end{cases}$$

What are the factors that determine the branching between the two possible (chemical-type) products? On an even more basic level, consider

$$F + H_2 \rightarrow \begin{cases} F + H_2 \\ H + HF \end{cases}$$

What is the fraction of collisions that are reactive?

Given that the outcomes of the experiment can be classified into two categories (or "bins") (say, reactive vs. nonreactive), we wish to determine the branching fraction into each. It has been shown[71] that a choice between two such alternatives can be reasoned out on thermodynamic-like grounds, i.e., that there is a "most reasonable" (i.e., most probable) choice, subject to the imposed constraints. It has been demonstrated[71] that this, most reasonable, branching fraction is just the thermodynamic probability (cf. Section 2.2) of the distribution of outcomes within the bin. In other words, given the (normalized) distribution of states for each reaction path separately, one can determine the branching ratios.

## 4.2. Analysis of Branching Ratios for Molecular Collisions

Given the (normalized) product state distribution for each reaction path, one can compute the entropy (and entropy deficiency) of the distribution within the path.

Let $P_a$ be the fraction of collisions that lead to products of type $a$. If the only additional constraint that is imposed is normalization,

$$P_a + P_b = 1 \tag{101}$$

then it has been shown[71] that*

$$P_a = \exp(S_a - \lambda_0) \tag{102}$$

Here $S_a$ is the entropy of the products *within* path $a$ and $\lambda_0$ is a normalization constant.

It is important to note, however, that the products' state distribution within path $a$ need not be a prior one. There may definitely be constraints that might give rise to a finite entropy deficiency $\Delta S_a = S_a^0 - S_a$, for any path $a$.

The prior branching fraction is, by definition, the branching fraction when there are no constraints at all (apart from normalization). Hence

$$P_a \propto \exp(S_a^0) \tag{103}$$

In terms of the entropy deficiency $\Delta S_a = S_a^0 - S_a$, for the branching ratio[71,72] we obtain

$$\begin{aligned}
\Gamma_{ab} = P_a/P_b &= \exp(S_a - S_b) \\
&= (P_a^0/P_b^0) \exp[-(\Delta S_a - \Delta S_b)] \\
&= \Gamma_{ab}^0 \exp[-(\Delta S_a - \Delta S_b)]
\end{aligned} \tag{104}$$

Here $\Gamma_{ab}^0 = (P_a^0/P_b^0)$ is the prior branching ratio.

If there are further constraints on the branching, over and above the constraints *within* each reaction path, then Eq. (104) needs to be modified in a suitable fashion.[71]

*A proof of Eq. (102) can be provided by using Eq. (67).

The prior branching ratio is given in terms of the (total) densities of states [Eq. (44)]:

$$\Gamma_{ab}^0 = \rho_a(E_a)/\rho_b(E_b) \tag{105}$$

Here $E_a$ and $E_b$ are the total (available) energies for the two reaction paths.

The total density of states can be readily calculated.[5,23,25] For an atom–diatom collision, in the range of energies where the diatomic molecule can be approximated as a (classical) rigid rotor–harmonic oscillator (RRHO), one obtains [5]

$$\rho(E) = \tfrac{4}{15} A_T A_I E^{5/2} \tag{106}$$

Here $A_T$ was previously defined [cf. Eq. (42), $A_T \propto \mu^{3/2}$] and $A_I = (\hbar \omega h c B_e)^{-1}$, where $\omega$ and $B_e$ are the vibrational frequency and rotational constant for the diatom.

The prior branching ratio can then be written in the RRHO approximation as [cf. Eq. (105)]

$$\Gamma_{ab}^0 = (\gamma_a/\gamma_b)(E_a/E_b)^{5/2} \tag{107}$$

The *energetics* factor $(E_a/E_b)$ will invariably favor the more exoergic reaction path, but as the total energy increases, this factor tends toward unity. (Of course, at higher energies the use of the RRHO density of states is no longer appropriate, becoming entirely misleading near the dissociation limit.) The structural factor can deviate markedly from unity, particularly if $H_2$ is a possible product.[72]

Figure 8 compares the prior[73] and the observed[74] branching fractions for an exoergic and an endoergic (reactive/nonreactive) branching process that proceeds via an intermediate complex:

$$M'X + M \rightarrow \begin{cases} MX + M' \\ M'X + M \end{cases}$$

It is seen that although the qualitative nature of the energy dependence is accounted for by the energetics term in the prior, there is no quantitative agreement. Other "statistical" approaches have also failed to account quantitatively for observed branching ratios for this class of reactions.[75–77]

When the distribution of states within each reaction branch is known one can compute the entropy deficiencies $\Delta S_a$ and $\Delta S_b$ and use Eq. (104) to "correct" the prior $\Gamma^0$ for the entropic bias. One can thereby allow for the fact that the product state distribution is not necessarily uniform within its phase space and that this "constrained" product phase space would necessarily imply a lower weight of the entire group of states [cf. Eq. (102)]. These considerations have been applied, say, to the

$$F + HI \rightarrow \begin{cases} HF(v') + I\,(^2P_{1/2}) \\ HF(v') + I\,(^2P_{3/2}) \end{cases}$$

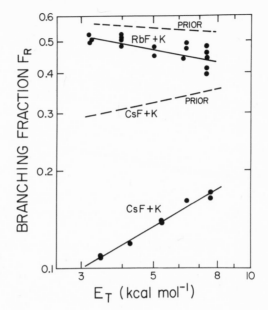

Fig. 8. Experimental[74] (dots) vs. prior[73] (broken line) reactive branching fractions $F_R$ for the M+M'F reactions for M = K and M' = Cs and Rb. Note the positive and negative energy dependencies for the endoergic (Cs) and the exoergic (Rb) reactions, respectively.

branching process.[78–80] Here the formation of a ground state iodine atom ($^2P_{3/2}$) leaves a considerable amount of energy ($\sim 60$ kcal mole$^{-1}$) to be partitioned between HF excitation and relative translation. The HF vibration population distribution would then be strongly "inverted," leading to a high $\Delta S[v']$. One then expects that the path leading to the formation of ground state I will have a high $\Delta S$. Thus, whereas on prior energetic grounds the formation of I ($^2P_{3/2}$) is strongly favored, it is disfavored on entropic grounds. Available experimental results[81] support this conclusion.[78]

When the experimental branching ratio deviates from the prior, one thus suspects a "constrained" phase space for one (or both) reaction path(s). For the reaction

$$F + HD \rightarrow \begin{cases} D + HF \\ H + DF \end{cases}$$

the measured[37,82] branching ratio is $1.4_5$, whereas $\Gamma^0 = 0.89$ (reflecting the lower zero-point energy of DF). Using Eq. (104), one infers that $\Delta S_{DF} - \Delta S_{HF} = 0.98$ e.u. From the measured[37] vibrational energy disposal $\Delta S_{DF}[v'] - \Delta S_{HF}[v'] = -0.73$ e.u. Hence,[83] there is a DF–HF difference of some 1.7 e.u. still to be accounted for. Most of this difference would be expected to arise from the rotational [$\Delta S[J|v]$; cf. Eq. (79)] contribution.[83]

Recent experiments[84] have confirmed these predictions. By use of $\Delta S[J', v']$ for the HF and DF distributions in Eq. (104) the computed branching ratio was found[84] to be 1.3 (vs. an experimental value[84] of 1.4).

The comparsion of experiment and theory for branching ratios has recently been reviewed.[85] Applications of the theory to reactive–nonreactive branching processes have also been reported.[86]

### 4.3. The Road Ahead

From a practical point of view, the most immediate goal, i.e., the analysis of the physical disequilibrium so characteristic of molecular dynamics, is essentially completed. Population inversion, preferred recoil and angular distributions, and selective energy consumption can be analyzed via the surprisal approach.

At this time, the road from Coulomb's law (via the Schrödinger equation) to the entropy deficiency has been traversed.[87] Using an *ab initio* potential energy hypersurface and exact quantal, 3D, collision dynamics, the rotational energy disposal for the reaction

$$H + H_2(J) \rightarrow H_2(J') + H$$

has been analyzed.[87] In accord with previous findings[5,39] the rotational surprisal is a linear function of the fraction, $g'_R$, of the available energy in product rotation (Fig. 9).

At the next level of sophistication there is also progress: correlating different aspects of the disequilibrium—in particular, the way microscopic chemical disequilibrium is related to the energy disposal. Another second-generation question is the relation between energy consumption and energy

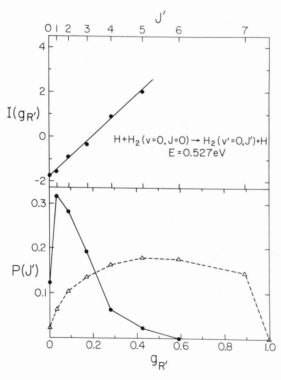

Fig. 9. Rotational energy disposal surprisal analysis[87] of the H+ $H_2(J) \rightarrow H_2(J') + H$ reaction. Lower panel: exact quantal computations[87] $P(J')$ (solid circles) and the prior expectation $P^0(J')$ [open triangles,

$$P^0(J') \propto (2J'+1)\rho_T(E'-E'_R)$$

$$\propto (2J'+1)E_T'^{1/2}$$

where $E'$ and $E'_R$ are the product total and rotational energy, respectively] vs. $g'_R = E'_R/E'$. These computations pertain to an energy $E' = E = E_T = 0.527$ eV. Upper panel: rotational surprisal $I_R = -\ln[P(J')/P^0(J')]$ vs. $g'_R$.

disposal for the same reaction. This will require the application of the maximal entropy procedure to the entire $P$ matrix (or the matrix of detailed rate constants $[k(n \rightarrow n')]$).

Next to be dealt with is the problem of a priori synthesis. Missing is the disequilibrium analogue of the zeroth law for the equilibrium situation. We still lack a thermometer with which to "read" the disequilibrium values of the Lagrange parameters. Until such time, synthesis will be limited to determining the functional form of the distributions or special procedures will have to be invoked (e.g., experimental measurements of, say, $\langle \varepsilon \rangle$) to determine the magnitude of the constraint. Even so, the ability to synthesize an entire array of detailed rate constants from one (or a few) experimentally determined moment(s) is a worthwhile objective.

Ultimately, one will achieve not just a thermodynamic but an entire statisticomechanical approach to microscopic disequilibrium, where the constraints will follow directly from mechanical principles.

## Appendix

### The Gibbs Inequality

The maximal properties of the entropy are readily demonstrated through an inequality that was certainly known to both Gibbs[6] and Boltzmann.[88] Planck[20] provides his own proof, as have many authors since.

In its most essential form the inequality is

$$\ln x \leq x - 1 \tag{A.1}$$

with equality for $x = 1$. For application it is convenient to rewrite the inequality (A.1) by introducing two real numbers $p_i$ and $q_i$ (which may depend on an index $i$) such that $x = q_i/p_i$. Multiplying Eq. (A.1) by $p_i$, we have

$$p_i(\ln p_i - \ln q_i) - (p_i - q_i) \geq 0 \tag{A.2}$$

with equality only for $p_i = q_i$. Summing Eq. (A.2) over $i$, we find that

$$\sum_i p_i \ln(p_i/q_i) \geq \sum_i (p_i - q_i) \tag{A.3}$$

If $p_i$ and $q_i$ are (normalized) probabilities, then the inequality takes an even simpler form,

$$\sum_i p_i \ln(p_i/q_i) \geq 0 \tag{A.4}$$

Equation (A.4) is the proof of the statement that the entropy deficiency [cf. Eq. (72)] is nonnegative.

As an application we shall now prove that the extremum derived in Section 2.3 is indeed the maximal value of the entropy. Given the constraints (19), we need to prove that among all distributions $\{p_i\}$ that are consistent with Eq. (19) the distribution $\{x_i\}$ [given by Eq. (21)] has the maximal entropy.

The proof follows directly. We take $q_i = x_i$ when Eq. (A.4) gives

$$-\sum_i p_i \ln p_i \le -\sum_i p_i \ln x_i$$

$$= \sum_i p_i[\lambda_0 + \sum_{m=1} \lambda_m g_m(i)]$$

$$= \lambda_0 + \sum_{m=1} \lambda_m \langle g_m \rangle \qquad (A.5)$$

The left-hand side is the entropy of the distribution $\{p_i\}$, whereas the right-hand side is [cf. Eq. (25)] the entropy of the distribution $\{x_i\}$. Q.E.D.

Boltzmann[88] (and Planck) use the inequality to prove the $H$ theorem ("the increase of entropy with time during a relaxation process"). With our treatment of relaxation (Section 2.5) such a proof is immediate.

ACKNOWLEDGMENTS

This work was supported by the Air Force Office of Scientific Research under Grant 74-2725 and by the National Science Foundation under Grant MPS73-04940 A01.

## References

1. R. D. Levine and R. B. Bernstein, *Molecular Reaction Dynamics*, Clarendon Press, Oxford (1974).
2. D. R. Herschbach, Reactive scattering, *Faraday Discuss. Chem. Soc.* 55, 233–251 (1973).
3. J. C. Polanyi, Molecular beam scattering, *Faraday Discuss. Chem. Soc.* 55, 389–409 (1973).
4. R. D. Levine and R. B. Bernstein, Energy disposal and energy consumption in elementary chemical reactions: The information theoretic approach, *Acc. Chem. Res.* 7, 393–400, (1974).
5. R. B. Bernstein and R. D. Levine, Role of energy in relative molecular scattering: an information-theoretic approach, *Adv. At. Mol. Phys.* 11, 215–297 (1975).
6. J. W. Gibbs, *Collected Works*, Vol. II, Longmans, Green & Co. Ltd., London (1928).
7. E. T. Jaynes, Information theory and statistical mechanics, *Phys. Rev.* 106, 620–630 (1957); Information theory and statistical mechanics. II, *Phys. Rev.* 108, 171–190 (1957).
8. E. T. Jaynes, in: *Statistical Physics*, 1962 Brandeis Lectures, Vol. 3, p. 181, W. A. Benjamin, Inc., Reading, Mass. (1963).
9. M. Tribus, *Thermostatics and Thermodynamics*, Van Nostrand Reinhold Company, New York (1961).
10. A. Katz, *Principles of Statistical Mechanics*, W. H. Freeman and Company, Publishers, San Francisco (1967).
11. R. Kubo, *Statistical Mechanics*, North-Holland Publishing Company, Amsterdam (1965).

12. R. D. Levine and R. B. Bernstein, Analysis of energy disposal: Thermodynamic aspects of the entropy deficiency of a product state distribution, *Chem. Phys. Lett.* **22**, 217–221 (1973).

13. J. E. Mayer and M. G. Mayer, *Statistical Mechanics*, John Wiley & Sons, Inc., New York (1940).

14. A. I. Khinchin, *Mathematical Foundations of Information Theory*, Dover Publications, Inc., New York (1957).

15. R. D. Levine, to be published.

16. C. B. Moore, Laser studies of vibrational energy transfer, *Acc. Chem. Res.* **2**, 103–109 (1969).

17. R. G. Gordon and J. I. Steinfeld, *in: Molecular Energy Transfer* (R. D. Levine and J. Jortner, eds.), pp. 67–84, John Wiley & Sons, Inc., New York (1976).

18. R. M. Osgood, Jr., P. B. Sackett, and A. Javan, Measurement of vibrational–vibrational exchange rates for excited vibrational levels ($2 \le v \le 4$) in hydrogen fluoride gas, *J. Chem. Phys.* **60**, 1464–1480 (1974).

19. I. Procaccia and R. D. Levine, The populations time evolution in vibrational disequilibrium: An information theoretic approach with applications to HF, *J. Chem. Phys.* **62**, 3819–3820 (1975).

20. M. Planck, *Theory of Heat Radiation*, McGraw-Hill, Inc., Blakiston Division, New York (1914).

21. L. P. Wheeler, *Josiah Willard Gibbs*, p. 160, Yale University Press, New Haven, Conn. (1952).

22. T. S. Kuhn, *Structure of Scientific Revolutions*, 2nd ed., University of Chicago Press, Chicago (1970).

23. A. Ben-Shaul, R. D. Levine, and R. B. Bernstein, Entropy and chemical change. II. Analysis of product energy distributions; temperature and entropy deficiency, *J. Chem. Phys.* **57**, 5427–5447 (1972); Prior-expectation distribution functions for energy disposal and energy consumption in reactive molecular collisions, *J. Chem. Phys.* **61**, 4937–4938 (1974).

24. R. D. Levine, *Quantum Mechanics of Molecular Rate Processes*, Clarendon Press, Oxford (1969).

25. J. L. Kinsey, Microscopic reversibility for rates of chemical reactions carried out with partial resolution of the product and reactant states, *J. Chem. Phys.* **54**, 1206–1217 (1971).

26. R. B. Bernstein and R. D. Levine, The relation between collinear and 3-D dynamical calculations of reactive molecular collisions, *Chem. Phys. Lett.* **29**, 314–318 (1974).

27. C. Rebick, R. D. Levine, and R. B. Bernstein, Energy requirements and energy disposal: Reaction probability matrices and a computational study of a model system, *J. Chem. Phys.* **60**, 4977–4984 (1974).

28. I. Procaccia and R. D. Levine, Vibrational energy transfer in molecular collisions: An information theoretic analysis and synthesis, *J. Chem. Phys.* **63**, 4261–4279 (1975).

29. J. Manz and R. D. Levine, The effect of reagent energy on chemical reaction rates: An information theoretic analysis. *J. Chem. Phys.* **63**, 4280–4303 (1975).

30. M. Rubinson and J. I. Steinfeld, Entropy analysis of product energy distributions in nonreactive inelastic collisions, *Chem. Phys.* **4**, 467–475 (1974).

31. U. Dinur and R. D. Levine, On the entropy of a continuous distribution, *Chem. Phys.* **9**, 17–28 (1975).

32. O. Kafri and R. D. Levine, Thermodynamic analysis of chemical laser systems, *Chem. Phys. Lett.* **27**, 175–179 (1974).

33. R. D. Levine, Analysis of Molecular Collisions: An Information-Theoretical Approach, Invited papers, VIII International Conference on the Physics of Electronic and Atomic Collisions, Belgrade (1973), pp. 567–582.

34. A. Ben-Shaul, On entropies and temperature parameters characterizing product distributions in chemical reactions and corresponding thermodynamic quantities, *Mol. Phys.* **27**, 1585–1600 (1974).

35. K. L. Kompa, Chemical lasers, *Top. Curr. Chem.* **37**, 1–92 (1973).

36. A. Ben-Shaul, G. L. Hofacker, and K. L. Kompa, Characterization of inverted populations in chemical lasers by temperaturelike distributions: Gain characteristics in the $F + H_2 \rightarrow HF + H$ system, *J. Chem. Phys.* **59**, 4664–4673 (1973).

37. M. J. Berry, $F + H_2$, $D_2$, HD reactions: Chemical laser determination of the product vibrational state populations and the $F + HD$ intramolecular kinetic isotope effect, *J. Chem. Phys.* **59**, 6229–6253 (1973).

38. O. Kafri, A. Ben-Shaul, and R. D. Levine, Chemical lasers: A thermodynamic analysis of a surprisal in disequilibrium, *Chem. Phys.* **10**, 367–392 (1975).

39. R. D. Levine, B. R. Johnson, and R. B. Bernstein, On the product rotational state distribution in exoergic atom–diatomic molecule reactions, *Chem. Phys. Lett.* **19**, 1–6 (1973).

40. A. Ben-Shaul, Product state distributions in chemical reactions: Vibrational temperature and rotational distributions, *Chem. Phys.* **1**, 244–255 (1973).

41. R. D. Levine, *in: Proceedings of the International Symposium on Chemical and Biochemical Reactivity*, pp. 35–50, Academy of Science, Jerusalem (1973).

42. D. L. King, H. J. Loesch, and D. R. Herschbach, Comments on extremely inelastic energy transfer, *Faraday Discuss. Chem. Soc.* **55**, 222–225 (1973).

43. M. J. Berry, Golden rule calculation of reaction product vibronic state distributions, *Chem. Phys. Lett.* **27**, 73–77 (1974); Vibronic surprisal analysis of the dynamics of photo-dissociation and related reactions, *Chem. Phys. Lett.* **29**, 323–328 (1974); Golden rule calculation of product vibronic population inversions in photodissocation and related reactions, *Chem. Phys. Lett.* **29**, 329–336 (1974).

44. N. C. Blais and D. G. Truhlar, Monte Carlo trajectories: The reaction $H + Br_2 \rightarrow HBr + Br$, *J. Chem. Phys.* **61**, 4186–4203 (1974).

45. J. C. Polanyi and K. B. Woodall, Energy distribution among reaction products. VI. $F + H_2, D_2$, *J. Chem. Phys.* **57**, 1574–1586 (1972).

46. G. L. Hofacker and R. D. Levine, Diabatic transition state theory and the concept of temperature, *Chem. Phys. Lett.* **15**, 165–170 (1972).

47. G. L. Hofacker and R. D. Levine, A non-adiabatic model for population inversion in molecular collisions, *Chem. Phys. Lett.* **9**, 617–620 (1971).

48. A. Ben-Shaul and G. L. Hofacker, *in: Handbook of Chemical Lasers* (J. F. Bott and R. W. Gross, eds.), John Wiley & Sons, Inc., New York (1976), in press.

49. J. P. Toennies, *in: Molecular Energy Transfer* (R. D. Levine and J. Jortner, eds.), pp. 16–52, John Wiley & Sons, Inc., New York (1976).

50. R. D. Levine and R. B. Bernstein, Energy disposal and energy requirements for elementary chemical reactions, *Faraday Discuss. Chem. Soc.* **55**, 100–112 (1973).

51. C. Rebick and R. D. Levine, Collision induced dissociation: A statistical theory, *J. Chem. Phys.* **58**, 3942–3952 (1973).

52. W. A. Chupka, J. Berkowitz, and M. E. Russell, A study of some reactions of $H_2^+$ in selected vibrational states, *in: Abstracts VI International Conference on the Physics of Electronic and Atomic Collisions*, pp. 71–72, The M.I.T. Press, Cambridge, Mass. (1969).

53. I. W. M. Smith, *in: Molecular Energy Transfer* (R. D. Levine and J. Jortner, eds.), pp. 85–113, John Wiley & Sons, Inc., New York (1976).

54. J. I. Steinfeld, *in: Molecular Spectroscopy: Modern Research* (K. N. Rao and C. W. Mathews, eds.), pp. 223–230, Academic Press, Inc., New York (1972).

55. I. Procaccia and R. D. Levine, Vibrational energy transfer in non–reactive molecular collisions: An information theoretic analysis, *Chem. Phys. Lett.* **33**, 5–10 (1975).

56. D. L. Thompson, Monte Carlo classical dynamical study of the $Cl + Cl_2$ and $I + I_2$ systems: Vibrational relaxation and atom-exchange reactions, *J. Chem. Phys.* **60**, 4557–4567 (1974).

57. I. Procaccia and R. D. Levine, From bulk vibrational relaxation data to the detailed (microscopic) rate constants, *J. Chem. Phys.* **62**, 2496–2497 (1975).

58. F. F. Crim and G. A. Fisk, unpublished; F. F. Crim, Ph.D. thesis, Cornell University, Ithaca, New York (1974).

59. J. C. Polanyi and K. B. Woodall, Mechanism of rotational relaxation, *J. Chem. Phys.* **56**, 1563–1572 (1972); R. B. Bernstein, Note on the Polanyi–Woodall Equation for rotational relaxation, *J. Chem. Phys.* **62**, 4570 (1975).

60. A. B. Callear and J. D. Lambert, *in: Comprehensive Chemical Kinetics* (C. H. Bamford and C. F. Tipper, eds.) Vol. 3, pp. 182–273, American Elsevier Publishing Company, Inc., New York (1969).

61. C. B. Moore, Vibration–vibration energy transfer, *Adv. Chem. Phys.* **23**, 41–83 (1973).

62. I. Oppenheim, K. E. Shuler, and G. H. Weiss, Stochastic theory of multistate relaxation processes, *Adv. Mol. Relaxation Processes* **1**, 13-68 (1967).

63. C. A. Parr, J. C. Polanyi, and W. H. Wong, Distribution of reaction products (theory). VIII. Cl + HI, Cl + DI, *J. Chem. Phys.* **58**, 5–20 (1973).

64. D. H. Maylotte, J. C. Polanyi, and K. B. Woodall, Energy distribution among reaction products. IV. X + HY (X $\equiv$ Cl, Br; Y $\equiv$ Br, I), Cl + DI, *J. Chem. Phys.* **57**, 1547–1560 (1972).

65. A. Kafri, E. Pollak, R. Kosloff, and R. D. Levine, Translational energy disposal in molecular collisions, *Chem. Phys. Lett.* **33**, 201–206 (1975).

66. D. R. Herschbach, Molecular beam studies of internal excitation of reaction products, *Appl. Opt. Suppl.* **2**, 128–144 (1965); *in: Molecular Beams, Advances in Chemical Physics* (J. Ross, ed.), Vol. X, pp. 319–393, John Wiley & Sons, Inc. (Interscience Division), New York (1966).

67. J. C. Polanyi and J. L. Schreiber, Distribution of reaction products (Theory). Investigation of an *ab initio* energy-surface for F + H$_2$ $\rightarrow$ HF + H, *Chem. Phys. Lett.* **29**, 319–322 (1974).

68. K. G. Anlauf, P. E. Charters, D. S. Horne, R. G. McDonald, D. H. Maylotte, J. C. Polanyi, W. J. Skrlac, D. C. Tardy, and K. B. Woodall, Translational energy-distribution in the products of some exothermic reactions, *J. Chem. Phys.* **53**, 4091–4092 (1970).

69. P. J. Kuntz, M. H. Mok, and J. C. Polanyi, Distribution of reaction products. V. Reactions forming an ionic bond, M + XC (3D), *J. Chem. Phys.* **50**, 4623–4652 (1969); P. J. Kuntz, The K + ICH$_3$ $\rightarrow$ KI + CH$_3$ reaction: Interpretation of the angular and energy distributions in terms of a direct interaction model, *Mol. Phys.* **23**, 1035–1050 (1972).

70. A. M. Ding, L. J. Kirsch, D. S. Perry, J. C. Polanyi, and J. L. Schreiber, Effect of changing reagent energy on reaction probability and product energy distribution, *Faraday Discuss. Chem. Soc.* **55**, 252–276 (1973).

71. R. D. Levine and R. Kosloff, The well-reasoned choice: An information-theoretic approach to branching ratios in molecular rate processes, *Chem. Phys. Lett.* **28**, 300–304 (1974).

72. R. D. Levine and R. B. Bernstein, Branching ratios in reactive molecular collisions, *Chem. Phys. Lett.* **29**, 1–6 (1974).

73. R. D. Levine, R. B. Bernstein, and M. Tamir, Information-theoretic analysis of reactive/non-reactive branching ratios for triatomic collision complexes: K + CsF, RbF, *in: Abstracts IX International Conference on the Physics of Electronic and Atomic Collisions*, pp. 333–334, University of Washington Press, Seattle (1975).

74. S. Stolte, A. E. Proctor, and R. B. Bernstein, Energy dependence of the branching fraction and cross sections for the decay of collision complexes: K + CsF, RbF, *J. Chem. Phys.* **61**, 3855–3856 (1974).

75. W. B. Miller, S. A. Safron, and D. R. Herschbach, Molecular beam kinetics: Four-atom collision complexes in exchange reactions of CsCl with KCl and KI, *J. Chem. Phys.* **56**, 3581–3592 (1972).

76. R. A. White and J. C. Light, Statistical theory of bimolecular exchange reactions: Angular distribution, *J. Chem. Phys.* **55**, 379–387 (1971).

77. G. H. Kwei, B. P. Boffardi, and S. F. Sun, Classical trajectory studies of long-lived collision complexes. I. Reaction of K atoms with NaCl molecules, *J. Chem. Phys.* **58**, 1722–1734 (1973).

78. U. Dinur, R. Kosloff, R. D. Levine, and M. J. Berry, Analysis of electronically non-adiabatic chemical reactions: An information theoretic approach, *Chem. Phys. Lett.* **34**, 199–205 (1975).

79. U. Dinur and R. D. Levine, Does the H + ICl reaction form electronically excited I ($^2P_{1/2}$) atoms?, *Chem. Phys. Lett.* **31**, 410–415 (1975).

80. M. A. Nazar, J. C. Polanyi, and W. J. Skrlac, Energy distribution among reaction products, H + NOCl, H + ICl, *Chem. Phys. Lett.* **29**, 473–479 (1974).

81. N. Jonathan, C. M. Melliar-Smith, S. Okuda, D. H. Slater, and D. Timlin, Initial vibrational energy level distributions determined by infra-red chemiluminescence. II. The reaction of fluorine atoms with hydrogen halides, *Mol. Phys.* **22**, 561–574 (1971).

82. A. Persky, Kinetic isotope effects in the reaction of fluorine atoms with molecular hydrogen. II. The F + HD/DH intramolecular isotope effect, *J. Chem. Phys.* **59**, 5578–5584 (1973).

83. R. B. Bernstein and R. D. Levine, Information-theoretic analysis of the kinetic intramolecular isotope effect for the F + HD reaction, *J. Chem. Phys.* **61**, 4926–4927 (1974).

84. D. S. Perry and J. C. Polanyi, An experimental test of the Bernstein–Levine theory of branching ratios, *Chem. Phys.* **12**, 37–43 (1976).
85. R. B. Bernstein, Branching ratios in reactive collisions: Theory vs. experiment, *Int. J. Quantum Chem., Symp.* **9**, 385–395 (1975).
86. H. Kaplan and R. D. Levine, Translation energy dependence of the reaction cross section: An information theoretic synthesis with applications to $M + CH_3I$, *J. Chem. Phys.* **63**, 5064–5066 (1975); Systematic trends in the alkali metal-alkyl iodide reactions: The translational energy dependence of the reaction cross section, *Chem. Phys. Lett.* **39**, 1–7 (1976).
87. R. E. Wyatt, Information-theoretic analysis of quantum mechanical reaction cross sections, *Chem. Phys. Lett.* **34**, 167–169 (1975).
88. L. Boltzmann, *Lectures on Gas Theory,* University of California Press, Berkeley (1964).

# Author Index

The suffixes A and B on the page numbers indicate the volume (Part A or Part B, respectively) in which the citation appears. Boldface page numbers indicate a chapter in one of these two volumes.

# Subject Index

The suffixes A and B on the page numbers indicate the volume (Part A or Part B, respectively) in which the citation appears.

*Ab initio* methods, 55B, 57B, 62B, 64B, 67B, 112B, 114B
Absorption cross section, 88A, 117A, 119A
Absorptive sphere, 106A
Accommodation coefficient, 240A, 242A
Action–angle variables, 18B, 48B, 172B, 187B, 189-192B, 197B, 202B, 206B
Action
  classical, 172B, 193, 212B
  internal, 14B
Action functional, 7B
Activated complex, 269B
Activation energy, 6B
Adams–Moulton method, 44B
Adiabatic process, 217B, 218B, 219B, 223B, 224B, 226B, 227B, 228B, 231B, 232B, 235B-238B *passim,* 241B, 245B, 246B, 250B, 257B, 258B, 259B
Adiabatic representation, 220B, 223B, 224B, 227B, 228B, 231B, 237B, 238B, 239B, 247B
Airy function, 175B, 178B, 182B, 195B, 197B
Alkali atom scattering, 108A
Amplitude of probability density, 10A
Amplitude of probability flux, 10A
Angle representation, 193B, 198B
Angular coupling, 230B
Angular distributions, 54B, 57B, 73B, 90B, 97B, 99B, 100B, 101B, 107B, 109B, 110B, 111B, 115B
Angular momentum, 30B, 37B
  rotational, 47B, 48B
Angular momentum barrier, 312-320B
Angular scattering, *see* Angular distributions
Anharmonicity, 32B
Antithreshold, 3B
Associative ionization, 233B
Asymmetric top, 18B, 30B

Asymptotic approximation, 104A
Atom–diatomic collisions, 123A
Atom–molecule scattering, 72-74A
Atoms-in-molecules, 229B
Attractive energy release, 79B, 80B, 81B, 90B, 95B, 97B, 103B, 107B, 110B, 112B, 113B, 115B
Autoionization, 258B
Avoided crossing, 224B, 226B, 227B, 230B, 243B, 244B, 251B, 257B
Axis of symmetry, 30B

Beam crossing angle, 25B
Bessel function, 183B, 197B
Bixon–Jortner model, 283-285A
Body-fixed coordinate frame, 14A, 35A, 36A, 48A, 49A, 50A, 59A
Body-fixed scattering amplitude, 19A
Bohr quantization condition, 179B
Boltzmann equilibrium, 15B, 18B
Born approximation, 231A, 241A, 242A, 236B, 241B
Born–Oppenheimer approximation, 219B
Born–Oppenheimer separation, 219B, 223B
Bound state, 46B
Bound state resonance, 221-226A *passim,* 231A
Boundary value problem, 203B
Branching ratios, 324B, 355B, 356B, 357B

Canonical transformation, 10B, 183B, 186B, 191B
Cartesian coordinates, 8B, 36B, 37B, 38B
Caustic, 172B, 197B
Center of mass system, 6A
Central potential, 2A
Centrifugal decoupling, 34A, 47-57A, 67A, 72-77A
Charge transfer, 218B, 244B

373